Industry 4.0 with Modern Technology

Editors Biography

Srinivas Sethi

Professor Srinivas Sethi is a Professor in Computer Science Engineering & Application, Indira Gandhi Institute of Technology, Sarang (IGIT, Sarang), India, and has been actively involved in teaching and research since 1997. He did his Ph.D., in the area of routing algorithms in mobile ad hoc network and is continuing his research work in the wireless sensor network, cognitive radio network, and cloud computing, Big-Data, BCI, Cognitive Science. He is a Member of Editorial Board for different journal and Program Committee Member for different international conferences/workshop. He is a Book Editor of four international conference proceedings published in Springer and Taylor & Francis. He has published more than 80 research papers in international journals and conference proceedings. He completed eight research and consultancy projects funded by different funding agencies such as DRDO, DST, AICTE, NPIU, and local Govt. office. He has delivered more than 125 talks as keynote speaker, resource person in different international and national conferences/workshops, conducted by different important national reputation institutions such as SAG group of DRDO, New Delhi, JNTU GV College of Engineering Vizianagaram, Pt. Rabishankar Sukla University, Raipur, CSIT, Raipur, Utkal University, Bhubaneswar, etc.

Mufti Mahmud

Dr Mahmud's research vision is to contribute toward a secure, smart, healthy, and better world to live in. In todays digitized world, converting the ever-expanding amount of raw data to smart data, and building predictive, secure and adaptive systems aiming personalized services are essential and challenging, which require cross-disciplinary and multi-stakeholder collaborations. Toward these goals, Dr Mahmud conducts problem-driven `Brain Informatics research where he works with problem domain experts to find multidisciplinary solutions to real-world problems. Dr Mahmud's research involves Computational, Health, and Social Sciences, and uses Neuroscience, Healthcare, Applied Data Science, Computational Neuroscience, Big Data Analytics, Cyber Security, Machine Learning, Cloud Computing, and Software Engineering and plans to develop secure computational tools to advance healthcare access in low-resource settings.

Rabinarayan Sethi

Dr. Rabinarayan Sethi is affiliated as Assistant Professor in the Department of Mechanical Engineering, IGIT, Sarang, India. He received a PhD degree. His research interests are broadly in the area of Mechanical Vibration, Machinery Condition Monitoring, Engineering Mechanics, M/c Dynamics, Strength of Materials, and Machine Design. He has 22 years of teaching and research experience in different capacities in the Department of Mechanical Engineering. He has more than 22 research publications in various International & National journals/conferences. He received a Fellowship from Global Institute of Science and Technology, Australia, in 2012. He is Fellow of the Institution of Engineers (India) (MIE), a registered member of other national and international professional societies, like, lifetime member of the Indian Society for Technical Education, Member of the Society for Automobile Engineers India, New Delhi. He participates as an expert in a weekly Science Programe *Gyana Vigyana, Doordarshan-6* (DD-6), Odia. He chaired a national and international conference session. He delivered an expert lecture on various short-term courses.

Sujit Kumar Pradhan

Sujit Kumar Pradhan is currently serving as an Assistant Professor in the Department of Civil Engineering, Indira Gandhi Institute of Technology Sarang, India. He holds PhD in Transportation Engineering from Indian Institute of Technology Bhubaneswar. His area of research includes pavement material characterization, pavement analysis & design, performance evaluation of asphalt binder and recycling, and geotextile and porous concrete. He has about 13 years of academic and research experience. He has published reputed international journal and conference papers and successfully completed TEQIP-III funded research and development projects. Also, he has reviewed many technical papers for various international and national journals.

Proceeding of International Conference on Emerging Trends in Engineering
and Technology-Industry 4.0(ETETI-2023), IGIT Sarang, Odisha, India

Industry 4.0 with Modern Technology

First edition

Edited By

Srinivas Sethi

Mufti Mahmud

Sujit Kumar Pradhan

Rabinarayan Sethi

CRC Press
Taylor & Francis Group
Boca Raton London New York

CRC Press is an imprint of the
Taylor & Francis Group, an **informa** business

First edition published 2024
by CRC Press
4 Park Square, Milton Park, Abingdon, Oxon, OX14 4RN

and by CRC Press
2385 NW Executive Center Drive, Suite 320, Boca Raton FL 33431

CRC Press is an imprint of Informa UK Limited

British Library Cataloguing-in-Publication Data
A catalogue record for this book is available from the British Library

ISBN: 9781032586472 (pbk)
ISBN: 9781003450924 (ebk)

DOI: 10.1201/9781003450924

Typeset in Sabon LT Std
by HBK Digital

Contents

Preface   xxi

Chapter 1   Design of integrated visible light communication-free space
optics under diverse climate conditions   1
Meet Kumari

Chapter 2   Exploration of strength properties of stabilized black cotton soil
using bamboo leaf as biomass   5
*Bernard Oruabena, Okoh Elechi, Okiridu M. Ugochukwu
and Blessing C. Odoh*

Chapter 3   Response of asymmetric building under different soil-foundation
conditions incorporating soil structure interaction   10
Arnab Chatterjee and Heleena Sengupta

Chapter 4   Laboratory investigation of aggregates in Gardez city for road
construction suitability   16
Naqibullah Zazaï, Karanjeet Kaur and Hazratullah Paktin

Chapter 5   Optical absorption and FTIR analysis of ZnS/PVK
nano-composites   20
Durgesh Nandini Nagwanshi

Chapter 6   Load sharing between group of two floating granular piled
rafts - an analytical study   25
Ashish Solanki and Jitendra Kumar Sharma

Chapter 7   An integrated approach to explore the suitability of surface
water for drinking purposes in Mahanadi basin, Odisha   31
Abhijeet Das, Krishna Pada Bauri and Bhagirathi Tripathy

Chapter 8   A comparative performance study on fly ash based geopolymer
by varying activator and class of base material   38
Darsana Sahoo and Suman Pandey

Chapter 9   A smart scheduling: an integrated framework under cloud
computing environment   44
*Biswajit Nayak, Bhubaneswari Bisoyi, Biswajit Das and
Prasant Kumar Pattnaik*

Chapter 10   Studying wind effect on high-rise building with different
cross-section   51
*Dhruv Chaudhary, Dhawal Tayal, Devesh Kasana, Ritu Raj,
S. Anbukumar and Rahul Kumar Meena*

Chapter 11   Modelling of preforming design criteria: closed-die forging of
axi-symmetrical cylindrical aluminum preforms   55
Saranjit Singh

Chapter 12   The influence of carbon content on low-carbon steel corrosion
              rates in water cooling synthesis   62
              Ahmad Royani and Sundjono

Chapter 13   Effect of cross anisotropy of pavement materials on flexible
              pavement response subjected to various subgrade modulus   68
              Dipti Ranjan Biswal and Brundaban Beriha

Chapter 14   Performance evaluation of four-lobed powder lubricated
              bearing   74
              Jijo Jose and Niranjana Behera

Chapter 15   Comparative seismic assessment of existing structures using
              P-delta effect in Indian context   80
              Manish Chakrawarty, Pradeep K. Goyal and Yaman Hooda

Chapter 16   Innovative cement-free concrete utilizing 100% recycled
              aggregates for sustainable construction   86
              Gopalakrishna Banoth and Dinakar Pasla

Chapter 17   Study on effect of damage location and severity on dynamic
              behavior of multi-storey steel framed structure   92
              Mahalaxmi S. Sunagar, Jayanth K and Naveen B.O.

Chapter 18   Performance analysis of internal fin concentric pipe heat
              exchangers   97
              Kailash Mohapatra and Ashutosh Mohapatra

Chapter 19   Fault detection of rolling bearing using artificial neural
              network and differential evolution algorithm   103
              Brahma Bibhutibhusan and Sethi Rabinarayan

Chapter 20   Application of machine learning models in Mahanadi River
              basin, Odisha for prediction of flood   108
              *Shradhanjalee Pradhan, Janhabi Meher and Bibhuti
              Bhusan Sahoo*

Chapter 21   Hydraulic conveying through the slurry pipeline:
              environment-friendly safe disposal of Indian coal fly ash   113
              Vighnesh Prasad, Anil Dubey and Snehasis Behera

Chapter 22   Study and prediction of built-up land use for Bhubaneswar
              and its corresponding effects   118
              Abhayaa Nayak and Anil Kumar Kar

Chapter 23   Streamflow prediction in a river using twin support vector
              regression and extreme learning machine   123
              Jagdish Mallick, Safalya Mohanty and Abinash Sahoo

Chapter 24   Seismic stability analysis of a small scaled hill slope reinforced
              with geogrid   129
              Rasmiranjan Samal and Smrutirekha Sahoo

Chapter 25	Role of additive and fine particle addition in the rheology of coarse bauxite slurry *N.V.K. Reddy and J.K. Pothal*	135
Chapter 26	Experimental research on the flexural strength of strengthened composite beams *Vijay Kumar*	140
Chapter 27	Impact of marble powder, silica fume, and steel fiber on fresh and hardened attributes of cement concrete *Niharika Pattanayak, Malaya Ku. Sahu[b] and Sudhanshu Sekhar Dasc*	144
Chapter 28	Waste plastic management and utilization in road construction: a review *Niketan Rana and Dr. Siksha Swaroopa Kar*	149
Chapter 29	Seismic response analysis and evaluation of vibration control of high-rise structure *Ushnish Roy and S. Pandey*	154
Chapter 30	Heat transfer analysis during the turning process under an air-assisted water-cooling environment *Chittibabu Gaddem, Manoj Ukamanal and Sasmita Bal*	161
Chapter 31	Performance evaluation of various concrete mixes: A new mix proportioning methodology for high-volume fly ash based concrete design mixes *Asif Ahmed Choudhury, Saurav Kar and Anup Kumar Mondal*	166
Chapter 32	Investigation on compressive strength of ultra-high-performance concrete (UHPC) under autoclave curing *S. Revathi, Dr. D. Brindha and R.Harshani*	171
Chapter 33	Study on mechanical performances of polyester composites filled with pistachio shell particle *Deepak Kumar Mohapatra, Chitta Ranjan Deo and Punyapriya Mishra*	176
Chapter 34	Mechanical and tribological characterization of AA5083 reinforced with boron carbide and titanium diboride *Debasish Rout, Dikshyanta Sahoo, Chitta Ranjan Deo and Rabindra Behera*	182
Chapter 35	Buckling analysis of FG plates using FEA *Abhijit Mohanty, Sarada Prasad Parida and Rati Ranjan Dash*	188
Chapter 36	Skin conductance based mental stress analysis using cross validation *Padmini Sethi, Ramesh K. Sahoo, Ashima Rout and M. Mufti*	194
Chapter 37	An investigation study on water quality of few lakes of Bengaluru *Amita Somya*	200

Chapter 38   Design and implementation of a cost-effective smart dustbin
              for segregating wet and dry waste   205
              Dwarikanath Choudhury, Ranjit Kumar Behera and
              Dayal Kumar Behera

Chapter 39   A study on abnormal medical image classification   210
              Khirod Kumar Ghadai, Subrat Kumar Nayak, Biswa Ranjan
              Senapati and Binaya Kumar Patra

Chapter 40   Studies on direct drop mode of atomization in a spinning cup   216
              Brahmotri Sahoo, Kshetramohan Sahoo and
              Chandradhwaj Nayak

Chapter 41   A comparative analysis on energy efficient narrow band-IoT
              technology   222
              Sunita Dhalbisoi, Ashima Rout, Srinivas Sethi and
              Ramesh Kumar Sahoo

Chapter 42   Investigations of the physical and antibacterial properties of
              vitamin e-enriched lemongrass oil nano emulsions   227
              Veda Prakash, Anil Kumar Murmu and Dr Lipika Parida

Chapter 43   Artificial intelligence based double fed induction machine
              controlled pumped storage turbine – A Critical Review   232
              T.K. Swain, Meera Murali and Raghu Chandra Garimella

Chapter 44   Understanding the link between above ground biomass and
              vegetative indices: an artificial neural network and regression
              modelling approach   237
              Kumari Anandita, Anand Kumar Sinha and C Jeganathan

Chapter 45   Design and fabrication of BLDC motor operated portable
              grass cutter   243
              Kamala Kant Sahoo, Baisalini Sethi, Shambhu Kumar
              Mahato, Sibasis Harihar Sahu, Saroj Kumar Padhi and
              Sovan Prasad Behera

Chapter 46   Rheological and mechanical properties of different types of
              lightweight aggregate concrete   248
              Snigdhajit Mukherjee, Rajesh Kumar, A Sofi and
              Monalisa Behera

Chapter 47   On game theory integrated particle swarm optimization for
              crack detection in cantilever beams   254
              Prabir Kumar Jena, Rabinarayan Sethi and Manoj Kumar Muni

Chapter 48   Evaluating chloride resistance of high strength recycled
              aggregate concrete integrating metakaolin   260
              Uma Shankar Biswal and DinakarPasla

Chapter 49   Hydration kinetics and stability study of pure phases of
              cement clinker with the addition of SCMs and dopants   266
              Dipendra Kumar Das, Rajesh Kumar and A Sofi

Chapter 50   Performance evaluation of quality of service parameters of
              proposed hybrid TLPD-ALB-RASA scheduling algorithm in
              cloud computing environment   270
              Vijay Mohan Shrimal, Yogesh Chandra Bhatt and
              Yadvendra Singh Shishodia

Chapter 51   Design and development of concrete paver block using waste
              tyre rubber   275
              Kali P Sethy, M. Srinivasula Reddy, Sundaram Jena, Debasmita
              Malla and Arpan Pradhan

Chapter 52   Effect of permeability property of micro concrete on durability
              of concrete repair   280
              Dipti Ranjan Nayak, Rashmi R. Pattnaik and
              Bikash Chandra Panda

Chapter 53   Effect of wood dust filler on mechanical properties of polyester
              composite   285
              Chandrakanta Mishra, Harish Chandra Baskey, Chitta Ranjan
              Deo, Deepak Kumar Mohapatra and Punyapriya Mishra

Chapter 54   Assessment of conservation potential of sub-watersheds in the
              Baitarani River basin in Odisha by using morphometric analysis   290
              Swagatika Sahoo and Janhabi Meher

Chapter 55   Next generation communication network using NB-IoT   296
              Lakhmi Priya Das, Srinivas Sethi, Ramesh Kumar Sahoo,
              Sunita Dalbisoi and Ashima Rout

Chapter 56   Effect of agro waste ash on the mechanical properties of
              recycled coarse aggregate concrete   301
              M. Srinivasula Reddy, Kaliprasanna Sethy, G. Nagesh Kumar
              and K.V.S Gopala Krishna Sastry

Chapter 57   Evaluation of expansive soil stabilized with lime and silica fume   307
              Sushismita Tripathy and Pragya Paramita

Chapter 58   Latex Modified Steel Fiber Reinforced Concrete   312
              Tapas Ranjan Baral, Sujit Kumar Pradhan and
              Debakinandan Naik

Chapter 59   Study of indirect tensile strength and cyclic indirect tensile
              modulus of fiber reinforced cement stabilized fly ash, stone
              dust and aggregate mixture for pavement application   316
              Sanjeeb Mohanty, Dipti Ranjan Biswal, Brundaban Beriha,
              Ramchandra Pradhan and Benu Gopal Mohapatra

List of Figures

Figure 1.1   (a) Proposed design of hybrid FSO-VLC system, (b) FSO system design and (c) VLC system design   2

Figure 1.2   BER analysis of integrated FSO-VLC model for variable (a) VLC length for various LEDs, (b) FSO range rate for different VLC ranges, (c) data rate over various lengths of both VLC and FSO and (d) FSO range under different climate conditions   3

Figure 2.1   Comparing soaked and un-soaked CBR   8

Figure 2.2   OMC and MDD variations for expansive black cotton soil   8

Figure 2.3   Impacts of the BLA + PLC content on the marine clay soil's UCS   8

Figure 3.1   Plan view of bare frame structure   11

Figure 3.2   3D View of frame simulating pile-soil interaction   11

Figure 3.3(a) Base shear along X-direction   12

Figure 3.3(b) Base shear along Y-direction   12

Figure 4.1   (a) Los Angles value of Gul Wall and Milan sourced aggregates; (b) Impact value of Gul Wall and Milan sourced aggregates; (c) crushing value of Gul Wall and Milan sourced aggregates   18

Figure 5.1   Optical absorption spectra for ZnS/PVK nano-composites   21

Figure 5.2   FTIR spectra of ZnS/PVK nano-compositesand pure PVK   23

Figure 6.1   Forces and stresses operating on two floating GPR units   26

Figure 6.2   (a) Fractional load variation with reference to the total load due to K_{gp} and the outcome of s/d = 3, 5, and 7(b) The impact of different ratios L/d = 20, D/d, and s/d = 2, 3, 4, and 5 on the change in fractional load relative to the total load   29

Figure 7.1   Spatial distribution of WA WQI and SWARA WQI map   34

Figure 7.2   Variation and rating of all quality monitoring points based on OWA   35

Figure 7.3   Variation and rating of all quality monitoring points based on compromise programming   35

Figure 7.4   Variation and rating of all quality monitoring points based on combined compromise solution   35

Figure 7.5   The best-fit semi variogram models (a) WA WQI and (b) SWARA-WQI   36

Figure 8.1   Day compressive strength of fly-ash geopolymer   40

Figure 8.2   Strength gain of specimens with aging   41

Figure 8.3   Apparent porosity of fly ash based geopolymer   41

Figure 8.4   Water absorption of fly ash based geopolymer   42

Figure 9.1   Total completion time when condition is false   47

Figure 9.2   Total completion time when condition is true   47

Figure 9.3   Total completion time when condition is false   49

Figure 9.4   Total completion time when condition is true   49

Figure 10.1   Model and domain   52

Figure 10.2   Pressure contours for model H   52

Figure 10.3   Pressure contours for model A   53

Figure 10.4   Velocity streamlines   53

Figure 11.1   Forging of cylindrical aluminum preforms and deformation
               modes   56
Figure 11.2 (a) Intermediate stages of die-cavity fills   58
Figure 11.2 (b) Variation of average load vs die corner fills   59
Figure 11.2 (c) Variation of shape complexity factor vs die fills   59
Figure 12.1   The results of corrosion rate for two different carbon steel at
               different exposure times   64
Figure 12.2   The photograph results of microstructure carbon steel after
               exposure   65
Figure 12.3   The morphology of corrosion products on the metal surface 65
Figure 12.4   The pattern of X-Ray diffraction of corrosion product   66
Figure 13.1   Variation of vertical compressive strain due to change in CBR
               value under different DOA   69
Figure 13.2   Variation of rutting life due to change in CBR value under
               different DOA   70
Figure 13.3   Variation of εv due to change in DOA value under different CBR 70
Figure 13.4   Variation of rutting life due to change in DOA value under
               different CBR   71
Figure 13.5   Variation of et due to change in CBR value under different DOA 71
Figure 13.6   Variation of fatigue life due to change in CBR value under
               different DOA   71
Figure 13.7   Variation of fatigue life due to change in DOA value under
               different CBR   72
Figure 13.8   Variation of fatigue life due to change in DOA value under
               different CBR   72
Figure 14.1   Four-lobed bearing   75
Figure 14.2   Stiffness and damping coefficients   75
Figure 14.3   Comparison of predicted frictional coefficient with the previous
               results   76
Figure 14.4   Variation of static and dynamic properties for various ellipticities
               with respect to eccentricity ratio   77
Figure 15.1   Storey displacement for G+4, G+10 and G+22 model (without
               P-Delta)   83
Figure 15.2   Storey displacement for G+4, G+10 and G+22 model (with
               P-Delta)   84
Figure 16.1   Fraction of recycled aggregate   87
Figure 16.2   Workability of concrete at various NaOH concentration   89
Figure 16.3   The compressive strength of RAGPC at various concentration
               of NaOH   89
Figure 16.4   Different molarity based geopolymer concrete with 100% RA
               28 days CS   90
Figure 17.1   Types of failures in connections   93
Figure 17.2   Responses due to 50% reduction in cross sectional area   95
Figure 18.1   Illustrative sketch of annulus tube with internal fins   98
Figure 18.2   Cell number v/s tube outlet temperature   100
Figure 18.3   Relation between wall temperature and fin shape   100
Figure 18.4   Average Nusselt number vs fin shape   101
Figure 18.5   Wall temperature vs pipe length   101
Figure 18.6   Local Nusselt number vs fin number   101

Figure 18.7	Friction factor vs fin number	102
Figure 19.1	Showing vibration signal for a healthy bearing	106
Figure 19.2	Showing vibration signal for a faulty bearing	106
Figure 19.3	Showing error curve between simulation result and DE algorithm result	106
Figure 21.1	(a) Particle size distribution (b) SEM image 5000x (c) Chemical composition	114
Figure 21.2	Flow curves at varying Cw = 65 - 70 wt.% (a) η versus $\dot{\gamma}$ (b) τ versus $\dot{\gamma}$ curves	115
Figure 21.3	h_L calculation (a) varying C_w = 65 – 70 wt.% (b) varying D = 0.15 – 0.5 m	116
Figure 21.4	(a) C_w versus mass loss of MS specimens; (b) SEM images of MS specimens at 5000x (b) before erosion test when flow variables are absent (c) after erosion test when MS specimen is subjected to 70 wt.% CFA slurry flowing at v_m = 3.5 m/s	116
Figure 22.1	State water plan showing per capita availability (Department of Water Resources, Govt. of Odisha)	119
Figure 22.2	Model verification for Bhubaneswar	120
Figure 22.3	Year wise NDBI for Bhubaneswar	121
Figure 23.1	Location of specified soradagauge station	125
Figure 23.2	Scatter plot of applied ELM and TSVR models	127
Figure 23.3	Actual vs. predicted streamflow based on ELM and TSVR models	127
Figure 24.1	Geometry of hill slope along with meshing	130
Figure 24.2	Time-varying horizontal displacement along the slope's crest	131
Figure 24.3	Time-dependent horizontal displacement along slope face	131
Figure 24.4	Time-varying horizontal displacement along the slope's toe	132
Figure 24.5	Time-dependent horizontal displacement along slope body	132
Figure 24.6	Time-dependent vertical displacement along slope crest and face	133
Figure 24.7	Time-dependent vertical displacement along slope toe and body	133
Figure 24.8	Variation of horizontal acceleration with time	133
Figure 25.1	Cumulative PSD of bauxite samples of size fraction < 45 µm, (-150+45) µm, and (-300+150) µm	136
Figure 25.2	(a) Rheogram (τ vs $\dot{\gamma}$); (b) Apparent viscosity (η) vs shear rate ($\dot{\gamma}$) of bimodal bauxite slurry of <45 µm and +150-300 µm fraction at C_w = 50%	137
Figure 25.3	(a) Rheogram (τ Vs $\dot{\gamma}$); (b) Apparent viscosity (η) Vs shear rate ($\dot{\gamma}$) of bimodal bauxite slurry of <45 µm and (-300+150) µm fraction in 20:30 at total 50 wt%, with varying SHMP dosage	137
Figure 26.1	Comparison of different strengthened beams	142
Figure 26.2	Compression of load carrying capacity of different beams	142
Figure 27.1	Grading curve	146
Figure 28.1	Plastics use in Mt (OECD, 2022)	150
Figure 29.1	G+10 building (plan view) and G+10 building (3D view)	157
Figure 29.2	Story height vs. displacement height	158
Figure 29.3	Story drift	159
Figure 29.4	Response spectrum function	159
Figure 30.1	Spray assisted cooling during turning AISI 316 SS	163
Figure 30.2	Average heat transfer coefficient in the air-water SIC environment	164

Figure 30.3   Cutting temperature contour plot under air-water SIC for varying machining parameters   164

Figure 31.1   New mix proportioning technique adopted in this current research work   167

Figure 31.2   Various hardened concrete samples with varying fly ash content   167

Figure 31.3   Workability properties: (a) Variation of slump values at various w/b ratios. (b) Variation of compacting factor (C.F.) values for various concrete mixes at varied w/b ratios   169

Figure 31.4   Variation of flow spread values for various concrete mixes at varied w/b ratios   169

Figure 32.1   SEM image of Silica fume, quartz powder, and quartz sand   172

Figure 32.2   RPC mixing techniques, autoclave curing, and testing of the specimen   173

Figure 33.1   Tensile strength of PSP/polyester composites   178

Figure 33.2   Flexural strength of PSP/polyester composites   178

Figure 33.3   Impact strength of PSP/polyester composites   179

Figure 33.4   Micro-hardness of PSP/polyester composites   179

Figure 33.5   SEM image of fracture composites (PSP-5) under (a) tensile loading, (b) flexural loading   180

Figure 34.1   (a) Tensile strength, (b) Impact strength (c) Hardness value   185

Figure 34.2   Wear rate Vs Sliding speed at Normal load (a) 20 N (b) 40 N & (c) 60 N   185

Figure 34.3   Wear rate vs normal load at sliding speed (a) 1.5 m/s (b) 3.0 m/s, (c) 4.5 m/s and (d) Co-efficient of friction vs normal load   186

Figure 34.4   Wear rate vs normal load at sliding speed (a) 1.5 m/s (b) 3.0 m/s, (c) 4.5 m/s and (d) Co-efficient of friction vs normal load   186

Figure 35.1   Schematic diagram of assumed FG-plate   189

Figure 35.2   Schematic diagram of 2D nine noded element   190

Figure 35.3   FEA models of FGM; (a) Meshed model (b) buckling mode shape   191

Figure 35.4   Variation of buckling stress ;(a) with grading index, (b) a/h ratio   192

Figure 35.5   Comparison of error percentage; (a) mesh size (b) Aspect ratio   192

Figure 36.1   Data acquisition and observing node setup [9]   196

Figure 36.2   Comparative analysis of actual and predicted no. of records   197

Figure 36.3   Error report of various machine learning algorithms   197

Figure 36.4   Classification report of various machine learning algorithms   197

Figure 36.5   Accuracy report of various machine learning algorithms   198

Figure 38.1   Block diagram of the model   206

Figure 39.1   Artificial neural network model diagram   212

Figure 39.2   Model diagram of the image classification process   213

Figure 40.1   Free body diagram showing action of competing forces during drop detachment from a spinning cup   217

Figure 40.2   A schematic of the spinning cup atomizer used in this work   218

Figure 40.3   A schematic of the experimental setup (Liquid supply and Imaging). The spinning cup receives inlet flow of 100 mL/min   219

Figure 40.4   A representative histogram showing distribution of drop size (Water) produced from the cup spinning at 104 rad/s. The cup was continuously fed at 100 mL/min   219

Figure 40.5   Effect of cup speed on Sauter mean diameter (d32)   220

Figure 40.6   Plot showing Weber number-dimensionless drop size relationship   220
Figure 41.1   Architecture of NB-IoT   223
Figure 41.2   Operation modes of NB-IoT (Kanj et al., 2020)   223
Figure 41.3   NB-IoT Uplink resource grid with 15KHz subcarrier spacing   225
Figure 41.4   NB-IoT Downlink resource grid in operation mode of guard
              band   225
Figure 41.5   Downlink resource grid in NB-IoT operation mode of
              non-anchor In-band   225
Figure 41.6   Power allocation in NB-IoT   226
Figure 42.1   Effect of SOR on vitamin E-lemongrass nanoemulsions'
              a) droplet size and b) PDI   229
Figure 42.2   a) Antibacterial activity of nanoemulsions against *S. aureus*
              (left) and *E. coli* (right); a-e represent SOR values from 0.25 to
              1.25, b) Effect of SOR of nanoemulsions on inhibition zone
              diameter   230
Figure 43.1   Typical layout of a hydropower plant with two surge tanks   233
Figure 43.2   Functional block diagram of the pumped storage turbine   233
Figure 43.3   Typical methodology of AI based control algorithm   235
Figure 44.1   Study area (Birla Institute of Technology, Mesra. Ranchi)   238
Figure 44.2   AGB results of different data sources (nntool). (a) Case 1; and
              (b) Case 2   240
Figure 44.3   Best regression model output for prediction of AGB for Case 2
              a) linear regression model b) support vector machine regression
              model   240
Figure 44.4   AGB map a) observed values b) predicted values Case 1
              c) predicted values Case 2   241
Figure 45.1   Design and assembled components of grass cutter   245
Figure 45.2   Fabricated components   245
Figure 45.3   Block diagram   246
Figure 46.1   Relation between slump flow and w/b ratio   250
Figure 46.2   Relation between J-ring and V-funnel for different LWA for
              different w/b ratio   250
Figure 46.3   Relation between torque and speed for mix M 2   251
Figure 46.4   Relation between CS and w/b ratio for mix M 2 different
              replacement percentage of LWA   251
Figure 47.1   Geometry of cantilever beam   255
Figure 47.2   Comparison of normalizednatural frequencies for three different
              modes for a definite non-dimensional crack depth   257
Figure 47.3   Comparison of convergence between GTPSO and PSO for crack
              location 600 and crack depth 2.4 mm   258
Figure 48.1   Grading of aggregates   261
Figure 48.2   Test setup for RCPT test   262
Figure 48.3   Influence of MK on total charge passed (TCP)   263
Figure 48.4   3D response surface plot of TCP for different levels of w/b and
              metakaolin (curing of 60 days)   263
Figure 50.1   Flowchart of proposed hybrid TLPD-ALB-RASA scheduling
              algorithm   272
Figure 50.2   Graphical representation of total processing time in different
              scenarios   273

Figure 51.1	Slump value with waste tyre rubber percentage	277
Figure 51.2	Comp. strength value with replaced rubber percentage	278
Figure 51.3	Flexural strength value with replaced rubber percentage	278
Figure 52.1	Typical arrangement for sorptivity test	282
Figure 52.2	Initial and secondary absorption of specimen M1 to M6	283
Figure 53.1	(a) Tensile strength (b) Flexural strength (c) Impact strength (d) Hardness strength	287
Figure 53.2	Analysis of TGA curves WD filled polyester composite	288
Figure 54.1	Delineation of sub-watersheds in Baitarani Basin	292
Figure 55.1	Layered structure of the NB-IoT architecture	297
Figure 55.2	Deployment of NB-IoT	298
Figure 55.3	frequency spectrum using gaussian filters method	299
Figure 56.1	7 day and 28 day compressive strength of various mixes	304
Figure 56.2	7 day and 28 day split tensile strength of various concrete mixes	304
Figure 56.3	7 day and 28 day flexural strength of the developed mixes	305
Figure 57.1	Atterberg's limit with different percentages of silica fume and life	309
Figure 57.2	Variation of shrinkage limit with different percentages of lime and SF	309
Figure 57.3	Compaction curve with different percentages of lime and SF	310
Figure 57.4	Bearing capacity curve with different percentages of lime and SF with unsoaked and soaked condition	310
Figure 57.5	UCS curve with different percentages of lime and SF	310
Figure 57.6	Variation of swelling pressure and permeability coefficient with additives	311
Figure 58.1	Volume of 1 m3 concrete	314
Figure 59.1	Grain size distribution of FA, 70FA-30SA and 60FA-40SA	317
Figure 59.2	Cyclic IDT modulus jig in servo hydraulic universal testing machine	318
Figure 59.3	(a) Effect of fiber content on UCS of 70FA+30SA various cement content	318
Figure 59.3	(b) Effect of fiber percentage on UCS of 60FA-40SA	318
Figure 59.4	Indirect tensile strength of 70FA-30SA and 60FA-40SA at 4% and 6% cement and 0% fiber, 0.25% fiber, 0.35% fiber and 0.5% fibers	319
Figure 59.5	Variation of cyclic IDT modulus of 70FA-30SA and 4% cement at various fiber percentage and stress ratio	319

List of Tables

Table 1.1:   Performance of the proposed work   4
Table 2.1:   Physicochemical assets of soil and those of treatment agents   6
Table 3.1:   Classification of soil parameters   12
Table 3.2:   % Increase in base shear values along x-direction   13
Table 3.3:   % Increase in base shear values along Y-direction   13
Table 3.4:   % cecrease in storey displacement at storey level 11 along x-direction   13
Table 3.5:   % Decrease in storey displacement at storey level 11 along Y-direction   14
Table 5.1:   Particle size evaluation from absorption edge using both models   22
Table 5.2:   Peak assignment of pure PVK and varing concentration of ZnS with PVK   23
Table 7.1:   Description of the surface water quality parameters' best-fitting variogram model   36
Table 8.1:   Composition of geopolymer paste with specimen ID   39
Table 8.2:   Results of workability test of the geopolymer paste   39
Table 8.3:   Class C fly-ash geopolymer test results on workability vs time activated in presence of sludge   40
Table 9.1:   Cloud configuration   46
Table 9.2:   Datacenter configuration   46
Table 9.3:   Resource specification   46
Table 9.4:   Task specifications   46
Table 9.5:   Tasks execution time on each resource   47
Table 9.6:   Datacenter configuration   48
Table 9.7:   Resource specification   48
Table 9.8:   Task specifications   48
Table 9.9:   Execution of tasks on each resource   48
Table 11.1:   Details of preform, forged component, and preforming stages   60
Table 12.1:   The chemical composition of the steel specimens in this study   63
Table 12.2:   The essential elements of solution media in this study   63
Table 12.3:   The water quality of the solution test   65
Table 13.1:   Pavement materials properties used for analysis   69
Table 15.1:   Summary of results   82
Table 16.1:   XRF analysis of fly ash   87
Table 16.2:   Chemical composition of NaOH and Na_2SiO_3 88
Table 16.3:   Various NaOH concentration-based mix design proportion   88
Table 17.1:   Variation of natural frequency (Hz): 50% reduction in cross-sectional area   95
Table 17.2:   Variation of lateral displacement (mm): 50% reduction in cross-sectional area   95
Table 20.1:   Prediction results on training and test datasets   111
Table 21.1:   Flow coefficients during fitting of power-law model in Figure 21.2b   115
Table 23.1:   Performance assessment results   126
Table 24.1:   Material properties   130

Table 25.1: List of model parameters and corresponding flow type for varying types of slurry   138
Table 26.1: Number of samples   141
Table 26.2: Properties of GFP obtained from tensile test   142
Table 26.3: Woven wire mesh properties   142
Table 27.1: Mix proportions for different designations   146
Table 27.2: Comparison of properties of different concrete mixes   147
Table 28.1: Comparison of projections with the existing literature   151
Table 28.2: Asphalt modification in previous studies   152
Table 29.1: Elements of the building's geometry   155
Table 29.2: Damper's properties   156
Table 29.3: Displacement of G+10 building   156
Table 31.1: Mix proportioning stipulations: a new methodology adopted HVFAC   168
Table 32.1: Mix proportions of UHPC   173
Table 32.2: Compressive strength of RPC   174
Table 33.1: Combination of PSP/polyester composites   177
Table 34.1: Composition of fabricated HAMMC   183
Table 37.1: Physico-chemical parameters of collected water samples from four lakes   202
Table 38.1: Data captured by different sensors   208
Table 39.1: Performance metrics of different classifiers   214
Table 40.1: Properties of the atomized liquids measured at 20°C   219
Table 42.1: Zeta potential of formulated nanoemulsions at different SOR   230
Table 44.1: Details of vegetative indices calculated   239
Table 44.2: Significant linear correlations between vegetation indices and AGB   240
Table 46.1: Material component per cubic meter   249
Table 47.1: Setting of control parameters for PSO and GTPSO algorithms   257
Table 48.1: Properties of Materials considered for mix design   261
Table 48.2: Mix design details   262
Table 48.3: ANOVA on total charge passed of MK-based RACs   264
Table 49.1: Mix constituents for the synthesis of C_3S 267
Table 49.2: Mix constituents for the synthesis of C_2S 268
Table 51.1: Specification details of aggregate (fine and coarse)   276
Table 51.2: Mix details for 1 m^3 of rubber concrete   276
Table 52.1: Grading of micro-concretes from M1 to M6   281
Table 52.2: Compressive strength of the micro-concretes specimens from M1 to M6   282
Table 53.1: Combination of WD-polyester composites   286
Table 54.1: Importance of parameters on soil erosion susceptibility   293
Table 54.2: Compound factor based prioritization of sub-watersheds   294
Table 56.1: Binder materials oxide composition   303
Table 56.2: Mix proportioning of different mixes   303
Table 57.1: Geotechnical properties of expansive soil   308
Table 58.1: Materials quantity used in experimental program   313
Table 58.2: Mix proportion (Mix design data was decided as per Shetty 2006)   313
Table 58.3: Test results   314

List of Contributors

Patron-in-Chief

Satyabrata Mohanta (Director) IGIT Sarang

Patron(s)

B. D. Sahoo IGIT Sarang
T. K. Nath IGIT Sarang
S. Mishra IGIT Sarang

General Chair

Radhakant Padhi HAL Chair Professor, Department of Aerospace Engineering & Associate Faculty (DCC Chair), Centre for Cyber-Physical Systems, Indian Institute of Science, Bangalore, India

Programme Chairs

Srinivas Sethi IGIT Sarang
Mufti Mahmud Nottingham Trent University UK

Convenors

Sujit Kumar Pradhan IGIT Sarang
Rabinarayan Sethi IGIT Sarang

Advisory Board

Sukanta Kumar Senapati IGIT Sarang
Pratap Kumar Pani IGIT Sarang
B. P. Panigrahi IGIT Sarang
Pitambar Das IGIT Sarang
Urmila Bhanja IGIT Sarang
Ashima Rout IGIT Sarang
B. B. Choudhury IGIT Sarang
Goutam Kumar Pothal IGIT Sarang
P. K. Mallik IGIT Sarang
D. Das IGIT Sarang
Sunil Kumar Tripathy IGIT Sarang
Binod Bihari Panda IGIT Sarang

Website Chair

Gaurab Kumar Ghosh IGIT Sarang
Sanjaya Ku. Patra IGIT Sarang
Rabinarayan Sethi IGIT Sarang

Publication Chair

Rabinarayan Sethi IGIT Sarang
Sujit Kumar Pradhan IGIT Sarang

Sponsorship Chair

P. R. Dhal IGIT Sarang

Finance Chair

Rabinarayan Sethi IGIT Sarang
Sujit Kumar Pradhan IGIT Sarang

Publicity Chair

P. R. Dhal IGIT Sarang
Sanjaya Ku. Patra IGIT Sarang
Gaurab Kumar Ghosh IGIT Sarang

Registration Chair

Ansuman Padhi IGIT Sarang
Trushna Jena IGIT Sarang

Logistic Chair

Manoj Kumar Muni IGIT Sarang
Sushant Kumar Sahoo IGIT Sarang
Ramesh Kumar Sahoo IGIT Sarang
Binaya Kumar Patra IGIT Sarang

Organising Committee

Debjyoti Mishra IGIT Sarang
Chitta Ranjan Sahoo IGIT Sarang
Rabindra Behera IGIT Sarang
D K Behera IGIT Sarang
Supriya Sahu IGIT Sarang
Sudhakar Majhi IGIT Sarang
July Randhari IGIT Sarang
Babita Singh IGIT Sarang
Jayashree Nayak IGIT Sarang
Ritesh Kumar Patel IGIT Sarang
Soumya Ranjan Pradhan IGIT Sarang
AnsumanPadhi IGIT Sarang
Subrat Kumar Nayak IGIT Sarang
Ritwik Patnayak IGIT Sarang
Anwesha Rath IGIT Sarang
Sandeep Kumar Sahoo IGIT Sarang
Brijesh Kumar IGIT Sarang
Manoj Kumar Choudhury IGIT Sarang
Maheswar Behera IGIT Sarang
Jogendra Kumar Majhi IGIT Sarang
Umakanta Mahanta IGIT Sarang
Biswanath sethi IGIT Sarang
Ashok Kumar Pradhan IGIT Sarang
Himansu Sekhar Dash IGIT Sarang
Deepak Suna IGIT Sarang
Dillip Kumar Swain IGIT Sarang
Niroj Kumar Pani IGIT Sarang

Dipesh Kumar Nayak IGIT Sarang
Padma Lochan Nayak IGIT Sarang
Sabyasachi Aich IGIT Sarang
Arbind Kar IGIT Sarang
Bibhu Prasad Panda IGIT Sarang
Ipsita Dhar IGIT Sarang
Sudhanshu Bhushan Panda IGIT Sarang
Surjyakant Panda IGIT Sarang
Rakesh Mohanty IGIT Sarang
Kali Charan Rath GIET University

International and National Technical Committee

Dayal R. Parhi NIT Rourkela
Sudhir Kumar Kashyap CSIR-CIMFR, Dhanbad
Mihir Kumar Das IIT Bhubaneswar
Debendra K. Das University of Alaska Fairbanks
K. G. Prasanth, Norway University of Science and Technology
Debamalya Banarjee Jadavpur University
R. N. Mahapatra NIT Meghalaya
Subhasis Bhaumik IIEST Shibpur
S. K. Sarangi NIT Patna
S. K. Kulkarni AGU Simla
C. N. Bhende IIT Bhubaneswar
Tapas Kumar Roy IIEST Shibpur
Subrat Kumar Roy IIEST Shibpur
A. Abraham, Machine Intelligence Research Labs, USA
Ahamed Zobaa Brunel University UK
N. P. Padhy IIT Roorkee
S. K. Panda NUS Singapore
Kenji Suzuki, University of Chicago
P. Mohapatra, University of California
Sukumar Mishra IIT Delhi
B. K. Panigrahi IIT Delhi
Abhisek Rajan NIT Sikim
Prakash Kumar Ray CET Bhubaneswar
Joymala Moirangthen NUS Singapore
Anus K. Chandrappa IIT Bhubaneswar
C. Chandra Sekhar IIT Madras
Ajay Kalamdhad IIT Guwahati
Partha Pratim Dey IIT Bhubaneswar
Anil Kumar Mishra IIT Guwahati
Tapan Kumar Rout TATA Steel Jajpur
A.K. Chaubey CSIR-IMMT Bhubaneswar
M. Loganathan NMIT Bengaluru
Ratnakar Das NIT Rourkela
S.K. Biswal CSIR-IMMT Bhubaneswar
Joykrishna Dash VSSUT Burla
R.B.V. Subramanyan NIT Warangal

Preface

This Taylor & Francis, CRC Press volume contains the papers presented at International Conference on Emerging Trends in Engineering and Technology-Industry 4.0 (ETETI-2023) being organized by the prestigious Indira Gandhi Institute of Technology, Sarang (An Autonomous institute of Govt of Odisha), India, during 6th and 7th May 2023. The theme of the conference included various field of Civil, Electrical, Mechanical, Metallurgy, Chemical, Computer science, and Electronics.

The conference draws some excellent technical keynote talks and papers. Seven keynote talks by Prof. Radhakanta Padhi, IISC Bangalore, Prof. B Krishna Prapoorna, IIT Tirupati, Prof. Umesh Chandra Sahoo, IIT BBSR, Dr Siksha Swaroopa Kar, CSIR-CRRI, Prof. Niranjan Sahoo, IIT Guwahati, and Prof. Deepak Tosh, University of Texas has been presented.

For the Conference, received of about 131 full paper and accepted only 59 papers. The contributing authors are from different parts of the globe. All the papers are reviewed by at least two independent reviewers and in some cases by as many as three reviewers. After review, papers were scrutinized by Editorial Board Member. All the papers are also checked for plagiarism and similarity index. We had to do this unpleasant task, keeping the Taylor & Francis guidelines and approval conditions in view. We take this opportunity to thank all the authors for their excellent work and contributions and also the reviewers, who have done an excellent job.

On behalf of the technical committee, we are indebted to Prof. Radhakanta Padhi, General Chair of the Conference, for his timely and valuable advice. We cannot imagine the conference without his active support at all the crossroads of decision-making process. The management of the host institute, particularly the Director Prof. Satyabrata Mohanta, HOD Mechanical, Prof. B. D. Sahoo, HOD, Civil, Prof. T. K. Nath, HOD CSEA Prof. Sasmita Mishra, and Coordinators Prof. S. K. Pradhan and Prof. R. N. Sethi, and have extended all possible support for the smooth conduct of the Conference. Our sincere thanks to all of them.

We would also like to place on record our thanks to all the keynote speakers, session chairs, authors, technical program committee members, website chair, publication chair, finance chair, publicity chair, registration chair, logistic chair, members of organising committee, print and digital media and above all to the volunteers. We are also thankful to Taylor & Francis publication house for agreeing to publish the accepted and presented papers.

Best Wishes,
Prof. Srinivas Sethi
Prof. Mufti Mahmud

1 Design of integrated visible light communication-free space optics under diverse climate conditions

Meet Kumari[a]

Department of ECE, Chandigarh University, Mohali, Punjab, India

Abstract

Integrated free space optics (FSO) with visible light communication (VLC) system under different climate conditions has been proposed. The results show that the proposed design offer faithful FSO and VLC range of 230 km and 650 m using white light emitting diode (LED) under clear air. Further, the system provides a faithful transmission rate of 80 Gbps under diverse climate conditions. It can be used for various indoor and outdoor applications. The work novelty is to enhance the proposed work for wireless access under diverse environment conditions.

Keywords: Free space optics, light emitting diode, visible light communication, wireless comunication

Introduction

Visible light communication (VLC) has illustrated an important interest or researchers due to its combined feature of communication and illumination. It employs light emitting diode (LED) based communication for transferring data and provides cost-effectiveness, energy-efficiency, privacy and security. Besides this, the fast spread of LED as wireless access various other technologies have been integrated with VLC for fast wireless communication. One of them is free space optics (FSO) which can be integrated with VLC to provides high bandwidth long reach space transmission (Kumari et al., 2019a; Wang et al., 2020). The FSO is an effective and efficient communication using LASER light in free space. It provides high bandwidth, security, less mass, less power requirements, and high spatial confinements. Unlike VLC communication, which is used for short or indoor communication, FSO offers the transmission over earth-to-satellite, satellite-to-ocean etc. outdoor communication successfully. The integration of hybrid VLC-FSO system offers both indoor and outdoor communication. However, FSO transmission also includes external turbulence and environment affects. Recently, various researchers have worked on both technologies. From the comprehensive survey it is clear that most of the research is done for individual FSO and VLC technology but not on integrated FSO-VLC system. Thus here paper, a hybrid FSO-VLC under diverse climate conditions are designed and analyzed (Kumari et al., 2021). Section 2 depicts the system model followed by results and analysis in Section 3. Then it is progressed by Section 4, conclusion.

Proposed work

Figure 1.1 indicates the design of proposed hybrid FSO-VLC system employing OptiSystem software (Kumari et al., 2020; Smirani et al., 2022). Figure 1.1(a) indicates

[a]meetkumari08@yahoo.in

DOI: 10.1201/9781003450924-1

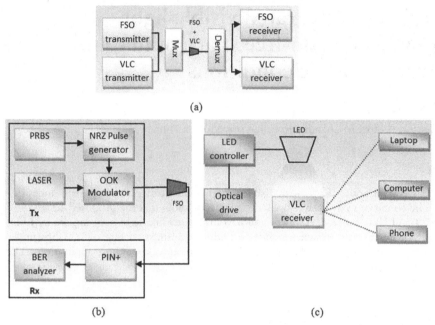

Figure 1.1 (a) Proposed design of hybrid FSO-VLC system, (b) FSO system design and (c) VLC system design
Source: (Chang, S H (2015); Kumari, M (2022))

the two transmitters i.e., FSO and VLC with LASER and LED input source respectively. Multiplexer and de-multiplexer are used to send the interrelated FSO-VLC link signals from transmitter to two different receivers. Figure 1.1(b) indicates the proposed design of FSO system used in the proposed work. The transmitter consists of a pseudo random bit sequence (PRBS) generator and non-return to zero (NRZ) generator to generate the random sequences. On-off keying modulation is used to modulate input LASER signal (1550 nm) with NRZ pulse patterns at 10 dBm input power. The FSO transmission link work under different climate conditions. Receiver section involves PIN photodetector and low pass filter to filter the original electrical signal. Bit error rate i.e., BER analyzer is utilized to presents the received signal (Kumari et al., 2021). Figure 1.1(c) shows the VLC system design used in the proposed work.

The transmitter consists of LED controller to control the input white LED signals to transfer the data from VLC receiver to different equipments like phone, computer and laptop (Chang, 2015; Kumari, 2022; Kumari et al., 2019a, 2019b).

Results and discussion

Proposed work performance is evaluated considering link noise as well as interference. Figure 1.2(a) depicts the measured BER analysis of hybrid FSO-VLC link for several VLC range with FSO range of 10 km at 10 Gbps traffic rate. It is depicted that as FSO range improves, the model performance diminishes for all LEDs. Out of all LEDs, white LED shows best performance followed by red, blue and green. Thus, the maximum archived VLC range is 650, 500, 350 and 300 m for white, red, blue and green LED respectively at 10^{-3} BER.

Figure 1.2(b) depicts the observed BER analysis of integrated FSO-VLC for variable FSO range for varied VLC ranges at 10 Gbps. It is analyzed that as FSO range

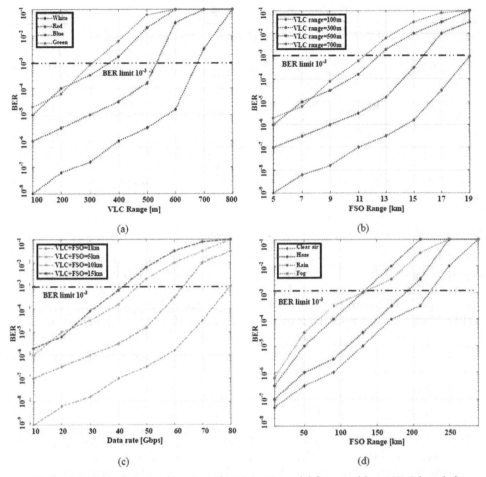

Figure 1.2 BER analysis of integrated FSO-VLC model for variable (a) VLC length for various LEDs, (b) FSO range rate for different VLC ranges, (c) data rate over various lengths of both VLC and FSO and (d) FSO range under different climate conditions

improves, the model performance diminishes. Besides this as VLC range enhances, again the model performance degrades. Thus, the maximum archived FSO range is 10, 15, 12 and 11 km with VLC range of 100, 300, 500, and 700 m respectively at 10^{-3} BER. Table 1.1 depicts the maximum obtained FSO range for different VLC distance. Figure 1.2(c) depicts the BER analysis of FSO-VLC for variable traffic rate over distinct ranges of combined VLC and FSO. This is noted that as traffic rate enhances, the performance of system diminishes and with increase in VLC + FSO length again system performance decreases. Therefore, the maximum obtained data rate over VLC+FSO ranges of 1, 5, 10 and 15 km is 80, 65, 45, and 40 Gbps respectively at 10^{-3} BER. Table 1.1 depicts the obtained data rate for variable transmission range. Figure 1.2(d) depicts the calculated BER analysis of hybrid FSO-VLC for variable FSO range and fixed VLC range of 10 m at 10 Gbps data rate under different climate conditions. It is observed as FSO range enhances; the model performance diminishes. Also, the model performs best considering clear air (0.2 dB/km) subsequence by haze (2.8 dB/km), rain (6.5 dB/km), and fog (14.7 dB/km) respectively. Hence, the maximum obtained FSO range is 230 km considering clear air, 180 km for haze, 130 km for

Table 1.1: Performance of the proposed work.

VLC range (m)	FSO range (km)	LED	VLC range (m)	VLC+FSO range (km)	Data rate (Gbps)	Climate condition	FSO range (km)
100	10	RED	300	1	80	Clear air	230
300	15	GREEN	350	5	65	Haze	180
500	12	BLUE	500	10	45	Rain	130
700	11	WHITE	650	15	40	Fog	125

rain and 125 km for fog at 10^{-3} BER. Table 1.1 indicates the summary of the proposed work.

Conclusion

An integrated free space optics (FSO) with visible light communication (VLC) system at 80 Gbps data rate under different climate conditions has been designed. It is concluded that the proposed out of al light emitting diodes (LEDs) white LED offers better performance than red, green and blue. The faithful transmission 230km FSO and 650 m VLC can be obtained successfully at high data rate under clear air. Also, FSO provides transmission ranges of 180, 130 and 125 km under haze, rain and fog conditions respectively.

References

Chang, S H (2015). A visible light communication link protection mechanism for smart factory. IEEE 29th International Conference on Advanced Information Networking and Applications Work WAINA 2015. pp. 733–737. doi: 10.1109/WAINA.2015.41.

Kumari, M (2022). Development and investigation of 5G Fiber-wireless access network based hybrid 16 × 10 Gbps 2048 split TWDM/DWDM Super PON for IoT applications. *Optical and Quantum Electronics*. 54(4), 1–20. doi: 10.1007/s11082-022-03616-9.

Kumari, M, Sharma, R, and Sheetal, A (2019a). A review on provision of cloud services over passive optical network. *International Journal of Control and Automation*.12(5), 437–442.

Kumari, M, Sharma, R, and Sheetal, A (2019b). Comparative analysis of high speed 20/20 Gbps OTDM-PON, WDM-PON and TWDM-PON for long-reach NG-PON2. *Journal of Optical Communications*. 1–14.

Kumari, M, Sharma, R, and Sheetal A (2020). Performance analysis of high speed backward compatible TWDM-PON with hybrid WDM–OCDMA PON using different OCDMA Codes. *Optical and Quantum Electronics*. 52(11). doi: 10.1007/s11082-020-02597-x.

Kumari, M, Sharma R, and Sheetal, A (2021). Performance evaluation of symmetric 8 × 10 Gbps TWDM-PON incorporating polarization division multiplexed modulation techniques under fiber-impairments. *Wireless Personal Communications*. 121(3), 1995–2010. doi: 10.1007/s11277-021-08751-2.

Smirani, L K, Kumari, M, Ahammad, S H et al. (2022). Signal Quality enhancement in multiplexed communication systems based on the simulation model of the optimum technical specifications of raman fiber optical amplifiers. *Journal of Optical Communications*. doi: 10.1515/joc-2022-0129.

Wang, Z, Han, S, and Chi, N (2020). Performance enhancement based on machine learning scheme for space multiplexing 2x2 MIMO VLC system employing joint IQ independent component analysis. *Optics Communications*. 458, 124733. doi: 10.1016/j.optcom.2019.124733.

2 Exploration of strength properties of stabilized black cotton soil using bamboo leaf as biomass

Bernard Oruabena[a], Okoh Elechi, Okiridu M. Ugochukwu[b] and Blessing C. Odoh[c]

Department of Civil Engineering, Federal Polytechnic, Ekowe, Nigeria

Abstract

This study evaluated the stabilizing properties of bamboo leaf ash (BLA) and Portland lime cement (PLC) as stabilizers for black cotton soil. This soil is typical of some regions of southern Nigeria. The soil was tested with different weight percentages (0, 5, 10, 15, 20, and 25) of the BLA mixture against the weight of the black cotton soil. Results from multiple tests included an increase in maximum dry density (MDD) from 1.65 to 1.72 KN/m^2. The California bearing ratio (CBR) test results increased from 8.26% at 0% to 13.22% at 20% admixture with the no-soak condition. Also, they rose from 2.5% at 0% admixture to 9.11% at 20% under wet conditions. After 28 days of curing, the unconfined compressive strength (UCS) value obtained by combining cement with 20% by weight BLA was 393.20 KN/m^2. The results indicated that stabilizing the expansive black cotton soil with PLC and BLA improved the soil's strength.

Keywords: Bamboo leaf ash, California bearing ratio, marine clay, stabilization, unconfined compressive strength

Introduction and background

Transportation plays a role in a country's economic development. Roads act as feeders for railroads, rivers, and airports, which supports the growth of these modes of transport. Transport affects the populace's economic health, cultural development, social, customs, and recreational habits. A nation, state, or local government area's high or poor standard of life can be accurately predicted by the presence or absence of suitable transportation infrastructure (Oguara, 2005). The primary obstacle to creating a comprehensive road network is money. If the local soil stability is insufficient to handle wheel loads, the qualities are increased by soil stabilization methods (Dumessa and Verma, 2018). Choosing local materials, including local soil, can lower the cost of road building. When coalescing materials cannot mechanically stabilize soil, stabilization can be accomplished by adding lime, cement, bituminous materials, or specific additives (Oguara, 2006). Cement stabilization is the most common technique used in road building when the subgrade has a high moisture content. Calcium hydroxide is the stabilizing material that is most frequently utilized. Although calcium oxide could be more efficient in some situations, it can corrode equipment and seriously burn or injure working people's skin.

The benefits of stabilization on the indexing and compacting features, UCS, CBR, and features swelling of black cotton soil have been discussed by several researchers. Stabilizing black cotton soil using solid wastes enhances the soil's geotechnical features. Most studies (Emesiobi and Aitsebaomo, 2002) have not thought enough

---

[a]boruabena2@gmail.com, baokiridu@gmail.com, cchidi.chinaemere@gmail.com

DOI: 10.1201/9781003450924-2

about how stabilization affects the consolidation characteristics, shear and splitting tensile strength, stiffness, and hydraulic conductivity of black cotton soil. Researchers seldom employ bamboo leaf ash as stabilizing agent for black cotton clay soil. To create flexible pavement, this study primarily uses bamboo leaf ash as a sterilizer on black cotton soil. The bamboo tree produces debris called bamboo leaf ash, and on many occasions, problems with disposal result in mounds that could be hazardous to the environment. Intensifying studies into using the waste product to sterilize black cotton soil may help to lessen the environmental risk this waste causes.

Materials and methods

Materials

Black cotton soil

At a deepness of 1.5 meters, the marine soil was gathered. After physically eliminating undesirable elements collected during excavation, the soil was sun-dried, burned, sterilized, and sieved accordingly. According to BS1377 (1990) standard, the soil index parameters, such as its moisture content, density, and ideal moisture content, were calculated.

Bamboo leaf ash

The waste bamboo leaf was taken from a pile, dried in the sun, and then heated to 600°c without the use of fuel. The ash from the burning process was ground on a clean, dry surface, sieved over 0.75 µm to create a consistent fine particle size, and then placed in an air-tight bottle for subsequent usage. Table 2.1 shows a list of the physicochemical characteristics of the materials used.

Water

Water that is fit for drinking is often acceptable for use in construction. It is assumed that the water used for the experiments in this study is free from any harmful pollutants and is of good drinkable worth.

Table 2.1: Physicochemical assets of soil and those of treatment agents.

Properties	Portland cement (42.5R)	Bamboo leaf ash	Marine soil
SG	3.02	2.61	2.44
pH	12	7.16	11
SiO_2 (%)	18.92	51.99	62.96
Fe_2O_3(%)	3.04	6.85	3.57
Al_2O_3 (%)	6.11	10.10	17.18
CaO (%)	65.02	12.51	0.16
MgO (%)	1.33	2.10	1.05
SO_3(%)	1.93	2.74	0.76
K_2O (%)	1.12	3.39	2.09
P_2O_3(%)	0.19	-	-
TiO_2 (%)	0.29	0.20	-

Source: Author

Method

Physicochemical assets of bamboo leaf ash, cement, and marine clay soil

The BLA, cement, and black cotton soil were analyzed for physicochemical properties. Testing was done to establish the essential elements of the soil, in combination with BLA and cement, to assess their average particle sizes and particle absorbance.

The UV/VIS spectrophotometer apparatus was employed for the following test (BS1377-1, 1990); the outcomes are displayed in Table 2.1.

Mix proportion

The black cotton soil was mixed with a 0%, 5%, 10%, 15%, 20%, and 25% proportion of BLA and subjected to the following tests: Grain size analysis test (wet and dry), specific gravity, liquid limit, plastic limit, proctor compaction, CBR, UCS, and indirect tensile strength tests. The Atterberg limit tests were done immediately and 24 hours after mixing. In addition, for the CBR test, the un-soaked conditioning test was done immediately after mixing and 3 days after for the soaked condition.

Methodology

Index properties

All laboratory examinations and approaches, including: (BS1377-1 1990, ASTMD6913-04 2009, ASTMD2488-09a 2015, BS5930 2015), were conducted according to the standard operating procedures outlined in the relevant I.S. codes. Additionally, a thorough analysis of the constituents utilized was conducted and stated.

CBR tests

The collected coastal soil was subjected to drying, homogenizing, sieving, and combining with BLA to create various CBR samples. Under both moist and dry conditions, the CBR readings at different BLA percentages (5, 10, 15, 20, and 25) were taken under both wet and dry conditions.

Compaction tests

The typical heavy compaction test evaluated the expansive marine clay's MDD and ideal moistness levels with changing amounts of BLA addition. For each test, the OMC and MDD were assessed.

Results and discussions

soaked/unsoaked

It can be seen from the CBR test results (Figure 2.1) for the soil that there is a drop in soaked CBR value when compared to the unsoaked CBR values due to the increased percentages of both cement and BLA. CBR values decrease from 8.26% (unsoaked) to 2.50% (soaked) at 0% replacement to 8.85% (unsoaked) to 4.11%, 9.47 to 6.86%, 11 to 6.92%, 13.22 to 9.11%, and 14.27% (soaked) to 9.76% (unsoaked) at 0%, 5%, 10%, 15%, 20%, and 25% replacements, respectively. According to the graph (Figure 2.2), which compares the CBR values of wet and unsoaked soil with respect to

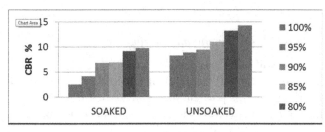

Figure 2.1 Comparing soaked and un-soaked CBR
Source: Author

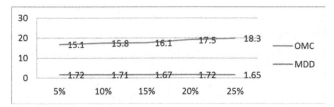

Figure 2.2 OMC and MDD variations for expansive black cotton soil
Source: Author

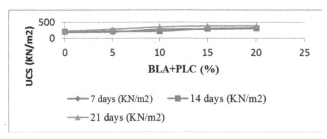

Figure 2.3 Impacts of the BLA + PLC content on the marine clay soil's UCS
Source: Author

0% replacement, the CBR value for 0% replacement (100% chikoko soil) is 8.26%. However, a consistent increase was seen at 5 to 25% compared to 0% replacement.

Compaction test

Variations in OMC and MDD were observed during the compaction test. The maximum OMC of 18.3 was achieved with a 75% soil substitution. A maximum MDD of 1.72 g/mL was reached at 80% and 95% soil replacement with BLA.

UCS

The UCS study for marine clay will enable construction engineers to determine the best improvement method for this soil type. The UCS at varying proportions of the mixture was also examined, and the results are shown in Figure 2.3. As the curing time increases, the UCS increases with increasing BLA and cement content. The rise in strength is due to adding the binder (cement and BLA), which had cementitious properties solidifying the soil matrix, thereby increasing the strength values of the marine clay soil samples.

Observation from the result indicates that a 20% PBA gave the highest UCS of 393.20KN/m² at a curing time of 28 days. This result tallies the requirements for stiff consistency for a sub-base material in civil engineering construction work (Onyelowe 2017, Gopal and Rao 2011). For an effective pozzolanic reaction, an admixture of (5%) gave a UCS value of 200.23KN/m^{2}, thus satisfying the requirements of firm reliability for usage as a subgrade and subbase course material (Onyelowe 2017, Gopal and Rao 2011). The results are comparable to that of (Otoko, et al. 2016), showing that BLA is the natural, highly reactive type of pozzolana and is very effective when combined with PLC.

Conclusion

The outcomes of the laboratory tests show that adding bamboo leaf ash (BLA) and cement as stabilizers to the black cotton soil reduced the volume variations that occur with the variation in soil's moisture content. The expansive black cotton soil treated with BLA and cement has a higher California bearing ratio (CBR). Therefore, cement and BLA as stabilizers in the flexible pavement design will impact layer thickness and fatigue performance. This study has also shown that BLA can potentially be used as stabilization materials in road construction and a potential solution to issues related to solid waste disposal, which will minimize environmental deterioration.

Acknowledgments

The authors would like to acknowledge the Departments of civil engineering, Niger Delta University Amassoma, and the Federal Polytechnic Ekowe, who made their laboratories available for the soil stabilization experiments.

References

ASTMD2488-09a. (2015). Standard Practice for Description and Identification of Soils (Visual Manual Procedure). Retrieved from: *www.astm.org*.

ASTMD6913-04. (2009). Standard Test Methods for Particle Size Distribution (Gradation of Soils using Sieve Analysis. Retrieved from: *www.astm.org*.

BS1377-1, 2. (1990). Method of test for soil for civil engineering purposes. *General Requirements and Sample Preparation*.

BS5930. 2015. Code of Practice for Site Investigation. Retrieved from: *www.bsigroup.com*.

Dumessa, G L, and Verma, R K (2018). Compacted behaviour of cement stabilized lateritic soil & its economic benefit over selected borrow material in road constr: a case study in Wolayita Sodo. *World Journal of Engineering Research and Technology*. 4(2), 1-34.

Emesiobi, F C, and Aitsebaomo, F O (2002). Testing of cement stabilized lateritic soil using compressive strength and various indirect tensile strength testing methods. *Journal of Construction and Materials Technology*. 2(1), 93-99.

Gopal, R and Rao, A S R (2011). Basic Applied Mechanics. New Delhi, India: New Age International Publishers.

Oguara, T M (2006). Highway Engineering: Pavement Design, Construction, and Maintenance. Lagos, Nigeria: Malt House Press Limited.

Oguara, T M (2005). Highway Geometric Design. 1st. Yenagoa, Bayelsa State: Treasure Communications Resource Limited.

Onyelowe, K C (2017). Nanostructured waste paper ash treated lateritic soil & its california bearing ratio optimization. *Global Journal of Tech. & Optimization*. 8(2), 1-6.

Otoko, R G, Fubara-Manuel, I, Chinweike, S I and Oyebode, J O (2016). Soft soil stabilization using palm oil fibre ash. *Journal of Multidisciplinary Engineering Science and Technology*. 3 (5), 4954-4958.

3 Response of asymmetric building under different soil-foundation conditions incorporating soil structure interaction

Arnab Chatterjee[a] and Heleena Sengupta[b]

Department of Civil Engineering, Techno India University, Kolkata, India

Abstract

The performance of any structure during an earthquake event is affected not just by the reaction of the superstructure but in addition to the reaction of the soil-foundation medium underneath the structure, which alters ground motion. Preceding significant structural failures have highlighted the significance of soil-structure interaction (SSI) consequences in seismically active regions or places with soft soil, which is the primary area of research here. Distinct models simulating three different soil-foundation medium classified upon N (standard penetration test) value using spring-dashpot mechanism. Also, the effect of shear wall under same conditions and a fixed-base condition for comparison purposes, are used to integrate and study the effect of SSI. Variation of seismic parameters such as base shear and storey displacement analyzed by the evaluation from diverse SSI models as well as comparing with the prevailing method assuming rigidity at the structure's base is studied. The outcomes clearly indicate the effect of SSI in the structure's dynamic characteristics that are significant in altering the seismic response. A major increase in the seismic parameters such as time period and other parameters in comparison with the fixed-base model is observed, which does not consider the supporting soil.

Keywords: Soil-structure interaction, shear wall, soil flexibility, base shear

Introduction

Structural failures encountered during 2001 Bhuj earthquake and in 1995 Kobe earthquake demonstrated the importance of considering soil-structure interaction (SSI) during the designing of structures. Since, seismic behavior of any structure is highly influenced not only by the response of the superstructure but also by the response of the relative rigidity of the soil-foundation medium underneath. Research by Bagheri et al. (2018) examined the parameters of a long–short, coupled raft-pile foundation system based on a knowledge of the interaction mechanics, which represented a fair correlation between structures shaking intensity rates (SIRs) and higher inter-story drifts. Further, Dey et al. (2020) focused on the variation in seismic behavior primarily elastic and inelastic on a RC frame and showed that with the inclusion of shear wall, it modifies the inelastic global and local response of the frame system depending on several parameters along-with the effect of SSI. Saha et al. (2015) studied the impact of SSI on the determination of forces used in seismic design on a one-storey structural system using soil-pile raft-structure system. The author showed that the fundamental period of structural system exhibits lengthening, which seems to be maximum in case of stiff structure system supported by slender piles under very soft soil and with the involvement of lesser number of piles such system tends to further extend the time period even more than its softer equivalents.

---

[a]arnab6138@gmail.com, [b]HS.square4@gmail.com

DOI: 10.1201/9781003450924-3

Methodology and model specifications

To conduct this research, 24 different models of asymmetric twelve-storey residential building satisfying the plan irregularity provisions affirmed in Cl.7.1 3B(Re-Entrant Corners) of IS:1893-2016(Part-1)with $A/L_2>0.15$ as shown in Figure 3.1 is taken up. Besides the effect of addition of shear wall at core compared to bare frame within the same study is also considered. Soil parameters classified as per IS:1893-2016 listed in Table 3.1 are given the due consideration for this study. The soil-raft interaction model of thickness 650 mm, modelled with springs and dampers to replicate the bi-axial translational and rocking stiffness and damping parameters according to the formulation from Richart and Lysmer (1970). While soil-pile interaction depicted in Figure 3.2 is modelled using the equations laid down by Nakazawa Method (2000).

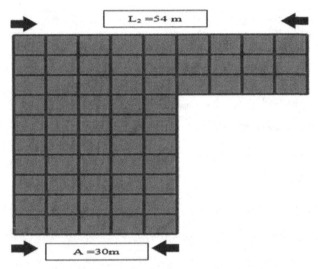

Figure 3.1 Plan view of bare frame structure
Source: SAP2000 software

Figure 3.2 3D View of frame simulating pile-soil interaction
Source: SAP2000 software

Table 3.1: Classification of soil parameters

Type of soil	N (Standard penetration test value)	Mass density (KN/m³)(ρ)	Shear wave velocity (m/s)(V_s)	Poisson's ratio(υ)	SBC KN/m²	Shear modulus (G) = ρ*Vs²
Stiff (type-I)	40	21	112	0.25	500	260000
Moderate-stiff (type-II)	20	18	84	0.33	230	131576
Soft (type-III)	9	17	56	0.48	150	50674

Source: Bowels, 1997 which is Foundation Analysis & Design Book by Joseph E. Bowels

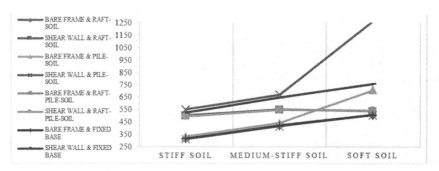

Figure 3.3(a) Base shear along X-direction
Source: SAP2000 software

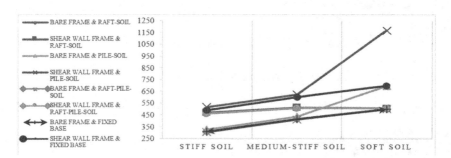

Figure 3.3(b) Base shear along Y-direction
Source: SAP2000 software

Response spectrum analysis corresponding to zone V of IS 1893:2016 (Part-1) design spectrum is used to study the finite element model of soil-structure system in SAP2000 software.

Results and discussions

Base shear

Values of base shear of the models analyzed is plotted in Figure 3.3(a) and (b) with the variation of structures modelled with shear wall in comparison with bare frame structures resting on different categories of soil-foundation medium besides comparing with fixed base is further tabulated in Tables 3.2 and 3.3.

Table 3.2: % Increase in base shear values along x-direction.

"Shear wall" structure in comparison to "bare frame" (considering SSI)			"Shear wall" structure in comparison to "bare frame" (considering fixed base)	
Raft soil Interaction	Stiff soil	37.24		
	Moderate-stiff soil	23.4	Stiff soil	40
	Soft soil	6		
Pile-soil Interaction	Stiff soil	39.86		
	Moderate-stiff soil	34	Moderate-stiff soil	34.6
	Soft soil	44		
Raft-pile-soil Interaction	Stiff soil	37.35		
	Moderate-stiff soil	24	Soft soil	33.23
	Soft soil	6.4		

Source: SAP2000 software

Table 3.3: % Increase in base shear values along Y-direction.

"Shear wall" structure in comparison to "bare frame" (considering SSI)			"Shear wall" structure in comparison to "bare frame" (considering fixed base)	
Raft-soil interaction	Stiff soil	37.2		
	Moderate-stiff soil	19	Stiff soil	37
	Soft soil	1.74		
Pile-soil Interaction	Stiff soil	36.8		
	Moderate-stiff soil	30.2	Moderate-stiff soil	31
	Soft soil	40.5		
Raft-pile-soil Interaction	Stiff soil	33.8		
	Moderate-stiff soil	19.5	Soft soil	28.85
	Soft soil	1.5		

Source: SAP2000 software

Table 3.4: % cecrease in storey displacement at storey level 11 along x-direction.

"Shear wall" structure in comparison to "bare frame" (considering SSI)			"Shear wall" structure in comparison to "bare frame" (considering fixed base)	
Raft-soil interaction	Stiff soil	19.6		
	Moderate-stiff soil	41	Stiff soil	16.4
	Soft soil	40		
Pile-soil Interaction	Stiff soil	17.6		
	Moderate-stiff soil	20.2	Moderate-stiff soil	18.8
	Soft soil	29.4		
Raft-pile-soil Interaction	Stiff soil	41.2		
	Moderate-stiff soil	44.4	Soft soil	19.2
	Soft soil	42		

Source: SAP2000 software

The consequence of SSI (building-foundation-soil system) is significantly recognized with the increase in soil flexibility as there is increment of base shear. Since, base shear is depended on the fundamental time period which also increases accordingly due to the descending curve of design response spectrum of design acceleration coefficient corresponding to 5% damping.

Table 3.5: % Decrease in storey displacement at storey level 11 along Y-direction.

"Shear wall" structure in comparison to "bare frame" (considering SSI)			"Shear wall" structure in comparison to "bare frame" (considering fixed base)	
Raft-soil interaction	Stiff soil	26.8		
	Moderate-stiff soil	35.3	Stiff soil	23.8
	Soft soil	47.6		
Pile-soil Interaction	Stiff soil	24.7		
	Moderate-stiff soil	26	Moderate-stiff soil	25.2
	Soft soil	36.7		
Raft-pile-soil Interaction	Stiff soil	38.7		
	Moderate-stiff soil	43.6	Soft soil	25.3
	Soft soil	47.6		

Source: SAP2000 software

Storey displacement

The average value of maximum elastic lateral deflection of each storey with respect to ground level known as storey displacement is also seen to be significantly influenced by the soil flexibility. Besides, the variation in storey displacement values of shear wall structure compared with bare frame at storey level 11 along X-direction is tabulated in Table 3.4 and along Y-direction in Table 3.5.

   Table 3.3 highlights that bare frame structure demonstrates more displacement values as compared to shear wall buildings where the deflection values are within the permissible limits in all the soil types. This is due to the additional mass concentration at core due to the inclusion of shear wall thereby showing the merits of including shear wall to reduce the lateral deflection of buildings. Maximum reduction in storey displacement stands at 47.64% for building modelled with shear wall with soil-raft-pile interaction at soft soil condition along Y-direction compared to the same modelling approach with bare frame.

Conclusion

The current study seeks to examine and appraise the influence of soil-structure interaction (SSI) on base shear and storey displacement on an asymmetric 12-storey RC building frame with shear wall positioned at core and bare frame while taking soil-foundation medium flexibility into account. It is observed that when soil flexibility rises, so does base shear. Maximum value is seen in soil-pile interaction, since, slender piles fixed under pile cap supported beneath asymmetric structures attract more base shear thereby showing raft-pile foundation a safer option while designing asymmetrical structures under soft soil as parameters. Yet when both the impact of SSI and fixed base are considered, the addition of a shear wall at the core causes the base shear value to significantly increase when compared to bare frame structures. This is because the inclusion of a shear wall tends to reduce the fundamental period of the structure due to the mass concentration at the core and increasing spectral acceleration coefficient than bare frame structural models. Shear wall provision is observed to be more significant in case of asymmetric building as the increase in storey displacement is effectively controlled. This results in improving the seismic performance of structures since storey displacement in the range of 18-48% reduces depending on soil-foundation parameters thereby making the structural design safe. Finally, it can be inferred that typical design procedure without SSI is unprogressive, thus considering SSI is required to ensure the structural safety of buildings resting over soft soil or at high seismic zones due to lateral deflection. However, addition of shear walls counterbalances the SSI effect by providing additional stiffness to resist the lateral earthquake forces.

References

Bagheri, M, Jamkhaneh, E M, and Samali, B (2018). Effect of seismic soil–pile–structure interaction on mid and high-rise steel buildings resting on a group of pile foundations. *ASCE.* 18(9), 1-27.

Dey, A, Sharma, N, and Dasgupta K (2020). Influence of shear wall on seismic response of RC frame buildings on pile foundation considering SSI. 17th World Conference on Earthquake Engineering, Sendai, Japan, pp. 1-12.

Saha, R, Dutta, S C, and Haldar, S (2015). Seismic response of soil-pile raft structure system. *Journal of Civil Engineering and Management.* 21, 2, 144–164.

4 Laboratory investigation of aggregates in Gardez city for road construction suitability

Naqibullah Zazai[a,1], Karanjeet Kaur[b,2] and Hazratullah Paktin[c,1]

[1]Paktia University, Gardez, Afghanistan

[2]University College of Engineering and Technology, Bikaner, India

Abstract

Aggregates must transmit the applied load to the underlying layers and must withstand to wearing, polishing, abrasion, impact, and crushing action due to the tires of vehicles during the different cycle of hot mix asphalt (HMA), starting from its production to its placement and compaction. The use of high quality of aggregate is necessary in order to have durable roads. Finding economically viable and physically suitable sources of aggregates for the roads construction in Paktia province of Afghanistan is one of the challenges which local government authorities have been facing. Therefore, the current comparative study is conducted to evaluate the physical properties of two different sourced aggregates (Milan and Gul Wall). Los Angles, impact value, and crushing value test were performed. The result shows that the aggregates from Milan source is highly suitable than Gul Wall source against wearing action of traffic, the aggregates from Milan source are exceptionally strong, and those from Gul Wall source are very strong against impact action, the source of Milan is better than Gul Wall source in crushing against loads. Aggregates from Milan source are more suitable than Gul Wall source to be used for construction of future roads in the area; however, both sourced aggregates fulfil the requirement for road construction.

Keywords: Abrasion, aggregates, crushing, impact, pavement, polishing, wearing

Introduction

Due to the rising extension of new roads, high quality of natural aggregates is necessary to be used for the constructions of highways for durable life. Aggregates are mainly investigated on mechanical-properties test methods, and how the properties would be improved during the production. Aggregates for road construction are mainly obtained from local supplies from the naturally available rocks. Natural rocks are typically found either as exposed formations at or near the surface, or as accumulations of gravel located primarily in former riverbeds. Apart from the aggregates obtained from the natural rocks, other types of aggregates are also sometime used in HMA (Kandhal et al., 2023).

The properties of aggregates which are used in pavement are vital for the performance of the pavement. Different specifications such as (AASHTO, 1990; IRC 37, 2001; and (MoRTH, 2013) are available in Afghanistan for the materials used in pavement layers. Many studies have been conducted to investigate the properties of different modified asphalt binder to minimize the distresses of flexible pavements (Gulzar and Underwood, 2020a; Gulzar and Underwood, 2019a; Gulzar and Underwood, 2019b; Gulzar and Underwood, 2020b; Gulzar and Underwood, 2020c). Distresses in pavement such as rutting and stripping often occur because of used aggregates. In order to achieve desired performance of the pavement, the usages of quality aggregate is necessary. Since, the natural rock materials can be found in many sources in the area, it is

[a]naqibzazai99@gmail.com, [b]karanjeetkaur@cet-gov.ac.in, [c]hzrtllh@gmail.com

DOI: 10.1201/9781003450924-4

important to sample and test the materials to ensure that which source is good and can be recommended for further use in construction of the highways. Aggregates should therefore be tested to ensure that it is good to be used in road construction. Different comparative studies have been conducted worldwide to evaluate the suitability of aggregates for road construction and also see the performance of different artificial aggregates with respect to natural aggregates (Thakur and Saklecha, 2019; Dhaqane, 2017; Manju et al., 2017) but no study have been found in context of Afghanistan to evaluate the local available sourced aggregates for the construction purpose.

The current study is therefore conducted to find suitable sources of aggregates for the construction of roads pavements in future. Two different sourced aggregates from Gardez city were selected for this study and the objectives of this study are to evaluating the Los Angles values, impact values, crushing values, and homogeneity level between these two sources.

Materials and testing

Material characterization

Two different sourced aggregates from Gardez city for the purpose of usage in road construction were investigated in this study. The names of the two sources are designated as Gul Wall and Milan which are the names of the areas from where the aggregates are produced.

Testing

The experimental plan for aggregates was prepared according to different worldwide accepted codes. Los Angles and impact values were found according to (ASTM, C131 1989) and (IS: 5640-1970) respectively while crushing value was found according to (IS: 2386 part IV).

Gradation 2 from (ASTM, C131, 1989) was chosen to perform Los Angles test in this study. The aggregates were kept in 100 110°C in the oven for four hours to be completely dry; and were kept in room temperature. Sample conditions were given to aggregates used for impact and crushing values tests. Eleven steel balls and (3033) revolutions per minutes were given. The tested aggregated were sieved on N.12 (1.68 mm) sieve to know the Los Angles value of the aggregates.

Impact value (IV) test was conducted according to (IS: 5640-1970). The mold was filled in three layers, each given 25 taps with the rod. After specified blows, the aggregates were sieved on the 2.36 mm until no further significant amount of aggregate passes. The IV percentage was found using the passing percentage with respect to original weight of the sample.

The crushing value test was conducted according to (IS: 2386 part IV). The mold of sized diameter 115 mm and height of 180 mm was filled in three layers, each given 25 taps with the rod. The load of 40 tons is applied at uniform rate of 4 tons per minutes. The aggregates are then sieved on the 2.36 mm sieve. The crushing value (CV) percentage was found from the materials passing 2.36 mm sieve.

Results and discussions

Los Angles value of the Gul Wall and Milan sourced aggregates were determined. Three samples of the aggregates were tested to avoid the uncertainty in the aggregates results. The result of the Los Angles value test is presented for all three samples of the two different sourced aggregates in the Figure 4.1(a). It is seen from the figure that the Los Angles value of the Gul Wall source is higher than the source of Milan. It is depicted that the difference in the results of the test is negligible for the same source of each sample, which shows that the rocks of each sources are homogeneous. The difference in the values of the Los Angles between both sources is about 7% which make the Milan sourced aggregates better than Gul Wall source.

The result of the IV test is presented for all three samples of the two different sourced aggregates in the Figure 4.1(b). It is seen from the figure that the IV of the Gul Wall source is higher than the source of Milan. The difference in the results of the test is negligible for the same sourced of each sample, which shows that the rocks of each source are homogeneous. The difference in the values of the Impact Values between both sources is about 4% which makes the Milan sourced aggregates better than Gul Wall source.

The result of the Crushing Value test is presented for all three samples of the two different sourced aggregates in the Figure 4.1(c). It is seen from the figure that the CV of the Gul Wall source is higher than the source of Milan. The difference in the results of the test is negligible for the same source of each sample, which shows that the rocks of each source are homogeneous. The difference in the values of the crushing values between both sources is about 6% which makes the Milan sourced aggregates better than Gul Wall source.

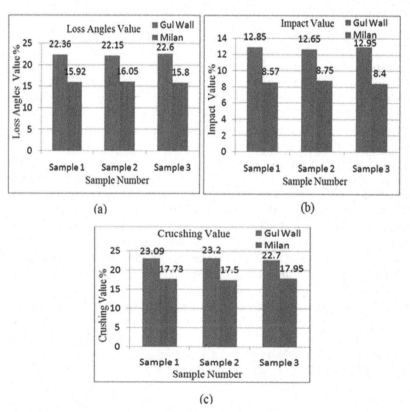

(a)   (b)

(c)

Figure 4.1 (a) Los Angles value of Gul Wall and Milan sourced aggregates; (b) Impact value of Gul Wall and Milan sourced aggregates; (c) crushing value of Gul Wall and Milan sourced aggregates

Source: Author

Conclusion

The Los Angles, impact, and crushing values of two different sourced aggregates (Gul Wall and Milan) in Gardez city, Paktia, Afghanistan were investigated in this study. The results show that aggregates from the Milan source are better than Gul Wall source. Drawn conclusions are explained as follows:

The rocks in each source are homogeneous as the different samples give almost the same results. Based on the results observations, the aggregates from Milan source are more suitable than Gul Wall source. The use of Milan source aggregates is highly recommended for the construction of future roads. The aggregates from Milan source are more suitable than Gul Wall source against wearing action of traffic. However, both sources of aggregates fulfil the requirements to be used for any kind of pavement. The aggregates from Milan source are exceptionally strong, and those from Gul Wall source are very strong against impact action. The source of Milan is better than Gul Wall source in crushing against loads. However, both sources can be used. There seems some correlation between the Los Angles test and CV test. The results of the two tests are quite same for the two sourced aggregates.

References

ASTM, C131. (2006). Standard test method for resistance to degradation of small-size coarse aggregate by abrasion and impact in the Los Angeles machine. *ASTM International, West Conshohocken, PA.*

Bureau of Indian Standards. (Reaffirmed 1998). Indian standard method of test for determining aggregate impact value of soft coarse aggregates. IS: 5640-1970. Indian Roads Congress.

Bureau of Indian Standards. (Reaffirmed 1997). Indian standard methods of test for aggregate for concrete. IS: 2386 part IV – 1963 Indian Roads Congress.

Dhaqane, E A I. (2017). Comparative study between natural and artificial aggregates. *Mogadishu University Journal.* 69–98.

Gulzar, S, and Underwood, B S (2020a). Nonlinear viscoelastic response of crumb rubber modified asphalt binder under large strains. *Transportation Research Record: Journal of the Transportation Research Board.* 2674(3), 139–149.

Gulzar, S, and Underwood, S (2019a). Nonlinear rheological behavior of asphalt binders. 91st Annual Meeting of The Society of Rheology, Raleigh, NC.

Gulzar, S and Underwood, S (2019b). Use of polymer nanocomposites in asphalt binder modification. Advanced Functional Textiles and Polymers, Wiley, (pp. 405–432).

Gulzar, S, and Underwood, S (2020b). Stress decomposition of nonlinear response of modified asphalt binder under large strains. Advances in Materials and Pavement Performance Prediction (AM3P 2020), San Antonio, Tx, USA: CRC Press, pp. 436–440.

Gulzar, S, and Underwood, S (2020c). Large amplitude oscillatory shear of modified asphalt binder. Advances in Materials and Pavement Performance Prediction (AM3P 2020), San Antonio, Tx, USA: CRC Press, pp. 432–435.

Kandhal, P S, Veeraragavan, A and Choudhary, R (2023). Bituminous Road Construction in India. PHI Learning Pvt. Ltd.

Manju, R, Sathya, S, and Sheema, K (2017). Use of plastic waste in bituminous pavement. *International Journal of ChemTech Research.* 10(08), 804–811.Thakur, N and Saklecha, (2019). Comparison of properties of steel slag and natural aggregate for road construction. Available at SSRN 3376488.

Transportation Officials. (1993). *AASHTO Guide for Design of Pavement Structures, 1993* (Vol. 1). Aashto

Indian Roads Congress. (2012). "IRC: 37-2012, Guidelines for the Design of Flexible Pavements." In Indian Roads Congress, New Delhi, India

Ministry of Road Transport and Highways (MoRTH). (2013). Specifications for road and bridge works. fifth revision.

5 Optical absorption and FTIR analysis of ZnS/PVK nano-composites

Durgesh Nandini Nagwanshi[a]

Department of Applied Physics, Jabalpur Engineering College, Jabalpur, India

Abstract

Currently, we studied and analyzed the Fourier Transform Infrared spectroscopy (FTIR) spectra and UV visible optical absorption of ZnS/PVK nano-composites. The composite was synthesized using a simple chemical process. In nano-composites, optical properties can be controlled through their particle size. The particle size varied with different ZnS concentrations. The estimated crystal size from X-ray diffraction (XRD) is 3-12 nm. The spectra of UV Visible optical absorption is obtain from 200 nm to 550 nm and the absorption increases at lower wavelengths for low concentrations of ZnS in PVK (poly N-vinyl carbazole). In ZnS/PVK nano-composites the absorption spectra show the superposition of ZnS nanoparticles and the absorption of PVK. It is observed that increasing the ZnS concentration in PVK than particle size increases. Particle size determined by the hyperbolic band model, effective mass approximation model, and XRD is approximately the same. FTIR help to investigate the chemical environment of the surface of the nano-composite. FTIR confirms the different composition of ZnS in PVK and various bond formation, stretching, bending and asymmetric stretching information in terms of broad and sharp peaks in the composites. On increasing the composition of ZnS other small intensity peaks are obtained and these new peaks and splitting of peaks confirm the formation of ZnS in PVK polymer. Hence all these results confirm the interaction and complexation between ZnS and PVK nano-composite.

Keywords: Fourier Transform Infrared spectroscopy, UV visible optical absorption, XRD, ZnS/PVK nano-composites

Introduction

Polymer nano-composites and inorganic nanoparticles, comprises of organic polymer, which has been extensively explored in the last few years. It has been used to develop advanced polymer material built on previously existing polymers (Okada and Usuki, 2006). The emerging polymer nano-composites exhibits significantly enhancement of some properties i.e., heats distortion, modulus, temperature, gas impermeability, hardness, and others. Wide band gap semiconductor nonmaterials attract much observation for their surprising electrical, chemical and physical properties because of their quantum confinement effects with high surface area (Chhowalla et al., 2013). Then a nocomposites have been also reviewed in defense applications (Kurahatti et al., 2010). Semiconductor nano particles confined in a polymer matrix exhibit different properties with respect to the corresponding bulk material form due to their quantum size effect (Wang and Herron, 1992). Zinc Sulfide is an II-VI group semiconductor of direct wide band gap range from 3.5eV to 3.7 eV at normal temperature. Due to this reason, ZnS semiconductor has been attract much attention for their excellent properties is compare to another chalcogenides which makes suitably for ZnS in various applications like biosensors, bioelectronics, phosphors, fluorescence, electronic and optical devices. Presently, we have used simple chemical

[a]nandininagwanshi2525@gmail.com

DOI: 10.1201/9781003450924-5

technique for synthesis of ZnS/PVK nano-composites and control their particle size and enhance their stability.

Synthesis of the ZnS/PVK nano-composites

For the preparation of ZnS/PVK nano-composites, dissolved the 400 mg poly N vinyl carbazole in 15 ml dimethyl formamide (DMF) at 80°C. In this solution, add the different concentrations of zinc acetate 10%, 20%, 30%, 40%, and 50% of 400 mg PVK. The obtained solution stirred was 30 minutes at room or normal temperature after applying H_2S gas for 30 second. So the solution was to convert to dirty white in color. This final solution was transferred to the plane glass plate and put in the oven for dry the sample in 6 hours at a temperature of 100°C. The chemical equation is given as:

$$Zn(CH_3COO)_2 + PVK + H_2S => 2CH_3COOH + ZnS/PVK$$

The surface studies of nano-composites were done using XRD, AFM, and SEM (Nagwanshi and Nigam, 2018) and here we have reported the studies and analysis of FTIR spectra and UV visible optical absorption spectra.

Results and discussion

UV visible optical absorption spectra

The optical absorption spectra are found within 200550 nm shown in Figure 5.1 for ZnS/PVK samples with different ZnS concentrations in PVK. It's observed that, there is a very minor absorption between 800-500 nm. Absorption slightly increases to a lower wavelength and then suddenly increases in the visible range. The absorption is obtained at 300 nm, 305 nm, 315 nm, 327 nm, 330 nm, and 340 nm for various concentrations of ZnS in PVK correspondingly. The absorption spectra of sample show the superposition of the absorption of ZnS and pure PVK. The energy band gap was obtained from the absorption edge through Eg = $\frac{hc}{\lambda}$. It is seen from Figure

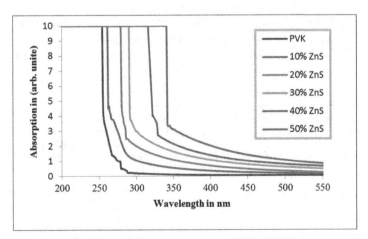

Figure 5.1 Optical absorption spectra for ZnS/PVK nano-composites
Source: Author

(5.1) that all the samples do not reveal any absorption in wavelength greater than 340 nm. When the concentration of ZnS nanoparticle is increases, then absorption shifts toward higher wavelengths (Masoud et al., 2009). The effective band gap of Eg is 4.13, 4.06, 3.93, 3.75, and 3.64 eV respectively for various ZnS concentration in PVK.

Determination of the particle size using absorption spectra

Different types of confinements are realized for based on crystalline size and shape of semiconductor nanoparticles. Effective mass approximation model (EMA) and hyperbolic band model (HBM) have been reported into their particle size with an effective band gap. From EMA and HBM model the radius of the nanoparticle (r) is given by equation (1) and (2). We have found the particle size as shown in Table 5.1.

$$r^2 = \frac{h^2}{8\left(E'_g - E_g\right)}\left(\frac{1}{m_e^*} + \frac{1}{m_h^*}\right) \tag{1}$$

$$r^2 = \frac{h^2 Eg}{2m_e^*\left(Eg'^2 - Eg^2\right)} \tag{2}$$

From the Table 5.1, it is observed that increasing ZnS concentration in the composite than particle size increases. The particle size determined using the hyperbolic model is higher as compared to the effective mass approximation model. As with ZnS concentration increases, the effective band gap is decreases and particle size increases.

FTIR spectra: FTIR spectroscopy shows the essential investigation of polymer structure and gives information of the interactions and complexation between the different compositions in the polymeric material. FTIR spectrum revealed the formation of the polymer. The FTIR spectra of ZnS/PVK are shown in Figure (5.2). The absorption band is obtained about 2925 cm^{-1} because of C-H stretching vibration with sp^3 hybridization, at 2240 cm^{-1} nitrile group of aromatic rings is obtained, the strong absorption peak that evidenced at 17201740 cm^{-1} was attributed in presence of C = O stretching modes obtain due to absorption of atmospheric CO_2 on the nano crystals surface and additional asymmetric stretching of poly vinyl carbazol is obtained in region 1660-1675 and band shift in higher concentration from 1650 to 1680 cm^{-1}, and 1452cm^{-1}, 1355cm^{-1}, 1208cm^{-1}, 970 cm^{-1}, 657 cm^{-1} symbolize, CH$_3$ bending, C-H bending, C-C obtain, alkane out of the plane, ring deformation respectively. The peaks at 17271737 cm^{-1}, 16501681 cm^{-1}, 14561444 cm^{-1} became broad

Table 5.1: Particle size evaluation from absorption edge using both models.

Name of sample	Concentration of ZnS in PVK (%)	Absorption edge wavelength	Energy band gap Eg(eV)	Size of particles	
				EMA (nm)	HBM (nm)
ZnS/PVK - I	10%	305	4.06	3.34	3.84
ZnS/PVK – II	20%	315	3.93	3.86	4.58
ZnS/PVK – III	30%	327	3.79	4.62	6.08
ZnS/PVK – IV	40%	330	3.75	5.58	6.83
ZnS/PVK – V	50%	340	3.64	7.9	8.82

Source: Author

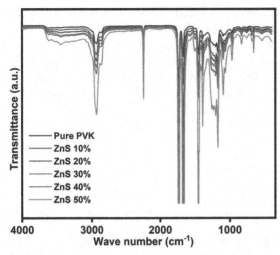

Figure 5.2 FTIR spectra of ZnS/PVK nano-compositesand pure PVK
Source: Author

Table 5.2: Peak assignment of pure PVK and varing concentration of ZnS with PVK.

Pure PVK	10% ZnS	20% ZnS	30% ZnS	40% ZnS	50% ZnS	Peak assignment
2935	2931	2933	2933	2931	2933	C-H stretching vibration sp^3 hybridization
2242	2240	2242	2240	2244	2242	Nitrile group of aromatic ring
1733	1735	1735	1729	1729	1726	C=O stretching modes
1668	1672	1670	1664	1662	1662	Asymmetric stretching of poly vinayl carbazol
1452	1452	1450	1452	1452	1446	CH$_3$ bending of vinyledine group
1386	1388	1384	1386	1386	1386	C-H bending of vinyledine group
1199	1199	1199	1201	1203	1199	C-C obtain of vinyledine group
1170	1168	1168	1168	1168	1168	C-N stretching
968	970	968	971	970	971	Alkane out of plane
657	659	659	659	659	659	Ring deformation

Source: Author

on higher concentrations of ZnS. To obtains the several characteristic bands at 1064, 1093, 1220, 1260, 1492 cm^{-1}, and 1527 cm^{-1} assigned for the stretching vibration of ZnS. The FTIR spectrum of PVK with ZnS reveals similar peaks as pure PVK. The peak arrived at 1201 cm^{-1} split into three peaks such as 1201 cm^{-1}, 1228 cm^{-1}, and 1255 cm^{-1}, indicating that the doped ZnS affected the structure of PVK. These results suggest the complexation and interaction between ZnS and PVK nano-composite (Menazea et al., 2020).

Conclusion

The composite was synthesized using chemical process. Increasing the concentration of ZnS in PVK the size of nanoparticle also increases. On increasing ZnS concentration the absorption edge gets shifted to red and larger crystal size for higher ZnS concentration because of quantum confinement effect. FTIR confirms the dopant incorporation of ZnS in PVK with different compositions and reveals various bond formation, stretching, bending and asymmetric stretching information in terms of broad and sharp peaks on pure PVK which are compared with the different compositions of ZnS. The other small intensity peaks are obtained and the splitting of peaks confirms the formation of ZnS in PVK polymer hence all these results confirm the interaction between ZnS and PVK nano-composite.

References

Chhowalla, M, Shin, H S, Eda, G, and Li, L J (2013). The chemistry of two-dimensional layered transition metal dichalcogenide nanosheets. *Nature Chemistry*. 5, 263.

Kurahatti, R V, Surendranathan, A O, Kori, S A, Singh, N, Kumar, A V R, and Srivastava, S (2010). Defence applications of polymer nano-composites. *Defence Science Journal*. 60, 551–563.

Masoud, S N, Fatemeh, D, and Mehdi, M (2009). Synthesis and characterization of ZnS nano-cluster via hydrothermal processing from [bis(salicylidene)zinc(II)]. *Journal of Alloys and Compounds*. 470, 502.

Menazea, A A, Ismail, A M, and Elashmawi, I S (2020). The role of Li4Ti5O12 nanoparticles on enhancement the performance of PVDF/PVK blend for lithium_ion batteries. *Journal of Materials Research and Technology*. 9(3), 5689–5698.

Nagwanshi, D N, and Nigam, R (2018). Synthesis and Surface Studies of ZnS/PVK Nanocomposites. *Nano Trends: A Journal of Nanotechnology and Its Applications*. 20(2), 0973-418X (Online).

Okada, K, and Usuki, A (2006). Twenty years of polymer-clay nano-composites *Macromolecular Materials and Engineering*. 291, 1449–1476.

Wang, Y, and Herron, N (1992). Semiconductor nanocluster. *Chemical Physics Letters*. 200, 71.

6 Load sharing between group of two floating granular piled rafts - an analytical study

Ashish Solanki[a] and Jitendra Kumar Sharma[b]

Civil Engineering Department Rajasthan Technical University, Kota Rajasthan, India.

Abstract

The best approach to enhance the quality of loose ground for the purpose of establishing a solid foundation structure such as bridge piers, and tall buildings etc. is ground stabilization utilizing granular piles along with raft. Three elements of the pile raft system i.e., raft, granular piles, and soil are used to distribute loads to the subsurface. In the current work, a group of two granular piles or stone columns (floating) with rigid rafts is studied using an elastic continuum technique. The current work examines the analytical examination of the load sharing between a group of two granular piled rafts, under various raft and pile parameters, including raft size, pile length, and spacing between the granular pile rafts. With regards to the fractional load distribution across the base, raft, and piles, the overall reaction of a rigid raft with a group of two granular piles is analyzed.

Keywords: Portion of the load transferred onto the pile, portion of the load transferred onto the base of pile, portion of the load transferred onto the raft

Introduction

The cost-effectiveness and affordability of the piled raft foundation system are enhanced in this study by replacing the conventional piled raft systems, which utilize concrete, steel, and wood piles, with granular piles. Madhav et al. (2009) used the continuum technique to investigate the load transfer mechanism between the Granular Pile and the raft on top. In order to compute the settlement of single and two floating GPs that have been considerably stiffened, Garg and Sharma (2019) developed analytical methods focused on an elastic approach. Utilizing the continuum method, Madhira et al. (2019) evaluated a partially stiffened granular piled raft (GPR). Settlement and interaction analysis of two floating GPRs were the focus of research by Solanki et al. (2022).

Methodology

Figure 6.1 provides a description of a research study that involves the use of two GPRs of length, 'L,' that are subjected to an axial force, 'P,' and placed at 's/d,' a certain distance from each other on a rigid raft of dia. 'D'. The study assumes that the soil is uniform, isotropic, and linearly elastic, and that the GPs have a hard and smooth base with a certain Poisson ratio (ν_s) and deformation modulus (E_s). In order to determine the vertical settlement of soil beneath the GP, the study utilizes Mindlin's equation, which consider the vertical displacements induced by a point load applied within the soil mass. Where 'n' shows the segments of the GP and the displacements for each segment are calculated at the nodes on the interface of GP-soil. The cylindrical segments are further divided into vertical and circumferential segments, with 'nz',

[a]ashish92.4ueverback@gmail.com, asolanki.phd19@rtu.ac.in, [b]jksrtu@gmail.com, jksharma@rtu.ac.in

DOI: 10.1201/9781003450924-6

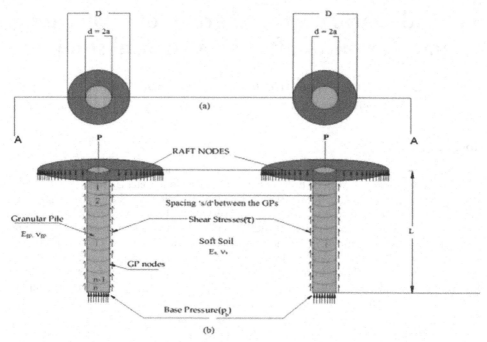

Figure 6.1 Forces and stresses operating on two floating GPR units
Source: Solanki et al. (2022)

and 'nt' segments, respectively. GP's relative stiffness is expressed by the ratio of the GP's elastic modulus to that of the surrounding soil, denoted by $K_{gp} = E_{gp}/E_s$.

Displacement of soil element

At the nodes of granular pile, soil displacements

Solanki et al. (2022) propose an assessment method where the displacement along the GP-soil interface is determined by integrating Mindlin and Boussinesq expressions. This evaluation considers the midpoint on each segment's side and at the base. The evaluation is presented using a matrix representation.

$$\{\rho^{ps}\} = \left\{\frac{S^{ps}}{d}\right\} = [I^{ppf}]\left\{\frac{\tau}{E_S}\right\} + [I^{prf}]\left\{\frac{p_r}{E_S}\right\} \tag{1}$$

The vectors $\{S^{ps}\}$ and $\{\rho^{ps}\}$ represent vertical soil displacements that have been normalized, respectively. $[I^{ppf}]$ – (n+1) × (n+1) sized matrix of the settlement influence coefficients (SIC). The (n+1) × kr $[I^{ppf}]$ – SIC matrix is utilized, where $\{\tau\}$ represents a column vector of size n+1, and $\{p_r\}$ represents a column vector of size kr."

At Raft Node Soil Displacements

According to Solanki et al. (2022), the raft is segmented into equal-sized 'kr' parts. At each element's node, displacements at the raft-soil interface are assessed and depend on the way the raft stresses interact with the elemental stresses of GP. The equation provided below presents the soil displacement at the raft node in matrix form:

$$\{\rho^{rs}\} = \left\{\frac{s^{rs}}{d}\right\} = [I^{rpf}]\left\{\frac{\tau}{E_s}\right\} + [I^{rrf}]\left\{\frac{p_r}{E_s}\right\} \tag{2}$$

The vectors $\{S^{rs}\}$ and $\{\rho^{rs}\}$ represent vertical soil displacements that have been normalized, respectively. The $[I^{rpf}]$ – matrix, with dimensions $kr \times (n+1)$, is analyzed to determine the SIC. $[I^{rrf}]$ – square dimension 'kr' matrix."

Displacements of pile

Elements of a granular pile's displacements are assessed as

$$\varepsilon_v = \frac{\sigma_v}{E_{gp}} \tag{3}$$

where
   E_{gp} = GP's elastic modulus of deformation and
   ε_v = Elemental strain in axial direction
   σ_v = Elemental stress in axial direction

Direct and shear stress relationship

The matrix shows the following relationship between shear and direct stresses on pile nodes:

$$\{\sigma_v\} = [A_1]\{\tau\} \tag{4}$$

where and are the $(n+1)$ shear and direct stress matrices applied to the pile nodes, respectively.

$$[A_1] = \begin{bmatrix} \frac{2(L/d)}{n} & \frac{4(L/d)}{n} & \frac{4(L/d)}{n} & \frac{4(L/d)}{n} & - & - & - & - & 1 \\ 0 & \frac{2(L/d)}{n} & \frac{4(L/d)}{n} & \frac{4(L/d)}{n} & - & - & - & - & 1 \\ 0 & 0 & \frac{2(L/d)}{n} & \frac{4(L/d)}{n} & - & - & - & - & 1 \\ - & - & - & - & - & 0 & \frac{2(L/d)}{n} & \frac{4(L/d)}{n} & 1 \\ - & - & - & - & - & - & 0 & \frac{2(L/d)}{n} & 1 \\ - & - & - & - & - & - & 0 & 0 & 1 \end{bmatrix}_{(n+1)\times(n+1)} \tag{5}$$

Granular pile displacements

Using Solanki et al. (2022) approach, displacements of a GP are assessed. Starting with the GP's top settlement 'ρ_t' by proceeding downward and taking into consideration the strain of each element progressively, the vertical displacements at every node of the granular pile are calculated."

$$\{\rho^{pp}\} = \rho_t\{1\} + [B_1]\left\{\frac{\sigma_v}{E_s}\right\} \tag{6}$$

where $[B_1]$ is a $(n+1)$ square matrix shown as

$$[B_1] = \frac{(L/d)}{nK_{gp}} \begin{bmatrix} -0.5 & 0 & 0 & 0 & - & - & - & 0 \\ -1 & -0.5 & 0 & 0 & - & - & - & 0 \\ -1 & -1 & -0.5 & 0 & - & - & - & - \\ - & - & - & - & - & - & - & - \\ -1 & -1 & - & - & - & - & -0.5 & 0 \\ -1 & -1 & - & - & - & \frac{34}{32} & -\frac{18}{32} & -\frac{6}{32} \end{bmatrix}_{(n+1)\times(n+1)}$$

(7)

GP nodes settlements in the form of shear stress is achieved by substituting the direct stress with shear stress using (Equation 4).

$$\{\rho^{pp}\} = \rho_t\{1\} + [D_1]\left\{\frac{\tau}{E_s}\right\}$$

(8)

where $[D_1] = [B_1][A_1]$.

Rafts displacements

According to Solanki et al. (2022) "raft is supposed to be rigid; all raft node settlements should be equal". The GP's top settlement (ρ_t), equals the settlement of the raft given as

$$\{\rho^{rr}\} = \rho_t\{1\}$$

(9)

where $\{\rho^{rr}\}$, is a vector of size 'kr' for the raft displacement."

Condition of compatibility

Utilizing the soil's and the GP's settlement compatibility
$\{\rho^{ps}\} = \{\rho^{pp}\}$
or

$$[AA_1]\left\{\frac{\tau}{E_s}\right\} + [I^{prf}]\left\{\frac{p_r}{E_s}\right\} = \rho_t\{1\}$$

(10)

Utilizing the soil's and the raft's settlement compatibility
$\{\rho^{rs}\} = \{\rho^{rr}\}$
or

$$\{\rho^{rs}\} = \left\{\frac{S^{rs}}{d}\right\} = [I^{rpf}]\left\{\frac{\tau}{E_s}\right\} + [I^{rrf}]\left\{\frac{p_r}{E_s}\right\} = \rho_t\{1\}$$

(11)

Standardized raft stresses and normalized interfacial shear stresses are calculated by solving Equations 10 and 11. The load distribution between the raft and pile is assessed using the shear stresses.
The portion of the load transferred onto the pile = $(P_P/P) \times 100$
The portion of the load transferred onto the base of pile = $(P_B/P) \times 100$
The portion of the load transferred onto the raft = $(P_R/P) \times 100$

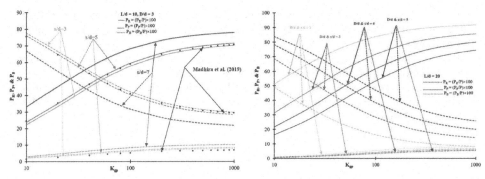

Figure 6.2 (a) Fractional load variation with reference to the total load due to K_{gp} and the outcome of s/d = 3, 5, and 7(b) The impact of different ratios L/d = 20, D/d, and s/d = 2, 3, 4, and 5 on the change in fractional load relative to the total load

Source: Author

Results and discussions

Results are derived for s/d =2-7, K_{gp}=10-1000, and D/d = 2-5. Results are obtained at K_{gp}=1000.

Figure 6.2 (a) and (b) depict the variations of "$(P_P/P) \times 100$, $(P_R/P) \times 100$, and $(P_B/P) \times 100$ with K_{gp}" for different ratios, including L/d values of 10 and 20, D/d values ranging from 2 to 5, and S/D values ranging from 2 to 7. Figure 6.2 curve distribution pattern illustrates that the "$(P_P/P)\times 100$ and $(P_B/P)\times 100$ increase while the K_{gp} rises, the $(P_R/P) \times 100$ generally decreases. There is an inverse relationship between the fractional load of a raft and that of a pile. Conversely, as expected, with an increase in the s/d ratio between the GPs, the $(P_R/P)\times 100$ value increases, while the $(P_B/P) \times 100$ value decreases. The results of this investigation were first compared to those of Madhira et al. (2019) to validate them. When the S/D, is increased the distance between the GPRs is increased relative to their dia. This can lead to a reduction in the interaction between the two GPRs, which can cause the system to behave more like a single GPR.

Conclusion

The load on the raft is distributed more accurately between the two granular piled rafts (GPRs), the portion of the load transferred onto the raft decreases. This mean that the GPRs system becomes more stable or more evenly loaded as the behavior of the system is better accounted. Also, as S/D, increases, the portion of the load transferred onto both GP and the base decreases, this is due to the fact that the load is spread out across a larger surface area, resulting in reduced concentration on both the GP and the base. Additionally, as the portion of the load transferred onto the raft increases, the load supported by the GP and base decreases, vice-versa also true. The specific details and implications of factors studied would depend on the specific context of the problem being studied.

References

Garg, V and Sharma,J K.2019. Analysis and settlement of partially stiffened single and group of two floating granular piles. *Indian Geotechnical Journal*, 49(2), 191203. https://doi. org/10.1007/s40098-018-0321-7

Madhav, M R, Sharma, J K, and Sivakumar, V (2009). Settlement of and load distribution in a granular piled raft. *Geomechanics and Engineering*, 1(1), 97-112. https://doi.org/10.12989/gae.2009.1.1.097

Madhav, M, Sharma, J K, and Sanadhya, R R (2019). Analysis of stiffened granular piled raft. *Geotechnical Engineering Journal of SEAGS-AGSSEA*, 51(2), 54-64.

Solanki, A, Sharma, J K, and Madhav, M R (2022). Interaction analysis of two floating granular piled raft units. *Geomechanics and Geoengineering*, 118. https://doi.org/10.1080/17486025.2022.2136409

7 An integrated approach to explore the suitability of surface water for drinking purposes in Mahanadi basin, Odisha

Abhijeet Das[1,a], Krishna Pada Bauri[1] and Bhagirathi Tripathy [2,b]

[1]Department of Civil Engineering, C.V. Raman Global University (C.V.R.G.U), Bhubaneswar, India

[2]Department of Civil Engineering, IGIT Sarang, Dhenkanal, India

Abstract

In this study, an effort has been made to comprehend the suitability of surface water in Mahanadi Basin, Odisha for human consumption. Nineteen samples consisting of 20 physicochemical parameters for 2019–2022 period, were evaluated in comparison to the recommended WHO standards to inspect water quality using two indexing protocols: weighted arithmetic water quality index (WA WQI) and stepwise weight assessment ratio analysis water quality index (SWARA-WQI). In order to resolve conflicts involving the WQI index and determine the best places along the river stretch where the water meets drinking requirements, multiple-criteria decision making (MCDM) models, such as compromise programming (CP), ordered weighted averaging (OWA), and combined compromise solution (CoCoSo), were adopted. To display the geographic distribution of water quality within each region, the spatial variation map was created using the inverted distance weighting (IDW) approach in the ArcGIS environment. The trends for the average anion and cation proportions were $Cl^->SO_4^{2-}>NO_3^->F^-$ and $Fe^{2+}>B^+$ respectively. However, the amounts of TC and TKN were quite high in comparison to WHO guidelines and gradually increased from upstream to downstream, indicating that the river waters were severely polluted. Further, the WQI results shows in the study area, which ranged from 23.78-96.09 (WA WQI) and 14.6-1065.2 (SWARA-WQI). The results of the WQI indicates that 15.79% (WA WQI and SWARA-WQI) of surface water samples had poor to very poor drinking water quality. The study area's total WQI shows that the water is potable and safe (around 84.21% good water) except few localized pockets in location SP-8, 9 and 19. A series of 20 in-situ inspections was used to validate each one of the modelled geographic maps. The interpretation of semi-variogram modelling also proves that Gaussian model finds the ideal fit for both WA WQI and SWARA-WQI dataset. Putting the above MCDM models into practice, it was clarified that SP-9 was most heavily polluted area compared to other places. This was obvious from the greater WA WQI and SWARA-WQI values at these survey points. Finally, this work offers suggestions for the future treatment of surface water contamination and the safeguarding of water resources.

Keywords: ArcGIS, combined compromise solution, compromise programming, Mahanadi basin, OWA, semi variogram, stepwise weight assessment ratio analysis water quality index, weighted arithmetic water quality index

Introduction

One of the world's most vital water resources is surface water, which is used for essential purposes like drinking, agriculture, and industry. In general, agricultural and industry development were primarily influenced by population increase and the

[a]das.abhijeetlaltu1999@gmail.com, [b]bhagirathitripathy@gmail.com

DOI: 10.1201/9781003450924-7

expansion of urbanisation, which led to water instability difficulties. There are, however, not many research, that describe how surface water irrigation and drinking affect the entire water cycle, from consumption to water to soil to crops. Consequently, there is a pressing need to create effective management plans for the sustainable use and preservation of essential surface water resources. To fill in this gap, earlier researchers have designed the framework of water quality index (WQI). In order to distinguish between different types of water quality, Horton created the WQI model in 1965. Since then, numerous indices have been put out, but there is no WQI that is universally recognised (Gao et al., 2020). The precision and objectivity of stepwise weight assessment ratio analysis (SWARA) weights are higher and stronger compared to those subjective appraisal methods, which can more fully explain the outcomes obtained. In the current study, three alternative multiple-criteria decision making (MCDM) approaches such as ordered weighted averaging (OWA), compromise programming (CP), and combined compromise solution (CoCoSo) were used to discuss the effectiveness of the WQI index. OWA uses a preference matrix to compare each of the discovered relevant criteria to one another in order to determine the necessary weighting factors (Yari and Chaji, 2012).

Study area

The study region is a segment of the Mahanadi River basin (MRB), which covers an area of about 141,600 Km² and is the third largest in the Indian Peninsula. It has a long history of providing irrigation for agricultural purposes and fisheries to the Indian states of Chhattisgarh and Odisha (Bastia et al., 2020). It is also regarded as the longest river in the state of Odisha, spreading over 494 km and flowing across 65,628 km² of land. The location is determined by geographic coordinates 80°30' E to 86°50'E and 19°20'N to 23°35'N. The basin has a tropical climate with 1200 to 1400 mm of annual precipitation on average. Approximately 54% of the MRB is covered by agricultural land. The basin's typical temperature fluctuates between 24-27°C, exhibiting the lowest temperature, highly fluctuating between 10-13°C in winter.

Sample collection, preservation, and analysis

To account for the worst-case scenario of pollution, sampling was done on an average yearly basis between 2019 and 2022. The Global Positioning System (GPS) captured the coordinates of the 19 monitoring locations. The sampling bottles were sealed, labelled, and delivered to the State Pollution Control Board's laboratory in Odisha, where they were kept at 4°C until additional examination. The research area took into account a total of 20 parameters, including, Alkalinity, TKN (total kjeldahl nitrogen), electrical conductivity (EC), pH, total coliform (TC), total suspended solids (TSS), chemical oxygen demand (COD), ammoniacal nitrogen (NH_3-N), dissolved oxygen (DO), biochemical oxygen demand (BOD), free NH_3, sodium adsorption ratio (SAR), chloride (Cl^-), sulphate (SO_4^{2-}), fluoride (F^-), boron (B), total dissolved solids (TDS), total hardness (TH), nitrate (NO_3^-), and iron (Fe^{2+}).

Methodology

WQI is a single arithmetic value that represents overall WQ and is based on a weighted average of specified characteristics. According to Shankar and Kawo (2019), the quantitative evaluation of WQ, utilizing the WA WQI, has been conducted, and it is further categorised into five classes. In order to calculate SWARA-WQI, each parameter is given a weight based on SWARA. It is an improvement over the currently used conventional WQIs.

Results and discussions

It is observed that the value was within the WHO standard limit (5 mg/l). The diversity of aquatic organisms may have diminished as a result of the increase in coliform, which indicates that the aquatic ecosystem's balance is upset. In the ongoing research, the readings fluctuate between 1212 and 42529 MPN/100 ml. Higher levels have been reported in the waters at SP-8, 9 and 19, which are near to factories, municipal sewers, or hospitals. Clay and silts, as well as biological solids like bacteria and algae cells, were the main sources of TSS. TSS value, in the current study, ranged from (28.6-74.9 mg/l), which is inside the threshold value of 100 mg/l. Higher alkalinity in water, and vice versa, increases its ability to neutralise acids. It should not be higher than 200 mg/l. Alkalinity at sampling sites was discovered to be between 70.4 and 100.90 mg/l. Because of the increased presence of these salts, the water at SP-9 was considerably more alkaline than water at other locations. COD is a measure of organic pollution coming from sources like home and industrial wastewater that has not been fully or properly treated in metropolitan regions. In the ongoing work, its value lies in between 0.41-16.59 mg/l. TDS readings that are relatively higher indicate that sewage from homes, runoff from farms, and industrial effluent have been discharged into rivers. If water hardness (TH) is too high, it can lead to human renal failure, scaling in pots and boilers, and other problems. In the investigated region, readings varied from 51-2195 mg/l were reported. Except for SP-9, all specimens were included inside the permitted level (300 mg/l) for drinking. Cl^- is a chemical that is used in water treatment to kill bacteria, parasites, viruses, and microorganisms by neutralising and oxidising them. Cl^- value was observed to range from 9.65 to 4904.91 mg/l. The permissible level set by the WHO is 250 mg/l. When Cl^- is present in excess at SP-9, it makes water taste salty, increases its corrosivity, and can have a laxative impact on people. High SO_4^{2-} concentration causes gastrointestinal discomfort in people and has a purgative impact. Except for SP-9, the range in the current study is 4.97–376.07 mg/l, which is substantially within the 250 mg/l acceptable limits. Significant quantities at SP-9 may also be caused by anthropogenic activity like industrial pollution. Due to geological processes, F^- is primarily found in water. The value of F^- was discovered to range from 0.26 to 1.00 mg/l. The WHO recommended limit was met at all sites, the concentration of F^- was acceptable, and the water could be drunk following disinfection. The principal ingredients of NO_3^- entering the aquatic environment are urban effluent discharge and livestock excretion of nitrogenous waste. The suggested NO_3^- concentration for drinking water is 45 mg/l. The findings showed that the nitrate concentration was between 1.29 and 2.70 mg/l. Since Fe^{2+} helps blood flow, it is not thought that the concentration found poses a health risk. Iron-rich water that has been concentrated can, however, cause turbidity to rise and turn reddish brown. The iron concentration fluctuates from 0.60 to 2.61 mg/l during the study period, which is within the permitted range of 1 mg/l. In the existing research, two indexing mechanisms namely WA WQI and SWARA WQI and three MCDM methods such as CP, OWA and CoCoSo were used to examine the WQI. The obtained WA WQI range is 23.78-96.09, suggesting categories ranging from 'good' to 'extremely poor'. These values were interpolated across the whole research to produce a map in ArcGIS 10.5

(Figure 7.1). For the study, SWARA-WQI was additionally used for the assessment of water quality. It values varied in the range 14.6-1065.2. These values signified that the water quality was excellent to extremely poor. The highest WQI value generated from

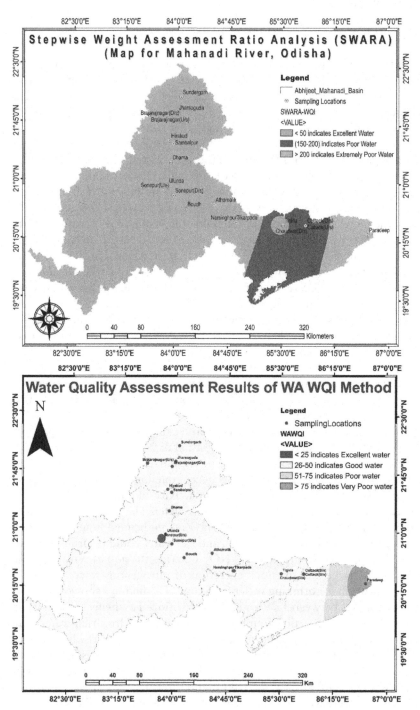

Figure 7.1 Spatial distribution of WA WQI and SWARA WQI map
Source: Author

both stated approaches was scrutinised at SP-9 which had elevated amounts of TC, TKN, EC, SAR, TDS, TH, Cl⁻, SO_4^{2-} and Fe^{2+}. According to WA WQI values, 15.78% and 68.42% of the selected locations had excellent to good water quality, whereas 15.79% had poor or very poor water quality. In case of SWARA WQI, approximately 84.21% of samples fall in excellent range, implying for drinking purposes. Around 15.79% ranged into poor or very poor category. SWARA-WQI variability thematic maps were created for the research region (Figure 7.1). Three locations (SP-8, 9 and 19) out of the total 19 samples reported poor or very poor quality WQI values; this observation may be the result of the area's careless handling of household, industrial, and industrial effluents. Although the river runs through the city's periphery, all sites had very high TKN and TC readings, which may indicate pollution from neighbouring home and agricultural operations. Three MCDM techniques, including OWA, CP, and CoCoSo, were used to resolve any inconsistencies between the basic WQI

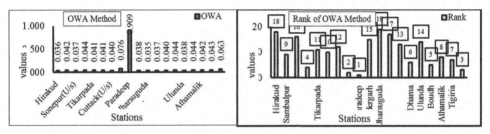

Figure 7.2 Variation and rating of all quality monitoring points based on OWA
Source: Author

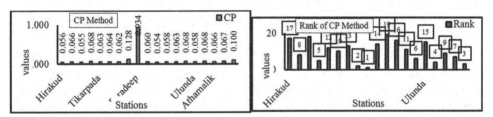

Figure 7.3 Variation and rating of all quality monitoring points based on compromise programming
Source: Author

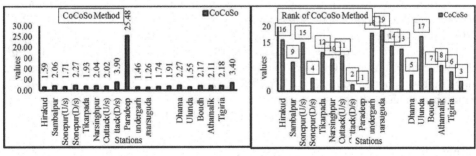

Figure 7.4 Variation and rating of all quality monitoring points based on combined compromise solution
Source: Author

Table 7.1: Description of the surface water quality parameters' best-fitting variogram model.

WQI models	Models outline	Nugget value	Sill value	Nugget/ sill	ASE	RMSE	MSE	RMSSE
WA WQI	Circular	95.67	91.71	1.04	11.08	13.63	-0.03	1.08
	Spherical	99.89	76.57	1.30	11.28	13.66	-0.03	1.07
	Exponential	115.28	83.01	1.39	12.44	13.86	-0.03	1.14
	Gaussian	43.60	361.50	0.12	7.85	12.07	-0.11	0.98
SWARA-WQI	Circular	5999.00	73663.00	0.08	124.06	224.41	-0.12	1.08
	Spherical	6342.30	66090.47	0.10	127.45	224.87	-0.12	1.05
	Exponential	7283.05	59900.06	0.12	146.86	229.99	-0.11	1.41
	Gaussian	2147.23	124816.00	0.02	68.81	193.45	-0.20	0.98

*ASE- Average standard error, RMSE- Root mean square error, MSE- Mean standardized error and RMSSE- Root mean square standardized error

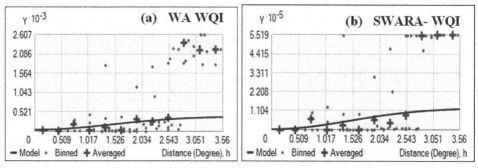

Figure 7.5 The best-fit semi variogram models (a) WA WQI and (b) SWARA-WQI
Source: Author

method, in order to comprehend the river's pollution levels. All the water quality parameters were subjected to these procedures, in order to create overall rankings, with the highest rank for each period denoting the most polluted sample location. Figures 7.2, 7.3, and 7.4 shows how OWA, CoCoSo, and CP vary among the sites as well as ranking. In compared to other places, the sampling location SP-9 was the most contaminated during the study period, followed by SP-8 and SP-19. It is evident that from the analysis at St. 9 that TC, TKN, EC, SAR, TDS, TH, Cl$^-$, SO$_4^{2-}$ and Fe^{2+} had high values relative to their desirable drinking water guidelines. The Gaussian model was regarded as the best-fit semi-variogram model for the WA WQI and SWARA-WQI datasets based on the results (Table 7.1). The Gaussian model's RMSE and ASE values, which were respectively 12.07 and 7.85 for the WA WQI and 193.45 and 68.81 for the SWARA-WQI, were the lowest of all the models. It is evident from the maps that were acquired in Figure 7.5, that shows majority of the samples in the SP-9 region had bigger values suggesting increased pollution. The activity of factories close to the river and the discharge of sewage into it may be the cause.

Conclusion

In conclusion, water pollution is a real issue that threatens aquatic life and causes water shortages as well as health issues. To reduce water pollution, it is crucial to monitor the quality of the water. In this study, the effectiveness of combined use of weighted arithmetic water quality index (WA WQI), stepwise weight assessment ratio analysis water quality index (SWARA-WQI), and multiple-criteria decision making (MCDM) such as compromise programming (CP), ordered weighted averaging (OWA), and combined compromise solution (CoCoSo) approach has been demonstrated with a case study. The outcomes of two indexing techniques, that seems to be WA WQI and SWARA-WQI, jointly reveal 84.21% of sites to come under the purview of excellent too good. The SWARA-WQI and WA WQI graded three Sampling locations (SP-8, 9 and 19) as poor/very poor and all other locations as excellent/good. With the exception of a few areas that need treatment, it is evident from the results of both indexing methods, that shows, bulk of the measured samples are in reasonably good state and hence suitable for drinking. In order to characterise sampling locations, CP, OWA, and CoCoSo were applied, which comprised of all meaningful data and offered an overall rating of the survey points, that primarily focused on their comparative levels of pollution. According to the findings, SP-9 was the most polluted location out of all the sites. This was apparent from the large WA WQI and SWARA-WQI values at this place. Additionally, it has high values of TC, TKN, EC, SAR, TDS, TH, Cl^-, SO_4^{2-} and Fe^{2+}, which were highest among all the areas and also higher than their desired concentration.

Acknowledgment

I would like to thank the research unit of C.V. Raman Global University (C.G.U), Bhubaneswar, India for financial assistance. Sincere, thanks are also due to the Editor and two anonymous reviewers for their constructive comments and suggestions that improved the original version of the manuscript.

References

Bastia, F, Equeenuddin, S, Roy, P D, and Hernández-Mendiola, E (2020). Geochemical signatures of surface sediments from the Mahanadi River basin (India): chemical weathering, provenance, and tectonic settings. *Geological Journal.* 55, 5294–5307. doi: 10.1002/gj.3746.

Gao, Y Y, Qian, H, Ren, W H, Wang, H K, Liu, F X, and Yang, F X (2020). Hydrogeochemical characterization and quality assessment of groundwater based on integrated-weight water quality index in a concentrated urban area. *Journal of Cleaner Production.* 260, 121006.

Shankar, K, and Kawo, N S (2019). Ground water quality assessment using geospatial techniques and WQI in north east of Adama Town, oromia region, Ethiopia. *Hydrospatial Analysis.* 3(1), 22–36. https://doi.org/10.21523/gcj3.19030103.

Yari, G, and Chaji, A (2012). Determination of ordered weighted averaging operator weights based on the m-entropy measures. *International Journal of Intelligent Systems.* 27(12), 1020e1033. https://doi.org/10.1002/int.21559.

8 A comparative performance study on fly ash based geopolymer by varying activator and class of base material

Darsana Sahoo[a] and Suman Pandey[b]

Techno India University, West Bengal, India

Abstract

Geopolymers have grabbed the attention of many researchers, environmentalists and material scientists as an alternative to OPC. This has shown effective results in reducing the associated emissions of greenhouse gas which is the key contributor to global warming. Earlier research-works have shown that Geopolymer is much more environment friendly unlike cement concrete which in turn causes less health hazards and pollution. Studies have also revealed its outstanding performance in terms of durability and suitability over other conventional cement.

Key words: Fly ash, geopolymer, sewage sludge, sodium silicate

Introduction

High demand of concrete as construction material and increasing necessity of production of cement has raised environmental degradation issues further. Emission of CO_2 and energy consumption associated with production of cement is another major concern and hence an eco-friendly substitute of this material is essential for sustainable development. On the way to develop green concrete which can impart high strength and durability and at the same time which can be produced with less emission of CO_2, geopolymer concrete (GPC) is gradually emerging as smart way of substitution of ordinary plain concrete.

Khale and Chaudhary (2007) addressed geopolymerization as a broad scope of research for utilizing solid waste products. The geopolymer paste is composed of sewage sludge (15% of total weight) and the base material, fly ash (Class F/ Class C). Van Jaarsveld et al. (2003) suggested in his research work that not every waste material is dissolved in alkali solution. Because of that the author mentioned that original structure of some waste particles remains intact and contribute to either quicken or toughen those developed frameworks. The work of sodium silicate is to initiate the polymerization at an earlier stage of reaction. Obonyo et al. (2011) studied the extent of polymerization of aluminosilicate. As metal hydroxide interacts with the reactive solid material in presence of silicate solution, it starts the generation of silicate and aluminate monomers in the mixture. The major drawback is that this silicate solution consists more than 65% of water which is responsible for porous character and semi crystalline phases in geopolymer. The scope of using alkaline sludge as an alternative to sodium silicate may be investigated. The present research work shows the comparative performance of geopolymer with fly ash by varying activator and class of base material. Different alkali combinations are also tried and the selection of natural alkali like sludge combination is made judiciously. Then the investigation is made to study the impact of sludge as natural activator to develop fly ash geopolymer

[a]darsanasahoo1999@gmail.com, [b]suman.p@technoindiaeducation.com

DOI: 10.1201/9781003450924-8

completely from waste. The performance and character of geopolymer in fresh and hardened state along with its micro structural changes are simultaneously analyzed.

Experimental investigation

First, the fly ash and sludge (15% of total wt.) were well mixed in partly wet condition in aluminium tray by hand mixing and Table 8.1 shows the details of the composition of the prepared geopolymer paste.

Workability test

Table 8.2 presents the results of workability of prepared specimen based on the combination of alkali and class of fly ash. It is again proved that geopolymer sample GPP2 failed to form gel. In absence of sodium silicate, the initiation of polymerization may not be possible for geopolymer with fly ash. But the addition of sludge in fly-ash geopolymer showed improved workability even in absence of sodium silicate in mix. In absence of sodium silicate, the fly ash based geopolymer sample GPP3 exhibits satisfactory performance.

It was reported earlier that there is a possibility of formation of calcium-based composites such as silicates, aluminate hydrates etc. during the time of synthesis of fly-ash geopolymer in presence of calcium. In earlier research-works it has been also suggested that the workability of fresh alkali activated fly ash composites decreases with the increase in the supplementary slag content. A similar nature of consistency was observed for class C fly ash (comprising calcium) based geopolymer activated with sludge. Paste samples were collected as shown in Table 8.3. It has indicated phase changes with time. After mixing the area factor dropped instantly at a constant

Table 8.1: Composition of geopolymer paste with specimen ID.

Specim-en ID	Class of fly ash	K_2O content acting as an activator [wt.%]*	SiO_2 content of activator (Silicate solution source) [wt.%]*	Slud-ge [wt.%]*	Water/fly ash ratio
GPP1	Class F	10	10	00	0.36
GPP2	Class F	10	00	00	0.36
GPP3	Class F	10	00	15	0.36
GPP4	Class C	10	00	15	0.36

*Content relative to the combined weight of sludge and the fly ash
Specimen size: 50 mm × 50 mm × 50 mm

Table 8.2: Results of workability test of the geopolymer paste.

Specimen ID	Final equivalent diameter (D_2) cm	Final area after flow (A_2) (cm^2)	Area factor (A_2/A_1)
GPP1	28	615.44	21.8
GPP2	-	-	-
GPP3	20	314	11.11
GPP4	16	200.96	7.11

Note: Initial diameter (D_1) cm = 6 cm, Initial area (A_1) cm^2 = 28.26

Table 8.3: Class C fly-ash geopolymer test results on workability vs time activated in presence of sludge.

Time elapse after mixing (mins)	Final equivalent diameter (D_2) cm	Final area after flow (A_2) cm^2	Area factor = A_2/A_1
5	16.0	200.96	7.11
10	15.0	176.79	6.26
15	15.0	176.79	6.26
20	12.0	113.14	4.00
25	10.0	78.57	2.78
30	6.0	28.26	1.00
35	6.0	28.26	1.00

Note: Initial diameter $(D_1$ cm = 6.0, Initial area (A_1) cm^2 = 28.26

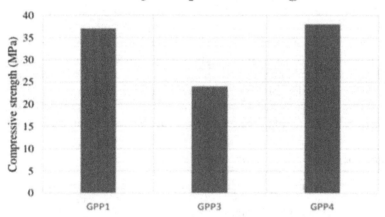

3 -Day Compressive strength

Figure 8.1 Day compressive strength of fly-ash geopolymer
Source: Author

rate of time. The area factor reached to unity between 30 and 35 minutes. A unit area factor may be considered as an end to the plastic state.

Compressive strength

To note the change in compressive strength with aging test specimens were tested for compressive strength after 3, 28, 60. and 90 days of curing. 3-day compressive strength for three typical series of specimens are presented in Figure 8.1. Alkali activation of fly ash based geopolymer could not be initiated without sufficient amount of sodium silicate which initiate the polymerization in earlier stage (Dutta et al., 2012). But for sludge blended geopolymer the scenario was different. Sludge may initiate the polymerization process by forming in-situ inorganic foam itself at the initial stage of activation like silica fume (Prud'homme et al., 2010). It is observed that compressive strength of class C fly-ash geopolymer blended with sludge remains almost unaffected by lowering content of sodium silicate. On contrary the presence of external sludge provided a better geopolymer where, maximum value of compressive strength was

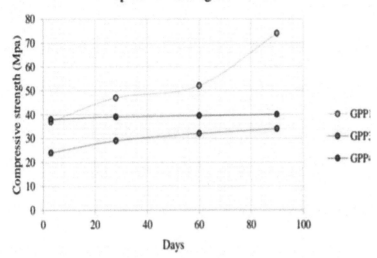

Figure 8.2 Strength gain of specimens with aging
Source: Author

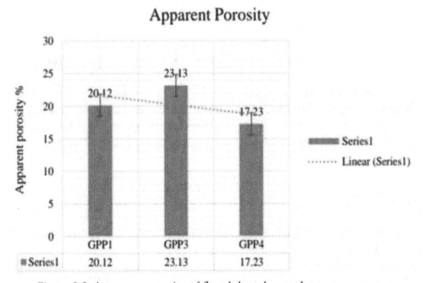

Figure 8.3 Apparent porosity of fly ash based geopolymer
Source: Author

obtained 38.0 MPa for specimen GPP4. Again, it was observed less change of strength value with age, for Class C fly ash based geopolymer.

Water absorption and porosity test

GPP1, GPP3 and GPP4 were selected to assess the apparent porosity and water absorption. These tests were conducted to find the preliminary idea about the permeable pore volume of fly-ash geopolymer. The apparent porosity and water absorption value were slightly increased for sample GPP3, with the replacement of silicate by

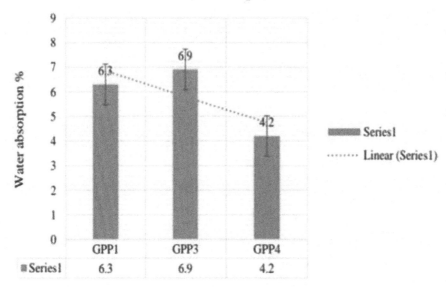

Figure 8.4 Water absorption of fly ash based geopolymer
Source: Author

sludge (for Class F based geopolymer). But minimum value of apparent porosity and water absorption were found for sample GPP4 (Figures 8.1 and 8.2). Earlier research indicated that the amorphous sodium aluminosilicate product is achievable in presence of calcium compound in alkaline environment (Van Jaarsveld et al., 1997). The replacement of sodium silicate by sludge is possible but the resultant product has a moderate level of performance for class F fly ash based geopolymer. Whereas class C fly-ash geopolymer performed tremendously well when activated with sludge.

Conclusions

Naturally available activator (sludge) and its impact on fly ash based geopolymer:

- In absence of sodium silicate, the initiation of polymerization may not be possible for fly ash based geopolymer (Example: Sample GPP2).
- On adding sludge in fly ash based geopolymer showed improved workability even if sodium silicate is not present in mix (Example: sample GPP3).
- Appreciable workability was also found for fly ash geopolymer activated with sewage sludge (waste) even in absence of sodium silicate. This phenomenon supports that the high alkaline sludge has equivalent potentiality like the sodium silicate in geopolymer mixture.
- Increasing trend of strength of fly-ash geopolymer (sample GPP1) with aging was found. This is mainly due to the internal pore pressure developed by the unreacted gel within the pore, may be responsible for the micro-cracking of the fly-ash geopolymer specimens with aging.
- Maximum 3-day compressive strength was obtained for specimen GPP4 i.e., 38.0 MPa.

- Replacement of sodium silicate by sludge is possible but the resultant product has a moderate level of performance for class F fly ash based geopolymer.
- Class C fly ash based geopolymer performed tremendously well when activated with sludge. It shows an improvement in pore distribution as confirmed by water absorption and apparent porosity test.

References

Dutta, D, Chakrabarty, S, Bose, C, and Ghosh, S (2012). Comparative study of geopolymer paste prepared from different activators. *STM Journals*. 2, 1–10.

Khale, D, and Chaudhary, R (2007). Mechanism of geopolymerization and factors influencing its development: a review. *Journal of Materials Science*. 42(3), 729–746.

Obonyo, E, Kamseu, E, Melo, U C, and Leonelli, C (2011). Advancing the use of secondary inputs in geopolymer binders for sustainable cementitious composites: a review. *Sustainability*. 3, 410–423.

Prud'homme, E, Michaud, P, Joussein, E, Peyratout, C, Smith, A, Arrii Clacens, S, Clacens, J M, and Rossignol, S (2010). Silica fume as porogent agent in geo-materials at low temperature. *Journal of the European Ceramic Society*. 30(7), 1641–1646.

Van Jaarsveld, J G S, Van Deventer, J S J, and Lorenzen, L (1997). The potential use of geopolymeric materials to immobilize toxic metals: Part I. *Theory and Applications Minerals Engineering*. 10, 659–669.

Van Jaarsveld, J G S, Van Deventer, J S J, and Lukey, G C (2003). The characterization of source materials in fly ash-based geopolymers. *Materials Letters*. 57(7), 1272–1280.

9 A smart scheduling: an integrated framework under cloud computing environment

Biswajit Nayak[1,a], Bhubaneswari Bisoyi[1,b], Biswajit Das[2,c] and Prasant Kumar Pattnaik[3,d]

[1]Faculty of Management Studies, Sri Sri University, Cuttack, India

[2]School of Management, KIIT, Bhubaneswar, India

[3]School of Computer Science.KIIT, Bhubaneswar, India

Abstract

Due to the concept of openness in the cloud computing environment there is an exponential increase in the data. The proper utilization of tasks plays a major role in solving this problem. Because task scheduling is a concept that schedule the task and existing resources in such a way that, performance of the system can be enhanced. Designed framework is tested and found the optimum performance of the system. An integrated framework is designed to execute or support variable length tasks. The framework is tested with different dataset and found optimum resource utilization.

Keywords: Algorithm, cloud computing, data centre, makespan, scheduling

Introduction

Task scheduling provides user satisfaction and stability of system by allocating resources in proper manner. Better throughput and improved response time can be achieved through several loads balancing algorithm available in cloud computing environment (Aggarwal and Gupta, 2017; Al-Qurishi et al., 2015).

The problem of scheduling tasks to the available resource can be solved by several task scheduling algorithms, because the concept behind the development of task scheduling algorithms is to increase the utilization of resources in such a way that, it will increase the efficiency by minimizing execution time, so that the efficiency of the system in the cloud computing environment can be boosted. An algorithm is required to be designed to execute or support variable length tasks and should also include the benefit of heuristic algorithms to optimize scheduling tasks with available resources to optimize system (Nayak et al., 2018a, 2018b).

Basic techniques

All the algorithms categorized into two different concepts based on the way they behave like static batch mode and dynamic batch mode. The concept of Traditional Min-Min and Max-Min algorithm are based on the minimum completion time and maximum completion time.

These methods execute the task in two different stages. First it finds the completion time then choose the task according to the minimum or maximum completion time

---

[a]biswajit.n@srisriuniversity.edu.in, [b]bhubaneswari.b@srisriuniversity.edu.in, [c]biswajit@ksom.ac.in, [d]patnaikprasantfcs@kiit.ac.in

DOI: 10.1201/9781003450924-9

based on algorithm it is using. If it is Min-Min algorithm, then it is good for small tasks at the same time produce starvation for large sized task. Whereas in case of Min-Min algorithm, it is idle for long sized task. But produces starvation for small sized task. So, because of the different issues of both the algorithms, it can be concluded that Min-Min algorithm can be suitable for smaller sized tasks whereas the Max-Min algorithm can be suitable for only tasks which are larger in size (Nayak et al., 2018a, 2018b).

Performance parameters

Performance parameters are used to evaluate the effectiveness of the system to satisfy the required requirements. Some of the parameters are responsible for such optimization of system (Kumari and Monika, 2015):

Makespan (MS): Makespan is a concept of time required to complete all tasks. It can also be defined as the maximum time needed to complete all the tasks over the datacentre. Like the resource utilization the makespan is also proportional to load balancing. A good task algorithm is that, which diminish makespan (Nayak et al., 2019).

A good scheduler means a technique that balance of load is optimum so the system will ensure optimum result. Let us consider 'N' virtual machine with 'n' application inputs and "T1, T2, T3, T4... Tn" are set of tasks. Cloud computing uses expected time to compute matrix for load balancing in the heterogeneous environment.

Makespan is the completion time of all machine and can be defined as:

$$MS = Max\ (T_{c_j}, \ldots\ldots\ldots T_{cm}) \tag{1}$$

Where: j =1, 2, …………m

One of the mostly used parameter to evaluate the performance in cloud computing environment is the makespan. Execution time always differs based on virtual machine. Better makespan means there exists a good load balance.

Methodology and model specifications

The algorithm aims to get rid of the existing difficulty present and enhance or maximize the performance. Some of the basic as well as heuristic algorithms used by different researchers for dealing with different problems but facing difficulties to achieve optimized solution (Rajput and Kushwah, 2016). A proposed hybrid model is designed to eliminate the problem and to resolve the existing starvation problem to optimize the performance of the system (Nayak et al., 2018a, 2018b). The proposed algorithm is also tested and found optimum result.

Result analysis

The developed algorithm first calculates completion time. Based on expected completion time the resources those are available, assigned to each task appropriately. Let us consider four tasks (C1, C2, C3, C4) with several parameters required for the execution. The table below (Tables 9.1 and 9.2) clearly shows cloud configuration used for execution of tasks and datacenter configuration respectively.

Table 9.3. represents the processing speed for all resources as well as the bandwidth. Bandwidth represented with mega bit per second (Mbps) and processing speed represented with million instructions per second (MIPS). Table 9.4 represents instructions and data volume of tasks. Instruction volume represented with millions of instructions (MI) and data volume represented with MB. From the execution of above available configuration parameter, the value at Table 9.5 can be found out for execution.

The expected execution-time (Equation. 2) and the completion-time (Equation. 3) on each resource can be evaluated by using the equation below:

Execution time:

Table 9.1: Cloud configuration.

Name of entities	Entity description
No. of dtacenter	1
No. of hosts	1
No. of virtual machines (VMs)	2
No. of Cloudlets	4

Source: Author

Table 9.2: Datacenter configuration.

Name of entities	Entity description
Architecture	x86
RAM (MB)	2048
Hypervisor	Xen
Storage (MB)	10000
MIPS	150, 300
Bandwidth (Mbps)	300, 15

Source: Author

Table 9.3: Resource specification.

Resources	R1	R2
Processing speed (MIPS)	150	300
Bandwidth (Mbps)	300	15

Source: Author

Table 9.4: Task specifications.

Tasks	Instruction volume (MI)	Data volume (Mb)
C1	257	87
C2	36	30
C3	328	95
C4	211	589

Source: Author

Table 9.5: Tasks execution time on each resource.

Resources	R1	R2
Tasks		
C1	2	6
C2	1	3
C3	3	8
C4	3	40

Source: Author

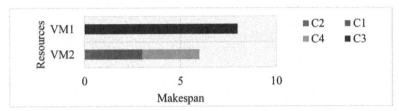

Figure 9.1 Total completion time when condition is false
Source: Author

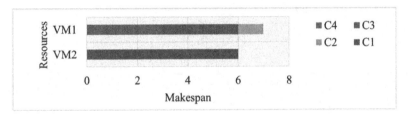

Figure 9.2 Total completion time when condition is true
Source: Author

$$ETij = \left(\frac{MIi}{MIPSj}\right) + \left(\frac{Mbi}{Mbpsj}\right) \tag{2}$$

Where: MIi - Millions of instructions, Mbi - Mega bit, $Mbpsj$ - Mega bit per second $MIPSj$ - Million instructions per second

The expected completion time can be calculated by adding ready time of resources because initially all the resources are in idle state.

$$CTij = ETij + RTij \tag{3}$$

Where: RT - Ready time

In Figure 9.1 tasks are executed based on their execution time and minimum completion time. The evaluation carries the makespan as 8. Whereas the Figure 9.2 process the given tasks based on maximum completion time. The evaluation carries a value of makespan as 7. As per the proposed algorithm the step 4 should give the optimum outcome as it consists of large numbers of small tasks and few numbers of large tasks only.

It also shows most of the task have the completion time more in case of execution through step 4 as compared to the execution through step 3 but still makespan of step 4 is less than process through step 3.

The above set of datasets can be challenged by taking opposite category of dataset that means a dataset that consists of larger tasks and few numbers of small tasks. The Table 9.9 shows the dataset description. It is carrying large numbers of large dataset

Table 9.6: Datacenter configuration.

Name of entities	Entity description
Architecture	x86
RAM (MB)	2048
Hypervizor	Xen
Storage (MB)	10000
MIPS	100, 350
Bandwidth (MBps)	70, 60

Source: Author

Table 9.7: Resource specification.

Resource	R1	R2
Processing Speed (MIPS)	100	350
Bandwidth (Mbps)	70	60

Source: Author

Table 9.8: Task specifications.

Tasks	Instruction volume (MI)	Data volume (MB)
C1	8175	137
C2	15295	1278
C3	12109	182
C4	6107	137

Source: Author

Table 9.9: Execution of tasks on each resource.

Resources	R1	R2
Tasks		
C1	81.78	23.36
C2	112.95	65.0
C3	121.09	34.59
C4	61.07	17.45

Source: Author

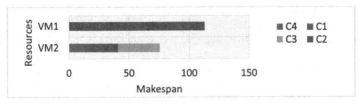

Figure 9.3 Total completion time when condition is false
Source: Author

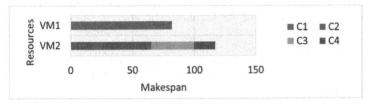

Figure 9.4 Total completion time when condition is true
Source: Author

and few numbers of small tasks. Table 9.6 represents the configuration of datacentre whereas Table 9.7 represents resource specification means the processing speed for all resources as well as the bandwidth. Bandwidth represented with mega bit per second (Mbps) and processing speed represented with million instructions per second (MIPS). Table 9.8 represents instruction and data volume of tasks. Instruction volume represented with MI and data volume represented with MB. Table 9.9 represents execution time for each task on each resource.

Again, the expected execution time and the completion time of each task on each resource can be evaluated by using the two equations: Equations (2) and (3).

In Figure 9.3 tasks are executed based on their execution time and minimum completion time and evaluation carry the makespan as 112.95(ms). Whereas Figure 9.4 process the given tasks based on maximum completion time and the evaluation carry a value of makespan as 117.08.

As per the proposed algorithm the step 3 should give the optimum outcome as it is a combination of larger tasks and few numbers of small tasks only. The Figure 9.4 showing the optimum outcome of step3 and step4.

Conclusion

The concept cloud computing offer to access unlimited resources that can be easily used any point of time on demand. That is the reason why it is required to handle the execution of data efficiently in the cloud computing environment. This can be solved by proper assignment of task to the available resources. The proposed integrated framework is tested with variable length data set. The result showing proper assignment of task and the optimized output as it is taking minimal time for execution of set of given tasks.

References

Aggarwal, R, and Gupta, L (2017). Load balancing in cloud computing. *International Journal of Computer Science and Mobile Computing*. 6(6), 180–186.

Al-Qurishi M, Al-Rakhami M, Al-Qershi F, Hassan M. M, Alamri A, Khan H. U, Xiang Y, (2015). A framework for cloud-based healthcare services to monitor non-communicable diseases patient. *International Journal of Distributed Sensor Networks.* 11(3), 1–11. http://dx.doi.org/10.1155/2015/985629.

Nayak, B, Padhi, S K, and Pattnaik, P K (2018a). Scheduling issues and analysis under distributed computing environment. *Journal of Advanced Research in Dynamical and Control Systems.* 10(02), 1475–1479. ISSN 1943-023X

Kumari, E, and Monika, A (2015). Review on task scheduling algorithms in cloud computing. *International Journal of Science, Environment and Technology.* 4(2), 433–439.

Nayak, B, Padhi, S K, and Pattnaik, P K (2018b). Static task scheduling heuristic approach in cloud computing environment. 5th Springer International Conference on Information System Design and Intelligent Applications. (vol. 862), pp. 473–480. DOI: 10.1007/978-981-13-3329-3_44.

Nayak, B, Padhi, S K, and Pattnaik, P K (2019). Optimization of cloud datacenter using heuristic strategic approach. 2nd International Conference on Soft Computing and Signal Processing, (vol 900). Springer, Singapore, pp. 473–480. https://doi.org/10.1007/978-981-13-3600-3_9.

Rajput, S S, and Kushwah, V S (2016). A genetic based improved load balanced min-min task scheduling algorithm for load balancing in cloud computing. 2016 8th International Conference on Computational Intelligence and Communication Networks (CICN). pp. 667–681. DOI:10.1109/cicn.2016.139.

10 Studying wind effect on high-rise building with different cross-section

Dhruv Chaudhary[1,a], Dhawal Tayal[1,b], Devesh Kasana[1,c], Ritu Raj[1,d], S. Anbukumar[1,e] and Rahul Kumar Meena[2,f]

[1]Civil Engineering Department, Delhi Technological University, Delhi, India

[2]Civil Engineering Department, Punjab Engineering College, Chandigarh, India

Abstract

International standard is helpful to design the tall building having regular shape while the irregular shape needs evaluation because wind load is governing the design of the tall building. With the development of technology, composite plan shaped buildings like pentagon, hexagon, etc. have been given consideration by many architects while keeping in mind the building's aesthetics. Hence pressure data is not presented in most of the standards for such shape hence investigation is must for such type of building. Pressure on the upstream side face is nearly same for both buildings while suction magnitude is different in the downstream side face of the high-rise building model having pentagon and square plan shape.

Keywords: ANSYS, CFD, flow patterns, high-rise building, pressure contours

Introduction

Infrastructure of high-rise building is changing the pattern in each decade. Now a days architects have limitations in developing regular plan geometry because of the reason of land shape. Regular shape structure design is possible through the available standards while in case of irregular shape investigation is must hence Computational Flud Dynamics or wind tunnel tests are used to evaluate the wind effects. With the increment in height of the structure wind load is also increases. Different research studies have already been conducted on different shape with different specification of the building. Some of the studies are as (Bhattacharyya and Dalui, 2018) on "E"-plan shaped; (Nagar et al., 2020) on "H" and square plan (Raj et al., 2020) on "+" shape building; (Pal et al., 2021) evaluated on fish and square plan shape models; and two identical shape building a remodeled triangle shape model; and (Nagar et al., 2021) evaluated the effect on two "Plus" shape model some of the important outcomes from their studies are as pressure contour is symmetrical for symmetrical wind angle. Wind ward face is not much dependent on the shape and size. Suction will mostly be observed in the downstream wake and sides of the buildings.

Wind effects are also investigated using the CFD such as on the rectangular (Paul and Dalui, 2021; octagonal Hajra and Dalui, 2016); rectangular with different corners (Meena et al., 2022); "Y" shape most of the study were validated with experimental studies or with international standards. However, the mesh sizing will significantly affect the results if grid independence test is not performed. K-epsilon model will perform best for the tall building problems.

[a]dhruvchoudhary_2k19ce044@dtu.ac.in, [b]dhawaltayal_2k19ce043@dtu.ac.in, [c]deveshkasana_2k19ce041@dtu.ac.in, [d]rituraj@dtu.ac.in, [e]sanbukumar@dce.ac.in, [f]rahulmeena@pec.edu.in

DOI: 10.1201/9781003450924-10

The present study investigates the wind effects on tall buildings with square and pentagon plan shape. However, the area is keeping constant for both types of buildings so that the architect can decide which shape is most effective in receptive scenario. ANSYS CFX student version is used for post CFD while for simulation purpose ANSYS academic version is used. Different results like pressure contours and streamlines are presented in this chapter.

Numerical model development

Geometric modelling

Present study is performed on two building model H and A having pentagon and square shape are depicted in Figure 10.1. The height of the model is kept constant as 600 mm while the plan area of the model is same for both the model. The geometric scale is 1:350 and buildings are assumed as non-deformable. Wind effects are investigated in virtual wind tunnels and modeling is performed in ANSYS and guidelines are used as available in ASCE report no. 67 wind tunnel manual.

Results and discussions

Pressure contours

The pressure contours for pentagon shape building model-H is presented in Figure 10.2 while for model-A, square shape is presented in Figure 10.3. The pressure on face which is under direct exposure to wind is in positive pressure which is more or less identical and same magnitude. The wind effect in the form of pressure on the side faces are found symmetrical for the symmetrical wind incidence angle and same can be verified with both the Figures 10.2 and 10.3. As the wind crossing the building from its

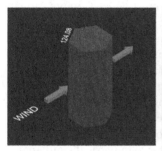 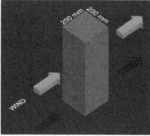 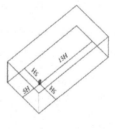

Figure 10.1 Model and domain
Source: Author

Face-1   Face-2   Face-3   Face-4   Face-5   Face-6   Top Face

Figure 10.2 Pressure contours for model H
Source: Author

| Face-1 | Face-2 | Face-3 | Face-4 | Top Face |

Figure 10.3 Pressure contours for model A
Source: Author

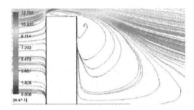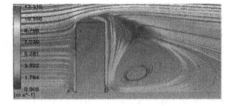

Model-H   Model-A

Figure 10.4 Velocity streamlines
Source: Author

height it is presented that upstream to downstream side wind effects are changing from pressure to suction it is because of the flow reattachment in the top of the building.

The average coefficient of pressure C_p on faces 1 is +0.54, face 2 is -0.13, face 3 is -0.35, face 4 is -0.30, face 5 is -0.35, face 6 is -0.13. On face 1, the maximum positive Cp is observed 0.54. Similarly negative Cp of 0.35 are observed for Faces 3 and 5. The negative pressure coefficient values for Faces 2 and 6 are also 0.13 for model H. It can be observed that the symmetrical face such as face-2 and face-6 has equal pressure coefficient similarly face-3 and face-5 is under the same or equal pressure distribution and lee ward side face which is face-4 is having less suction than the face-3 and 5 but more suction than the face 2 and 6 it is because of the reason that the wind is incident from face-1 to face-4.

Wind effects in graphical form of pressure contours is presented for building model-A which is having square plan shape. It is clearly presented in Figure 10.3 that the wind ward face is having pressure while rest three faces side face and lee ward face of the building are in suction. The side face 2 and face 4 is having similar pressure distribution also as the wind is passing the building initially face 2 some part on upstream is in positive then the pressure starts to decrease and before leaving the side face it is under the suction and the downmost part of building on down stream of wind, side face is under the more suction then the top of the building.

The average value of C_p on face 1 is 0.68, on face 2 is -0.66, on face 3 is -0.34, on face 4 is -0.66. The maximum positive pressure coefficient value is 0.68 occurred in face 1. Faces 2 and 4 are getting similar negative pressure coefficient values of 0.66.) for building model A.

Streamlines or flow patterns

Streamlines or flow lines are helpful in understanding the flow characteristics in the upstream and downstream, Figure 10.4 is clearly indicating the flow pattern along the height of the building for both model H and A. The flow in the upstream get increases wind speed in top and ground of the building model while from the top on third position the flow is in stagnation zone for both buildings. The flow reattachment is also presented in the top of the building and the flow reattaches the building at the top part earlier than the square building. The flow is reattaching the square building nearly at the central portion of building along the plan from upstream to downstream.

The wind speed clearly indicating different flow features such as upwash and downwash, upwash wind speed is slightly more than the downwash wind. Also, the vortex shedding in the downstream of wind for both the building is different like for square building strong vortices are observed than the pentagon shape building.

Conclusion

Due to its turbulent and unpredictable nature, wind can create various flow conditions when interacting with tall buildings, making it a complex phenomenon to study. Thus, investigating wind flow characteristics on tall buildings is crucial to understand their behavior. Important outcomes form the present research is summarized as below:

International standard is useful for the regular shape tall building while the designer needs to investigate the wind effect on irregular shape model.

Complex shapes such as hexagon, Model H, ANSYS CFX is employed to interpret wind flow characteristics on high rise structure and results are helpful to the architect in designing such type of tall building.

Effect of wind flow is presented in the form of pressure contours and vertical streamlines, are presented in vertical flow to understand different flow patterns.

The maximum positive pressure was observed on upstream face which is directly comes under the wind exposure and then the side faces as well as the downstream face are in suction.

References

Bhattacharyya, B, and Dalui, S K (2018). Investigation of mean wind pressures on "E" plan shaped tall building. *Wind and Structures, An International Journal*. 26(2), 99–114. Available at: https://doi.org/10.12989/was.2018.26.2.099.

Hajra, S, and Dalui, S K (2016). Numerical investigation of interference effect on octagonal plan shaped tall buildings. *Jordan Journal of Civil Engineering*. 10(4), 462–479.

Meena, R K, Raj, R, and Anbukumar, S (2022). Wind excited action around tall building having different corner configurations. *Advances in Civil Engineering*. 1-22. Available at: https://doi.org/https://doi.org/10.1155/2022/1529416.

Nagar, S K, Raj, R, and Dev, N (2020). Experimental study of wind-induced pressures on tall buildings of different shapes. *Wind and Structures, An International Journal*. 31(5), 441–453. Available at: https://doi.org/10.12989/was.2020.31.5.431.

Nagar, S K, Raj, R, and Dev, N (2021). Proximity effects between two plus-plan shaped high-rise buildings on mean and RMS pressure coefficients. *Scientia Iranica*. 0(0), 28. Available at: https://doi.org/10.24200/sci.2021.55928.4484.

Pal, S, Raj, R, and Anbukumar, S (2021). Comparative study of wind induced mutual interference effects on square and fish-plan shape tall buildings. *Sādhanā*. 46(2), 86. Available at: https://doi.org/10.1007/s12046-021-01592-6.

Paul, R, and Dalui, S K (2021). Shape optimization to reduce wind pressure on the surfaces of a rectangular building with horizontal limbs. *Periodica Polytechnica Civil Engineering*. 65(1), 134–149. Available at: https://doi.org/10.3311/PPci.16888.

Raj, R, Ark Rukhaiyar, Bhagya Jayant, Kunal Dahiya, Rahul Kumar Meena and Ritu Raj. (2020). Response analysis of plus shaped tall building with different bracing systems under wind load. *International Journal of Advanced Research in Engineering and Technology*. 11(3), 371–380. Available at: https://doi.org/10.34218/IJARET.11.3.2020.032.

11 Modelling of preforming design criteria: closed-die forging of axi-symmetrical cylindrical aluminum preforms

Saranjit Singh[a]

School of Mechanical Engineering, KIIT Deemed to be University, Bhubaneswar, India

Abstract

The preform design is a necessary step to achieve complete die-filling to accomplish a defect-free forged product as it involves steps for optimal design of preform shapes. The present paper investigates a case of a closed-die forging process using a modified preforming criterion for an optimal design of perform with an objective to reduce the forging load and die wear. The process involved deformation of cylindrical preforms into double-hub flange components. The criteria of preforming stages were validated by comparing the energy dissipated during forging of actual preforms and corresponding enveloping preforms. Three different closed-die forging stages were considered to validate the preforming criteria using the concept of shape complexity factors, which were examined for the proper die fills.

Keywords: Aluminum preform, closed-die forging, die fills, shape-complexity

Introduction

The preforming steps are extremely critical for the design and development of forging processes. Forging defects can be greatly reduced by adequate design of preforming steps, which leads to proper corner-fillings. The impression of preform shape influences distribution of metal within die-cavity. It is a vital factor influencing flash formation, forging load, die wear, and die life. Thus, the determination of intermediate forging stages required for transforming performs in multiple stages is fundamental to the design and must be included during the design of forging processes (Tomov and Wanheim, 1993; Zhao et al., 1996; Radev and Tomov, 2002).

The present paper presents a preforming criterion for an optimal preform design using a case of closed-die forging process. It involves deformation of cylindrical aluminum preforms into double-hub flange components. The deformations were carried out within closed-die sets having recess cavities on both the halves for proper preform location. The work is based on a novel concept of shape complexity factor (SF) (Tomov, 1997). The energy dissipations were theoretically computed based on the "upper bound" approach. Three different intermediate forging stages having specific SF values were examined for proper squeeze filling of die cavities. The criterion of performing stages and design was finally validated.

Forging of axi-symmetric cylindrical preform

Figure 11.1 shows the positioning of the preform with closed-die set and subsequent forging stages. The die punches have recess cavities for proper preform location, so that they can be located co-axially within a die-container axis. The entire deformation of preform is divided into two stages. The first stage is an upsetting or free barreling of

[a]ssingh@kiit.ac.in

DOI: 10.1201/9781003450924-11

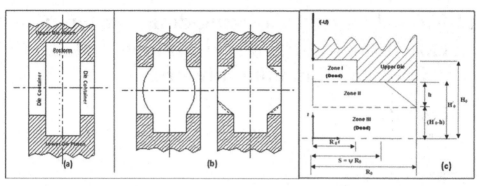

Figure 11.1 Forging of cylindrical aluminum preforms and deformation modes
Source: Author

the preform, whereas the subsequent second stage involves constrained deformation. The first stage of deformation is considered redundant with respect to the preforming step design and only the second stage of the deformation has been considered for the analysis. The die corners are formed and filled as shown in Figure 11.1(b). For the ease of analysis, only quarter portion of the preform has been considered. Further, the preform is divided into deforming and dead zones based on the concept of "Velocity Discontinuities". Figure 11.1(c) shows the deformation mode of the axi-symmetric preform along with the dimensions for deforming and dead zones. It may be noted that zones I and III are considered as dead zones, and deformation is considered only in zone II (Agarwal and Singh, 2021a, 2021b).

The present research considered few assumptions (Singh et al., 2007a, 2007b, 2007c):

- The compatibility condition during deformation was derived based on the "volume constancy" principle and is expressed as follows:

$$\dot{\varepsilon}_{rr} + \dot{\varepsilon}_{\theta\theta} + \dot{\varepsilon}_{zz} = 0 \tag{1}$$

- The interfacial frictional conditions at die-preform-container interfaces are considered to be hybrid, consisting of both sticking and sliding friction. The mathematical expression for same is as follows:

$$\tau = \mu\left\{P_{av} + \chi\left[1 - \left(\frac{R_m - r}{nR_0}\right)\right]\right\}; \text{ where } R_m = R_0 - \frac{H_0}{2\mu}\ln\left(\frac{1}{\mu\sqrt{3}}\right) \tag{2}$$

- The average forging load was computed during the conclusion or closing stages of the deformation stage towards (i.e., toward the end of squeeze die-cavity fills).
- The circumferential stress, strain and flow of the preform during the deformation was neglected (i.e., preform's deforming surfaces were considered to remain straight during the deformation).

As mentioned above, the total external energy (J^*) dissipations as supplied by the die platen during the deformation of axi-symmetrical cylindrical aluminum preforms was computed using "upper bound" approach. The mathematical expression is as follows:

$$J^* = (W_i + W_f + W_a) = \left[\frac{2\sigma_0}{\sqrt{3}} \int_v \sqrt{\frac{1}{2}\dot{\varepsilon}_{ij}\dot{\varepsilon}_{ij}} dV + \int_s \tau[\Delta U] dS + \int_v (a_i U_i) dV\right] \tag{3}$$

The individual expressions for the various energy dissipations were mathematically computed based on the compatibility condition, composite interfacial frictional law, boundary conditions, velocity field and the associated strain rates. The boundary conditions are given as follows:

$$U_z = 0 \text{ at } z = (H_0 - h); \ U_z = -U \text{ at } z = H_0 \tag{4}$$

The velocity field was derived based on the compatibility conditions and boundary conditions. The kinematically admissible velocity field along with the corresponding strain rates is given as follows:

$$U_r = \left(\frac{Ur}{2H_i}\right); \ U_z = -\left[\frac{z-(H_0-H_i)}{H_i}\right] U; \ U_\theta = 0 \tag{5a}$$

$$\dot{\varepsilon}_{rr} = \dot{\varepsilon}_{\theta\theta} = \left(\frac{U}{2H_i}\right); \ \dot{\varepsilon}_{zz} = -\left(\frac{U}{H_i}\right); \ \dot{\varepsilon}_{r\theta} = \dot{\varepsilon}_{\theta z} = \dot{\varepsilon}_{rz} = 0 \tag{5b}$$

According to "upper bound" approach, the energy dissipation due to internal deformation (W_i), frictional shear energy dissipations at die-preform (W_{f1}) and container-preform (W_{f2}) interfaces and inertia energy dissipation (Wa) were computed and mentioned as follows:

$$W_i = \left[\frac{\sqrt{3}\pi\sigma_0 U R_0^2(2\psi+1)}{3\sqrt{3}}\right] \tag{6}$$

$$W_{f_1} = \left[\frac{\pi\mu U R_0^3(\psi^3-\lambda^3)}{3H_i}\right]\left[P_{av} + \chi\left(1 - \frac{Rm}{6nR_0}\right) + \left(\frac{\phi_0}{8n}\right)\left(\frac{\psi^4-\lambda^4}{\psi^3-\lambda^3}\right)\right] \tag{7}$$

$$W_{f_2} = \left[\frac{\pi\mu U R_0(H_0-H_i)^2 U}{H_i}\right]\left[P_{av} + \chi\left(1 - \frac{Zm}{nH_0}\right) + \left[\frac{2\phi_0(H_0-H_i)}{3nH_0}\right]\right] \tag{8}$$

$$W_a = \left[\frac{\pi\rho_i(1-\psi)R_0^2 U}{10}\right]\left\{\begin{array}{l}\left[\frac{R(4\psi^3+3\psi^2+\psi+1)U^2}{8H_i}\right] + \\ \left[\frac{(1+4\psi)U^2}{3}\right] + \left[\frac{20H_i(2\psi+1)\dot{U}}{1-\psi}\right]\end{array}\right\} \tag{9}$$

The average forging load during the end of the deformation can be computed by substituting the above energy dissipation equations in the below equation:

$$F_{av} = J^*(U)^{-1}A_{av}, \text{ where } J^* = 4[W_i + W_{f_1} + W_{f_2} + W_a]; \psi = \left[\frac{S_i}{R_i}\right]; \lambda = \left[\frac{R_i}{R_0}\right] \tag{10}$$

Preforming criteria and analysis

The shape complexity factor (SF) is expressed as the ratio of total energy dissipated during forging of actual preforms and corresponding enveloping imaginary pancake preforms theoretically. Mathematically, it is defined as follows (Zhao et al., 1995):

$$SF = \left[\frac{2R_g P_F^2 A_C}{R_C P_C^2 A_F} \right] \tag{11}$$

The necessary condition for preforming impression stages to be present during the closed-die forging process depends upon the limiting condition, $SF_f \geq SF_p$. If the ratio of SFs of final forged component and initial preform is greater than or equal to unity; then the entire forging process need to be carried out in multiple stages. The SFs values are computed keeping the preform volume constant. The preform height during intermediate forging steps is determined by volume constancy principle, i.e., equality of volume transformed during closed-die forging represented as compatibility condition mentioned in Equation (2). It works on the assumption that total energy dissipation is computed when an arbitrary component is forged and squeezed into the die cavity at any instance of time. This consists of two parts, i.e., energy dissipated for upsetting preform and squeeze filling of preform into die cavity. In the present case, the energy dissipation for upsetting has been neglected.

To draw a criterion, the above equation is transformed into a dimensionless one and compared with the energy dissipation for squeezed filling of preform into die cavity. Thus, it is converted into a criterion to estimate the requirement of a blocker impression, i.e., bigger the energy dissipation for squeeze filling, higher is the necessity of a preform impression. Thus, the necessary condition for preforming stages depends upon (Tomov and Radev, 2001):

$$J^* = \emptyset_V + [1 - \emptyset_V]\emptyset_A \geq \emptyset_H; \text{ where } \emptyset_V = \left[\frac{V_{add}}{V_0}\right]; \emptyset_A = ln\left[\frac{A_0}{A_P}\right]; \emptyset_H = ln\left[\frac{H_0}{H_{add}}\right] \tag{12}$$

As stated above, three different intermediate forging stages were considered as shown in Figure 11.2(a). The details of the performance, forged component and

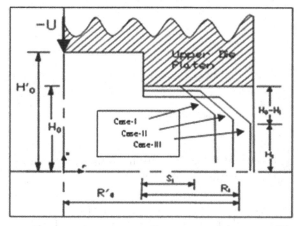

Figure 11.2 (a) Intermediate stages of die-cavity fills
Source: Author

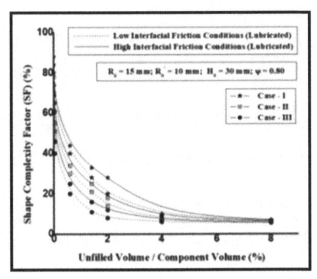

Figure 11.2 (b) Variation of average load vs die corner fills
Source: Author

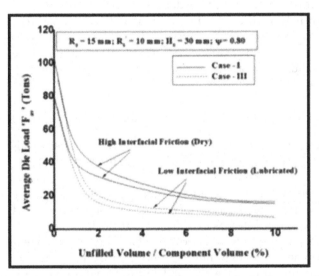

Figure 11.2 (c) Variation of shape complexity factor vs die fills
Source: Author

various factors affecting intermediate stages of forging and die-cavity fills are illustrated in Table 11.1. For each case, these parameters were computed, and criterion was tested. The values of SFs in all cases were found to be greater than one leading to conclusion that preforming steps is a necessity. The effect of deformation characteristics on the preforming step design are displayed graphically (Figures 11.2(b) and (c)). All forging experimentation was conducted both under dry and lubricated frictional conditions and corresponding die loads were recorded at different die cavity fills. It was found that the die load increases exponentially with die cavity fills and becomes asymptotically high at the end of forging. It can be seen that interfacial frictional

Table 11.1: Details of preform, forged component, and preforming stages.

	Preform and forged component			Factors affecting preforming stages						
	Preform	Forged component	Preforming stages	D_0	H_0	$\emptyset_V$	$\emptyset_A$	$\emptyset_H$	J^*	$\left(\dfrac{SF_f}{SF_p}\right)$
Shapes	Cylindrical	Double-Hub flange	Case-I	20	30	0.524	0.811	0.659	0.910	1.98
Dimensions	$D = 20$ mm $H = 40$ mm	$R_0 = 15$ mm $R'_0 = 10$ mm $H_0 = 30$ mm	Case-II	20	30	0.613	0.811	0.543	0.927	2.75
			Case-III	20	30	0.702	0.811	0.462	0.943	3.37

Source: Author

conditions dominate during the initial stages of the die cavity fills and the die load requirements are comparatively low for lubricated conditions. These findings suggest that all closed-die forging operations involving extreme squeeze fills, must be carried under lubricated frictional conditions.

Figure 11.2(c) depicts variation of SF for all the three cases of deformation modes under lubricated frictional conditions. It is found that the SF value increases with the die cavity fills and is highest for case-I followed by case-II except during the final stages of forging. The lubricated conditions exhibited better metal flow leading to lower SF values. Though, in all the three cases, SF values indicated the necessity of preforming steps, but case-I strongly advocate the proper design of preforming steps as compared to case-II and case-III. Also, if the initial preform shape is closer to shape of final component, the SF value is low and may not necessitate an intermediate preforming step. This finding validates the above theoretical modeling of preforming criteria as established by the authors.

Conclusion

The present article successfully demonstrated that the design of preforming steps leading to an optimal design of a preform during a complex closed-die forging process is a necessity. The validation of preform design criterion was done using a novel concept of shape complexity factor based on total energy dissipation. Different intermediate cases of preform deformation were considered as proof of concept wherein associated values of SF were computed and verified. It was established that the lower SF value corresponds to a preform initial shape which is closer to the final forged component and thus eliminates the preforming step design. But in all other cases, adequate preforming step design is a necessity for the manufacture of defect free final components. It is expected that the present research shall be extremely valuable to all the working professionals for the successful design and development of the defect-free forged industrial components.

References

Agarwal, M, and Singh, S (2021a). A study on aluminum metal matrix preforms forging into double-hub flange component. *Journal of Physics: IOP Conference Series*. 1950(012001), 1–14.

Agarwal, M, and Singh, S (2021b). Upper bound analysis of closed-die forging of eccentrically-located SiCp AMC preforms. *Journal of Applied Science and Engineering*. 25(2), 275–285.

Radev, R, and Tomov, B (2002). Preform design in hot die forging. Proceedings 11th International Scientific Conference on Achievements in Mechanical and Materials Engineering. pp. 1–4.

Singh, S, Jha, A K, and Kumar, S (2007a). Dynamic effects during sinter forging of axi-symmetric hollow disc preforms. *International Journal of Machine Tools and Manufacture.* 47(7-8), 1101–1113.

Singh, S, Jha, A K, and Kumar, S (2007b). Upper bound analysis and experimental investigations of dynamic effects during sinter-forging of irregular polygonal preforms. *Journal of Materials Processing Technology.* 194(1), 134–144.

Singh, S, Jha, A K, and Kumar, S (2007c). Upper bound analysis of flashless closed die sinter-forging of cylindrical preform into double-hub flange component. *International Journal of Manufacturing Technology Research.* 3(1&2), 68.

Tomov, B (1997). A new shape complexity factor. Proceedings Conference Advanced Materials Technologies. pp. 861–866.

Tomov, B, and Radev, R (2001). An example of determination of preforming steps for hot die forging. Proceedings 10th International Scientific Conference on Achievements in Mechanical and Materials Engineering. pp. 1–4.

Tomov, B and Wanheim, T (1993). Preform design based on model material experiments. Proceedings International Conference on Advanced Mechanical Engineering Technologies. pp. 147–156.

Zhao, G, Wright, E, and Grandhi, R V (1995). Forging preform design with shape complexity control in simulating backward deformation. *International Journal of Machine Tools and Manufacture.* 35, 1225–1239.

Zhao, G, Wright, E, and Grandhi, R V (1996). Computer-aided preform design in forging using the inverse die contact tracking method. *International Journal of Machine Tools and Manufacture.* 36, 755–769.

Nomenclature

P_F	perimeter of preform	$\emptyset_A$	logarithmic area strain	$\square_0$	preform relative density
P_C	perimeter of circumscribing cylinder	$\emptyset_H$	logarithmic height strain	R_0	flange radius
A	Preform c/s area	V_{add}	added volume	R_0	hub radius
R_g	distance between symmetry axis & CG	V_0	initial volume of forged component	r_m	sticking zone distance
R_C	radius of circumscribing cylinder	A_0	initial cross-sectional of forged	χ	specific cohesion of contact area
SF_f	shape-complexity of forged component	D_0	initial diameter of forged component	$\square$	interfacial frictional shear stress
SF_p	shape-complexity of preform	A_p	cross-sectional of forged component	H	preform height
U_{ij}	velocity field	W_i	internal energy dissipation	$\square$	coefficient of interfacial friction
ε_{ij}	strain rates	W_{f1}	die-workpiece frictional energy	P_{av}	average die pressure
U	die velocity	W_{f2}	die-container frictional energy	$\square_0$	flow stress

Subscripts

r	Radial	$\square$	circumferential	z	Vertical

12 The influence of carbon content on low-carbon steel corrosion rates in water cooling synthesis

Ahmad Royani[a] and Sundjono

Center of Corrosion Study, Research Center for Metallurgy, National Research and Innovation Agency – BRIN, Tangerang Selatan 15314, Indonesia

Abstract

Carbon steel is the most extensively used structural material in several industries, although this steel tends to corrode. For this reason, the immersion test of two low-carbon steel plates with different carbon contents was carried out in the synthesis water cooling. This study aimed to determine the influence of carbon content in steel on the corrosion behavior of the two steel plates. Weight loss was employed to analyze the corrosion performance of the steel. The physical characteristics of the resultant products of corrosion were evaluated using evaluated using scanning electron microscopy (SEM), energy dispersive spectroscopy (EDS), and X-ray diffraction (XRD). The result has shown that the corrosion rate of steel with higher carbon contents was more corrosive than the lower carbon contents could be because of intergranular corrosion. The steel corrosion rate decreases in aqueous conditions with constant corrosivity because of forming a thin protective layer of corrosion products on the metal surface. The corrosion products of the steel are uniform with the compounds $FeO(OH)$, $Fe_2O_3.H_2O$, and $Fe(OH)_3$. The corrosion performance, morphology, and corrosion products of carbon steel was influenced by carbon content.

Keywords: Carbon steel, corrosivity, immersion test, water quality

Introduction

Carbon steel is the most extensively utilized material in manufacturing and machining in various industries due to its availability, low cost, ease of production, and high strength (Tait, 2018). Low-carbon steel is mostly applied in the industrial and equipment sectors but is corrosive (Dwivedi et al., 2017). As a result, many corrosion researchers employ various techniques to produce effective steel corrosion control in aggressive environments. Carbon steels have been subjected to bases or acids for various industrial applications. The mechanical properties of carbon steel were determined by its microstructure and chemical composition. Corrosion is influenced by ferrite and pearlite grain size and ferrous carbide form, size, and distribution (Handoko et al., 2018)various approaches have been conducted through different heat treatment parameters to compare its microstructural engineering on chemical and mechanical properties. In this paper, correlation of different microstructure on corrosion resistance and hardness properties have been investigated. Three different heat treatment cycle have been applied on carbon steel with same composition to prepare dual-structure (DS). Thus, the chemical composition and the heat treatment of carbon steel are the major factors influencing iron carbide production.

Cooling towers were used to reduce heat from integrated unit cooling media. Because of the scarcity of fresh water, a water cooling selection is critical to avoid failure (Ahmed and Makki, 2020). Electrochemical processes and physicochemical

[a]ahmad.royani@brin.go.id

DOI: 10.1201/9781003450924-12

interaction between the metal and its surroundings cause metal corrosion in cooling water systems. Water quality parameters in cooling water systems must be investigated to reduce corrosion and deposit formation. This study evaluated the corrosion behavior of two low-carbon steel with different carbon content in water cooling synthesis. It aimed to obtain the corrodibility of metal and the corrosivity of liquid. The corrosion rates in this media are also determined to compare their effectiveness between two carbon steel plates. The morphology and characteristics of corrosion products are also described.

Materials and methods

Material and liquid test

Table 12.1 presents the chemical compositions of each of these specimens.
The test solution used was synthetic cooling water; its essential elements are stated in Table 12.2.

Preparation and immersion performance

The steel specimen surface must be cleaned following ASTM G-1 standards to prepare for the immersion experiment. The corrosion immersion test is based on a previous study (Royani et al., 2021). The reactor tube was filled with the test solution to submerge the test object and then placed in a water bath. A sensor probe evaluated water quality after measuring water level and temperature. The specimen was then threaded onto the holder and inserted into the test solution over the container side. The reactor tube was sealed and exposed for 728 days (ASTM G-31).

Weight loss, water quality parameters, and corrosion products analysis

The following equation was used to evaluate the performance of the steel corrosion rate based on the amount of weight loss:

$$Corrossion\ rate = k\frac{W}{(D\ A\ t)} \tag{1}$$

Table 12.1: The chemical composition of the steel specimens in this study.

Code	Element compositions (Wt. %)						
	C	Ni	Si	Mn	S	P	Fe
A	0.032	0.031	0.008	0.167	0.016	0.014	Bal.
B	0.176	0.143	0.134	0.717	0.041	0.012	Bal.

Source: Author

Table 12.2: The essential elements of solution media in this study.

Parameter	Conductivity	DO	Cal. hardness	Total hardness	Total alkalinity	pH
Unit	μmhos	Ppm	ppm, $CaCO_3$	ppm, $CaCO_3$	ppm, $CaCO_3$	-
Content	232	110.9	1.3	24.48	28	7.7

Source: Author

Where k is constant (3.45×10^6), W is weight loss (gram), D is the density of the specimen (g/cm^3), A is specimen area of exposure (cm^2), and t is exposure period (hours).

The water quality parameters were analyzed at the start of the operation, and each test specimen was collected with a multimeter portable (Hach HQ40d). At the same time, the product of corrosion was investigated using SEM (Jeol JSM 6390 A) equipped with EDS. The performance of the corrosion product was identified by X-ray diffraction (XRD), Shimadzu 700.

Results and discussion

The corrosion rate of steel

Figure 12.1 represents the carbon steel corrosion rate after 7 and 28 days of exposure. This data is provided as a comparison of the two different carbon steel.

The results indicated that specimen (A) had a higher level of corrosion resistance than specimen (B). The corrosion resistance performance of carbon steel could be affected by the amount of carbon, chromium, and copper present in the steel, among other elements (Katiyar et al., 2019). When added carbon, the pearlite phase ratio can increase, leading to intergranular corrosion (Soleimani et al., 2019). This type of corrosion could be caused by galvanic corrosion in grain boundaries. Some factors include the accumulation of impurities at grain boundaries (Jangir, 2018), the depletion of one of these elements at the grain-boundary areas, and or the enriching of one of the alloying elements (Jangir, 2018), which could be responsible for intergranular corrosion. Figure 12.2 shows metallographic photographs of the two types of steel examined.

Figure 12.2 shows that the grain size of the microstructure (pearlite) on carbon steel B is finer than that of carbon steel A. However, the content of the ferrite phase in carbon steel A is lower than in steel B. Carbon steel with ferrite phase has a lower corrosion rate than carbon steel with pearlite phase because no galvanic cell formation occurs.

The experimental results also reveal that the corrosion rate of specimen B in the initial steps of the reaction process increases with exposure time. In contrast, specimen A is relatively constant over time (Figure 12.1). That shows the development of surface film of corrosion products on surface metal inhibiting corrosion reactions.

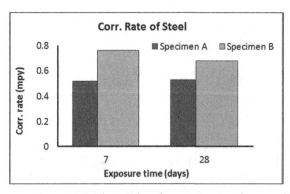

Figure 12.1 The results of corrosion rate for two different carbon steel at different exposure times

Source: Author

Specimen A   Specimen B

Figure 12.2 The photograph results of microstructure carbon steel after exposure
Source: Author

Table 12.3: The water quality of the solution test.

Sample	Time (day)	TDS (mg/L)	Cond. (µS/cm)	Salinity, (‰)	DO, (mg/L)	pH	Temp. (°C)
Blank	0	110.9	232	0.1	6.94	7.7	34.8
A	7	101.7	213.5	0.1	3.57	8.08	37.7
	28	99.2	208.4	0.1	3.95	7.57	37
B	7	80.2	169.1	0.08	3.64	7.58	37.1
	28	78.1	164.4	0.08	3.8	7.76	37.2

Source: Author

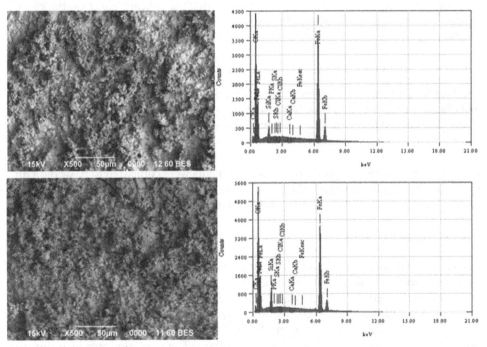

Figure 12.3 The morphology of corrosion products on the metal surface
Source: https://doi.org/10.17683/ijomam/issue8.52

Parameter of water quality

The parameter results of water quality measurements are presented in Table 12.3. The experimental results indicate that the corrosive characteristics of the test solutions did not significantly change as a result of the experiment (relatively constant).

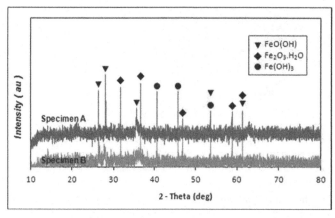

Figure 12.4 The pattern of X-Ray diffraction of corrosion product
Source: Author

There are shifts in the parameters of dissolved oxygen (DO). After a steel specimen has been immersed in the test solution, dissolved oxygen values in the solution have been shown to decrease. The corrosion process could have started in an anodic area and increased oxygen consumption, resulting in decreased DO in the solution media. The presence of DO in the liquid media has the most significant effect on the corrosion rate of carbon steel (Royani et al., 2019). This result is consistent with the results of other studies (Rybalka et al., 2018).

Analysis of corrosion products

Figure 12.3 presents the morphology and EDS data for corrosion products on the metallic surface for two different of carbon steel. As illustrated in Figure 12.3, the shapes are formed uniformly and evenly when seen from the perspective of the identified morphology and corrosion conditions.

The X-ray analysis results for the product of corrosion are presented in Figure 12.4. The results of the XRD diffractogram measurements were compared to basic data diffraction. The experiment results revealed that the oxides formed on the carbon steel surface were $FeO(OH)$, $Fe_2O_3.H_2O$, and $Fe(OH)_3$ compounds.

Conclusion

Immersing two steel plates with different carbon content has been successfully carried out in water cooling synthesis. The carbon content of carbon steel influenced its corrosion performance. Carbon steel with a high carbon content has a high corrosion rate due to intergranular corrosion between phases. Carbon steel, on the other hand, has a lower corrosion rate in the liquid corrosivity constant due to a thin layer protecting corrosion products on the steel surface. The corrosion products formed are homogenous, with constituents in iron oxide complexes such as $FeO(OH)$, $Fe_2O_3.H_2O$, and $Fe(OH)_3$ compounds.

Acknowledgment

This study was supported by the Research Center for Metallurgy, National Research and Innovation Agency (BRIN), by providing laboratory facilities and infrastructure.

References

Ahmed, S A, and Makki, H F (2020). Corrosion behavior of mild-steel in cooling towers using high salinity solution. In AIP Conference Proceedings. Dahham, O S, and Zulkepli, N N (eds.), 2213:020178–1–020178–6. Iraq: AIP Conference Proceedings. pp. 2213. https://doi.org/10.1063/5.0000274.

Dwivedi, D, Lepková, K, and Becker, T (2017). Carbon steel corrosion: a review of key surface properties and characterization methods. *RSC Advances*. 7(8), 4580–4610. https://doi.org/10.1039/C6RA25094G.

Handoko, W, Pahlevani, F, and Sahajwalla, V (2018). Enhancing corrosion resistance and hardness properties of carbon steel through modification of microstructure. *Materials*. 11(2404), 1–12. https://doi.org/10.3390/ma11122404.

Jangir, D K (2018). Influence of grain size on corrosion resistance and electrochemical behaviour of mild steel. *International Journal for Research in Applied Science and Engineering Technology*. 6(4), 2875–2881. https://doi.org/10.22214/ijraset.2018.4479.

Katiyar, P K, Misra, S, and Mondal, K (2019). Corrosion behavior of annealed steels with different carbon contents (0.002, 0.17, 0.43 and 0.7% C) in freely aerated 3.5% NaCl solution. *Journal of Materials Engineering and Performance*. 28(7), 4041–4052. https://doi.org/10.1007/s11665-019-04137-5.

Royani, A, Prifiharni, S, Nuraini, L, Priyotomo, G, Sundjono, S, Purawiardi, I, and Gunawan, H (2019). Corrosion of carbon steel after exposure in the river of sukabumi, west Java. *IOP Conference Series: Materials Science and Engineering*. 541, 1–10. https://doi.org/10.1088/1757-899X/541/1/012031.

Royani, A, Prifiharni, S, Priyotomo, G, and Sundjono, S (2021). Corrosion rate and corrosion behaviour analysis of carbon steel pipe at constant condensed fluid. *Metallurgical and Materials Engineering*. 27(4), 519–530. https://doi.org/10.30544/591.

Rybalka, K V, Beketaeva, , and Davydov, (2018). Effect of dissolved oxygen on the corrosion rate of stainless steel in a sodium chloride solution. *Russian Journal of Electrochemistry*. 54(12), 1284–1287. https://doi.org/10.1134/S1023193518130384.

Soleimani, M, Mirzadeh, H, and Dehghanian, C (2019). Effect of grain size on the corrosion resistance of low carbon steel. *Materials Research Express*. 7(1), 16522. https://doi.org/10.1088/2053-1591/ab62fa.

Tait, W S (2018). Controlling corrosion of chemical processing equipment. In Handbook of Environmental Degradation of Materials. Kutz, M (eds.), (3rd ed.), pp. 583–670. Oxford: William Andrew. https://doi.org/10.1016/C2016-0-02081-8.

13 Effect of cross anisotropy of pavement materials on flexible pavement response subjected to various subgrade modulus

Dipti Ranjan Biswal[a] and Brundaban Beriha[b]

School of Civil Engineering, KIIT Deemed to be University, Bhubaneswar, Odisha, Bhubaneswar, India

Abstract

Generally, layered elastic theory (LET) based pavement models considers the pavement layers to be linear elastic, homogenous and isotropic. But in reality, pavement materials exhibit cross-anisotropic behavior. Hence a study was undertaken to investigate the effect of anisotropy of pavement materials on fatigue strain and rutting strain which governs the fatigue and rutting life respectively. From the study it was observed that rutting is very sensitive to California Bearing Ratio value of subgrade. Hence, enhancement of subgrade CBR either by using lightly stabilized soil or good quality compacted granular materials should be used instead of higher thickness of base and subbase layer. Fatigue life has no sensitivity with the variation of subgrade CBR or subgrade modulus. The effect of degree of cross anisotropy (DOA) is not significant on fatigue life.

Keywords: Cross anisotropy, pavement analysis, rutting life, fatigue life

Introduction

In flexible pavement design, the fatigue life (Nf) is predicted using horizontal tensile strain at the bottom of bituminous layer (εt) and rutting life (Nr) is predicted using vertical compressive strain at the top of the subgrade respectively (εv). Generally Layered Elastic Theory (LET) based pavement models are used for the determination of these critical strain values for their simplicity. But most of the LET based model considers the pavement layers to be linear elastic, homogenous and isotropic (Biswal et al., 2019). But in reality, pavement materials exhibit cross-anisotropic behavior (Beriha and Sahoo 2020). In cross-anisotropy, the material possesses different modulus in horizontal (Eh) and vertical (Ev) direction. Hence a study was undertaken to investigate (1) the effect of cross-anisotropy of unbound materials on pavement response for different subgrade modulus; (2) to study the effect of subgrade modulus on εt , εv, rutting life and fatigue life.

Literature review

Researchers (Maina et al. 2017, Beriha and Sahoo 2020) have observed that the εt and εv value increases with decease in the degree of cross anisotropy (DOA=Eh/Ev) of unbound base and subbase. Whereas the εv value decreases with decrease in DOA. Saleh (2019) and Hu et al (2018) have observed that isotropic analysis overestimated the structural capacity of the pavement sections compared to the cross anisotropic analysis for thin bituminous layer (< 50mm) whereas for thick bituminous layer (100-240 mm) both analysis resulted in similar values. It can be observed from the

[a]dipti.biswalfce@kiit.ac.in, [b]brundaban.berihafce@kiit.ac.in

DOI: 10.1201/9781003450924-13

literature review that cross-anisotropic analysis of pavement is essential for accurate determination of pavement response parameters.

Pavement model

A typical flexible pavement having properties (Table 13.1) has been considered for the purpose of numerical study using the Crosspave Program (Beriha et al. 2020). The pavement sections were analyzed under the standard axle load of 80kN with tyre pressure of 0.56 MPa. The variation in the subgrade modulus from 50MPa to 120 MPa was achieved by changing the CBR value of compacted subgrade from 5 to 20. The rutting life and fatigue life of the pavement was estimated using the equation given for 90% reliability in IRC 37-2018.

Result and analysis

Effect of Subgrade Modulus on εv and Nr

It can be observed from Figure 13.1 that irrespective of DOA, εv decreases with the increase of CBR value of the subgrade. The εv for a hard subgrade of CBR 15 is decreased by 32% as compared to that of CBR 5. It can be concluded that εv is very sensitive to subgrade CBR value. It can be observed from Figure 13.2 that

Table 13.1: Pavement materials properties used for analysis.

Layer	Thickness (mm)	Elastic modulus (MPa)	Poisson's ratio	Density (kN/m3)	DOA
Bituminous layer	100	2000	0.35	23.5	
Granular base	250	280	0.35	19.5	0.251.0
Granular subbase	200	180	0.35	21.5	0.25-1.0
Compacted subgrade	500	50-120	0.45	18.0	0.25-1.0
Natural subgrade	1500	50	0.45	17.5	0.25-1.0

Source: Biswal et. al. (2019) and IRC: 37 (2018)

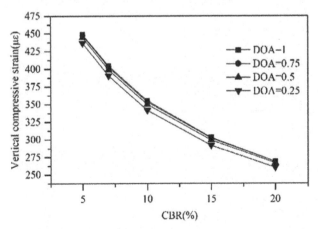

Figure 13.1 Variation of vertical compressive strain due to change in CBR value under different DOA

Source: Author

70

irrespective of DOA, rutting life increases at a high rate with increase in CBR. Rutting life increases by 5 times when the subgrade CBR is increased from 5 to 15. i.e., hard CBR results in drastic reduction of permanent deformation. Rutting life is very sensitive to subgrade CBR.

Effect of degree of anisotropy on εv and Nr

It can be seen from Figure 13.3 that irrespective of CBR, εv increases at a slow rate with increase DOA. It can be seen from Figure 13.4 that for subgrade of CBR 15, 20 the rutting life increases with decrease in DOA (decrease in horizontal modulus) whereas in the case of subgrade CBR of 5, 7 or 10, rutting life does not change significantly. It may be concluded that changes in rut life are more prominent at higher CBR than lower.

Effect of subgrade modulus on εt and fatigue life

Figure 13.5 and 13.6 shows εt and thereby the fatigue life decreases at a mild rate with the increase in subgrade CBR. εt decreases by 3% when the subgrade CBR is

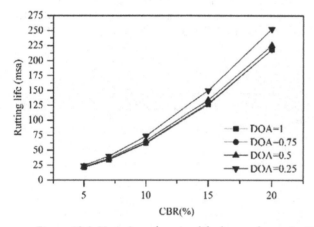

Figure 13.2 Variation of rutting life due to change in CBR value under different DOA
Source: Author

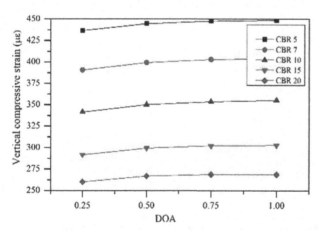

Figure 13.3 Variation of εv due to change in DOA value under different CBR
Source: Author

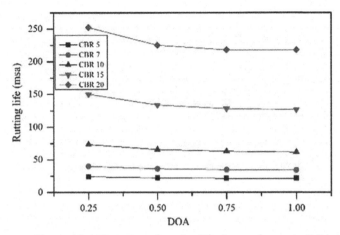

Figure 13.4 Variation of rutting life due to change in DOA value under different CBR
Source: Author

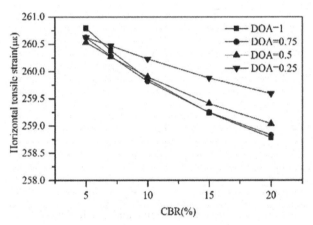

Figure 13.5 Variation of et due to change in CBR value under different DOA
Source: Author

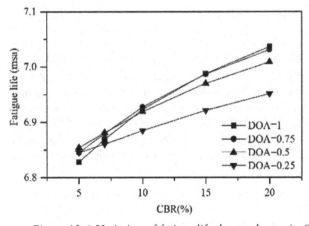

Figure 13.6 Variation of fatigue life due to change in CBR value under different DOA
Source: Author

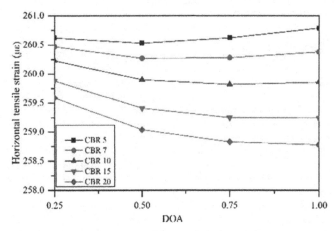

Figure 13.7 Variation of fatigue life due to change in DOA value under different CBR

Source: Author

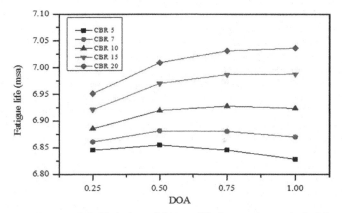

Figure 13.8 Variation of fatigue life due to change in DOA value under different CBR

Source: Author

increased from 5 to 20. Fatigue life has no sensitivity with the variation of subgrade CBR or subgrade modulus.

Effect of degree of anisotropy on εt and fatigue life

It can be noticed from Figure 13.7 that the εt decreases with increase in DOA from 0.25 to 0.5 and then increases upto DOA 1 for soft subgrade. However, εt decreases with increase in DOA from 0.25 to 1 for hard subgrade. Though the rate of change of εt is not significant, there is a need to study the effect for thinner pavements. Figure 13.8 shows that the fatigue life enhanced with increase in DOA from 0.25 to 0.5 and then lowered upto DOA 1 for soft subgrade. However, fatigue life increases with increase in DOA from 0.25 to 1 for hard subgrade but the rate of change is not significant.

Conclusion

In the present study, the sensitivity of the subgrade modulus on critical strains and rutting and fatigue life considering under isotropic and anisotropic material properties was investigated. From the observations it may be concluded that εv is very sensitive to subgrade CBR value of subgrade. Hence, enhancement of subgrade CBR either by using lightly stabilized soil or good quality compacted granular materials should be used instead of higher thickness of base and subbase layer. The εv increases at a slow rate with the increase of DOA value. Consideration of anisotropy of unbound layers results in reduction of rutting life for hard subgrade (CBR= 15, 20). However, it not so sensitive for subgrade CBR of 5, 7 or 10.

Fatigue life has no sensitivity with the variation of subgrade CBR or subgrade modulus. εt value goes higher with increase in DOA from 0.25 to 0.5 and then lowers upto DOA 1 for soft subgrade. However, εt increases with increase in DOA from 0.25 to 1 upto DOA 1 for hard subgrade. Though the rate of change of εt is not significant, there is a need to study the effect for thinner pavements.

References

Beriha, B and Sahoo, U C (2020). Analysis of flexible pavement with cross-anisotropic material properties. *International Journal of Pavement Research and Technology*, 13(4), 411–416.

Beriha, B, Sahoo, U C and Mishra, D (2020). Crosspave: a multi-layer elastic analysis programme considering stress-dependent and cross-anisotropic behaviour of unbound aggregate pavement layers. *International Journal of Pavement Engineering*, 23(6), 17231737.

Biswal, D R, Sahoo, U C, Dash, S R (2019). Structural response of an inverted pavement with stabilised base by numerical approach considering isotropic and anisotropic properties of unbound layers. *Road Material Pavement Design*, 21(8), 21602179.

Ghadimi, B, Nikraz, H (2017). A comparison of implementation of linear and nonlinear constitutive models in numerical analysis of layered flexible pavement. *Road Material Pavement Design*, 18(3), 550572.

Hu, Y., Yan, K, and You, L (2018). Mechanical response analysis of asphalt concrete overlay placed on asphalt pavement considering cross-anisotropic pavement materials. *In* Transportation Research Congress 2016: Innovations in Transportation Research Infrastructure (pp. 333340). American Society of Civil Engineers, Reston, VA, United States.

IRC: 37 (2018). Guidelines for the design of flexible pavements. Indian Roads Congress, New Delhi, India.

Maina, J W, Kawana, F, and Matsui, K (2017). Numerical modelling of flexible pavement incorporating cross-anisotropic material properties. Part II: Surface rectangular loading. *Journal of the South African Institution of Civil Engineering*, 59(1), 2834.

Saleh, M (2019). Effect of anisotropy and nonlinearity assumptions on the predicted surface deflections. *International Journal of Pavement Research and Technology*, 12, 472–477.

14 Performance evaluation of four-lobed powder lubricated bearing

Jijo Jose[1,a] and Niranjana Behera[2,b]

[1]Amal Jyothi College of Engineering, Koovappally, Kerala, India

[2]VIT University Vellore, Tamilnadu, India

Abstract

The effectiveness of a four-lobed hydrodynamic bearing which is lubricated using powder particles which are 2 microns in diameter is assessed in this paper using the finite difference method. Static and dynamic properties of four lobed bearing have been studied. The stability and friction coefficient of the bearing rise as the ellipticity ratio rises, but the load-bearing capacity decreases. Four-lobed bearings that have an ellipticity ratio of 0.25 perform better than those with ellipticity ratios with 0.5 and 0.75 because they can carry a greater load and are more stable. Circular journal bearings, which have an ellipticity ratio of zero, are shown to be less stable than four-lobed bearings with ellipticity ratios of 0.25, 0.5, and 0.75.

Keywords: Ellipticity ratio, four-lobe bearing, particle diameter, powder lubricant

Introduction

In hydrodynamic bearings that operate in harsh environments where conventional lubricants fail, such as at high temperatures and under difficult working circumstances, dry powdered particles have been suggested as viable lubricant substitutes (Jose and Behera, 2020a). Unlike liquid lubricants, which degrade at high temperatures, powder lubricants do not (Jose and Behera, 2020b). Rahmani et al. (2016) conducted the initial study on the powder lubrication of non-circular bearings. In this study, the dynamic and static performances of an elliptical journal bearing lubricated with solid granular particles of sizes 1 micron and 2 micron were assessed for a range of ellipticity ratios (δ) and L/D ratios. Later, Jose and Behera (2021) expanded this research to examine the static and dynamic behavior of three-lobed bearings. In this paper, the performance of a four-lobed bearing that is lubricated with powdered particles that are 2 microns in size is assessed using the finite difference technique.

Model developed from mathematical equations

In this section, the mathematical model that was developed to do the analysis is briefly described. A four-lobed bearing is depicted schematically in Figure 14.1. By using the macroscopic grain model, the equations for thermal diffusivity (K), viscosity (η), pressure (p) and energy dissipation (I) are obtained.

Modified expression for Reynolds equation applicable for normal circular journal bearings (Khonsari and Booser, 2008) is given below.

$$\frac{\partial}{\partial x}\left(\frac{h^3}{12\eta\prime}\frac{\partial p}{\partial x}\right)+\frac{\partial}{\partial z}\left(\frac{h^3}{12\eta\prime}\frac{\partial p}{\partial z}\right) = \frac{U}{2}\frac{\partial h}{\partial x} + \frac{\partial h}{\partial t} \tag{1}$$

[a]jijojose@amaljyothi.ac.in, [b]niranjanabehera@vit.ac.in

DOI: 10.1201/9781003450924-14

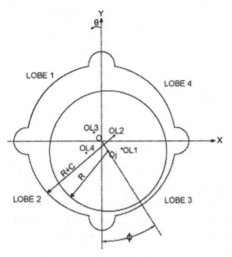

Figure 14.1 Four-lobed bearing
Source: Author

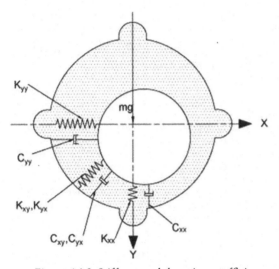

Figure 14.2 Stiffness and damping coefficients
Source: Author

p, h, U and □' are pressure, film thickness, surface velocity and average flow viscosity respectively.

Fluid film force components along the X and Y directions on the journal are

$$F_X = \int_0^L \int_0^{2\pi} pR\cos\theta d\theta dz \tag{2}$$

$$F_Y = \int_0^L \int_0^{2\pi} pR\sin\theta d\theta dz \tag{3}$$

R *and* L are the radius of journal and bearing width respectively. z and θ are longitudinal axes and angular coordinate respectively.

The force of friction acting on the bearing can be determined by the expression

$$F_s = \int_0^L \int_0^{2\pi} \left(\frac{h}{2R} \frac{\partial p}{\partial \theta} + \eta \frac{U}{h} \right) R d\theta dz \tag{4}$$

Frictional coefficient is evaluated using the equation

$$f = \frac{F_s}{W} \tag{5}$$

Figure 14.2 shows the cross coupled and direct stiffness and damping properties of the powdered fluid film, which are characterized using the Finite perturbation method.

The finite difference approach is used to find the numerical solution to Reynolds equation (1). The solution can be found by iteration using the Gauss-Seidel method with an over relaxation factor. Stability parameter (M_{cr}) is calculated as follows.

$$M_{cr} = \frac{Keq}{\lambda_{cr}^2} = m\omega^2 \frac{C_r}{W} \tag{6}$$

The system will be in stable condition if $M_{cr} > m\omega^2 \frac{C_r}{W}$

Each lobe in the current investigation had a mesh size of 90 × 90. The iteration terminates when the pressure value meets the following requirements.

$$\frac{\sum_{i=1}^m \sum_{j=1}^n \left| (p_{i,j})^N \right| - \left| (p_{i,j})^{N-1} \right|}{\sum_{i=1}^m \sum_{j=1}^n \left| (p_{i,j})^N \right|} \le 10^{-5} \tag{7}$$

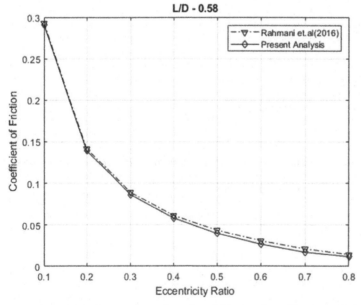

Figure 14.3 Comparison of predicted frictional coefficient with the previous results
Source: Author

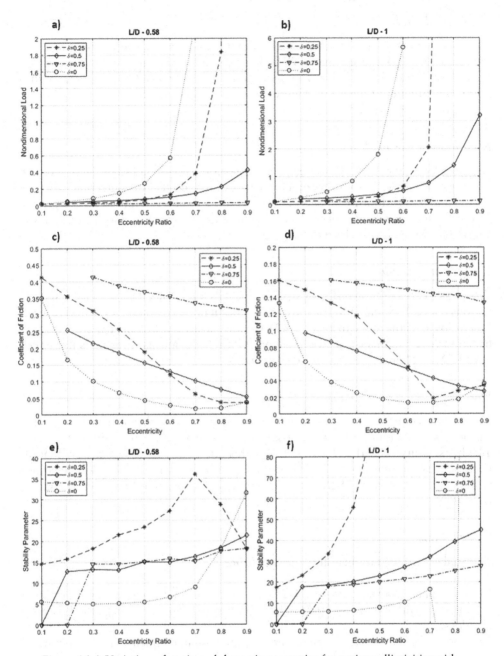

Figure 14.4 Variation of static and dynamic properties for various ellipticities with respect to eccentricity ratio

Source: Author

Where i, j indicate the grids considered in the θ and z directions. m,n indicate the total number of grids considered in θ and z directions. N indicates the total number of iterations.

Model validation

The coefficient of friction for a normal journal bearing case corresponding to d = 1 μm and L/D = 0.58 was compared to findings of Rahmani et al. (2016). The analysis was found to have L = 20 mm, D = 34 mm, Radial clearance C = 0.1 mm, speed = 2000 rpm, number of lobes = 4. Figure 14.3 shows that the acquired results showed a good degree of agreement.

Non-dimensional load

Figures 14.4(a) and 14.4(b) show how the non-dimensional load varies with increase in the eccentricity ratio for different ellipticity ratios and L/D ratios. In comparison to δ = 0.5 and 0.75, a four-lobed bearing with δ = 0.25 has been found to have a greater load carrying capacity. Powdered layer thickness grows as δ rises which causes a low bearing pressure and low load carrying capability.

Coefficient of friction

The plots of coefficient of friction with eccentricity ratio for L/D = 0.58 or 1 and for various δ = 0, 0.25, 0.5, and 0.75 are shown in Figure 14.4(c) and 14.4(d). For both L/D ratios 0.58 and 1, it can be inferred that normal circular bearing has a lower coefficient of friction. This is because the load for a normal journal bearing is more. For both L/D = 0.58 and 1, the coefficient of friction is maximum at δ = 0.75.

Stability

Figure 14.4(e) and 14.4(f) display variations in stability with eccentricity ratio for various ellipticity ratios and L/D ratios. Bearings with an ellipticity ratio of 0.25 have a higher stability value than bearings with other ellipticity ratios. When the eccentricity ratio was increased, the stability at δ = 0.25 increased faster. Four-lobe bearings with δ = 0.25, 0.5, and 0.75 have higher stability values than normal circular journal bearing.

Conclusions

The analysis resulted in the following conclusions.

(i) Four-lobed bearings with δ = 0.25 have a larger load carrying capability than bearings with δ = 0.5 and 0.75. The load carrying capacity rises as the L/D ratio does.

(ii) The coefficient of friction increases with increase in ellipticity ratio. The normal circular bearing has the lowest coefficient of friction.

(iii) Four-lobed bearings with δ = 0.25 are more stable than bearings with other ellipticity ratios. Bearing is stable at higher values of ellipticity ration and at lower eccentricity ratio.

(iv) A four-lobed bearing with δ = 0.25 is more stable and has high load capacity than δ = 0.5 and 0.75.

(v) Four-lobed bearings at δ = 0.25, 0.5, and 0.75 are more stable than normal circular bearings.

References

Jose, J and Behera, N (2020a). Study of wear and friction characteristics of powder lubricants under extreme pressure conditions. *Materials Today: Proceedings.* 22(4), 1306–1317.

Jose, J and Behera, N (2020b). Lubrication capability of solid powder lubricants in high pressure situations. *Jurnal Tribologi.* 26, 101–119.

Jose, J and Behera, N (2021). Static and dynamic performance characteristics of powder lubricated symmetrical three-lobed bearing. *Proceedings of the Institution of Mechanical Engineers, Part J: Journal of Engineering Tribology.* 235(5), 916–930.

Khonsari, M M and Booser, E R (2008). Applied Tribology: Bearing Design and Lubrication. New York: Wiley and Sons.

Rahmani, F, Dutt, J K, and Pandey, R K (2016). Performance behaviour of elliptical-bore journal bearings lubricated with solid granular particulates. *Particuology,* 27, 51–60.

15 Comparative seismic assessment of existing structures using P-delta effect in Indian context

Manish Chakrawarty[1,a], Pradeep K. Goyal[1,b] and Yaman Hooda[2,c]

[1]Department of Civil Engineering, Delhi Technological University, New Delhi, India

[2]Department of Civil Engineering, Manav Rachna International Institute of Research and Studies, Haryana, India

Abstract

Earthquake activity is considered to be the utmost catastrophic and erratic natural occurrence phenomenon, that results in major devastation to the overall society, including infrastructure and human lives. Also, urbanization is growing rapidly, existing land for building is becoming less and costlier. So, popularity of tall buildings is increasing. Based on guidelines of Indian Standard Code IS:16700-2017, the P-Delta analysis must be considered while analyzing tall buildings, known to be an iterative type of method. When a member is loaded axially and a lateral force act on it but it remains in place then moment is only produced by this lateral load, but when the member gets deflected by Δ, with initial moment still acting on it, a secondary moment starts acting on it due to this deflection which is called as P-Delta effect. In most of the structures with elements which are applied with axial load, the non-linear effect, which is called as P-Delta, occurs. More the height of building, more is lateral seismic load. The significance of P-Delta effect comes into consideration when the height of the structure starts increases. This study focused on building models of a residential complex located at East Delhi, which comprise of three different storey configurations as G+4, G+10 and G+22 storeyed – building. All the three models had the same dimensions of structural elements (beams, columns and slabs). The work is done in two phases. In one phase, the analysis of the structure was performed while neglecting the consideration of P-Delta effect, while in second phase, the considerations of the P-Delta effect on the structure was observed. The study reported the behavior of the structure against the seismic actions in terms of displacement and respective drift observed at different storeys. Results show significant difference in storey displacement and storey drift with and without P-Delta effect. For future considerations, a correlation analysis is done and correlation factors are determined.

Keywords: Correlation factor, P-delta analysis, seismic analysis

Introduction

As per United Nations Office for Disaster Risk Reduction (UNISDR), a disaster can be stated as "a serious disruption of the functioning of a community or society involving widespread human, material, economic or environmental losses and impacts, which exceeds the ability of the affected community or society to cope with using its own resources". The enhancement in technological sciences, forecast of various kind of calamities had become possible and outcome of these disasters can be greatly lowered by providing appropriate and intime warnings and cautions. From the year 1950 to 2011, it had been perceived that the events of the various calamities had

[a]manish1071995@gmail.com, [b]pkgoyal@dtu.ac.in, [c]yamanhooda@gmail.com

DOI: 10.1201/9781003450924-15

been increasing, especially of geophysical disaster. India is a massive country with unalike topographical features. The Indian Standard Codal Provision used for evaluating the earthquake on any structure is IS:1893(Part 1)- 2016. While contemplating the seismic effects in tall buildings, in last seismic zone (Zone V) horizontal shaking and vertical shaking both should be taken into account. A site- specific spectrum had to calculated and shall be used in design of buildings in Zone IV, and Zone V. When site-specific research yields excessive hazard assessment, site – specific research results had to be used.

Literature review

Dinar et al. (2103) observed 6 RC structures starting from G+5 to G+30 storey, showcased that increasing height under P-Delta analysis, displacement varied exponentially and the axial force also varied appreciably. Similar results were obtained by the researches conducted by Patil et al. (2013), Mallikarjuna (2014), Hooda et al. (2021), Verma and Verma (2019), and Dheeb and Abbas (2019). They had investigated P-Delta effects considering wind load on multi-storey steel buildings. Results recommend that the dynamic response of buildings with storeys more than 20 is highly influenced by P-Delta effect. Hooda and Goyal (2021) studied the behaviors of the hospital structure in Delhi under the action of seismic forces with the consideration of P – Delta effect and pushover analysis, suggested the best bracing retrofitting technique so as to with stand the seismic forces. Bhikshma et al. (2019) had studied the effect of wind and earthquake loads and showed that with greater wind speed in higher seismic zones, story displacements, story drifts also increased. Hooda and Goyal (2023) and Jadav and Desai (2020) studied P-Delta effects on tall structure resulted that its necessary to incorporate effects of P-Delta in buildings with 20 storey or more .

Methodology

The problem of this study is totally restricted to the Delhi/NCR Region of India. In this analysis, the software used is StaadPro-Connect edition V22 update 9, with guidelines as IS:1893(Part1)-2016. As per the IS codal provisions, P-Delta analysis is the best possible analysis for considering seismic effects on the tall buildings. The total design area is 15.770 m × 22.700 m. M25 grade of concrete was used for building up all the structural elements. The thickness of the slab was 150 mm, column size and beam size being 800 mm × 800 mm, and 600 mm × 350 mm in one direction and 400 mm × 300 mm in other direction respectively. Three models of varying storey heights are considered as G+4, G+10 and G+22. Storey height was 3.5 m; 230 mm, and 115 mm are the thickness of the external walls and internal walls respectively. The DL and LL were assigned on the models as per the guidelines of IS 875 – Part I and Part II.

Results and analysis

When any structure is subjected to seismic waves, the structural members get displaced from their original state. The basic parameters upon which the results can be compared are therefore the maximum displacement. Thus, in this study, the results include the observation of maximum displacement and maximum drift at storey levels Figure 15.1 and Figure 15.2.

The results are summarized in the following Table 15.1

Table 15.1: Summary of results.

S. No.	Model	Without P-Delta				With P-Delta			
		Max displacement (mm)		Max drift		Max displacement (mm)		Max drift	
		X	Z	X	Z	X	Z	X	Z
1	G+4	20.137	13.631	4.167	2.962	20.382	13.676	4.225	2.977
2	G+10	44.958	29.044	5.053	3.421	46.379	29.516	5.274	3.496
3	G+22	155.171	67.053	8.672	3.741	185.099	70.818	10.79	4.016

Source: Author

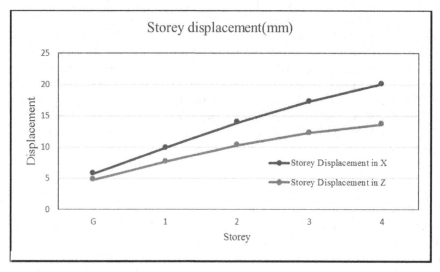

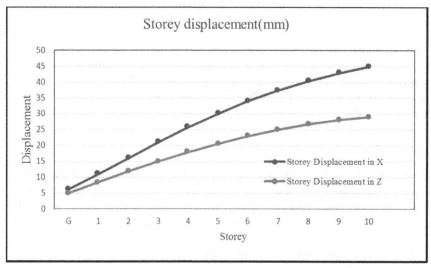

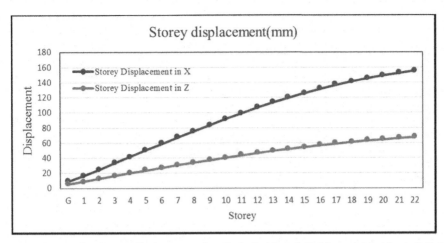

Figure 15.1 Storey displacement for G+4, G+10 and G+22 model (without P-Delta)
Source: Author

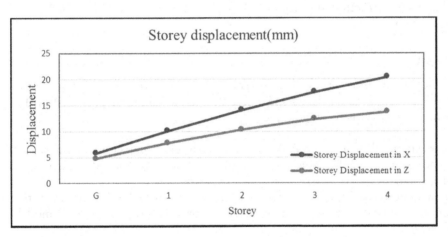

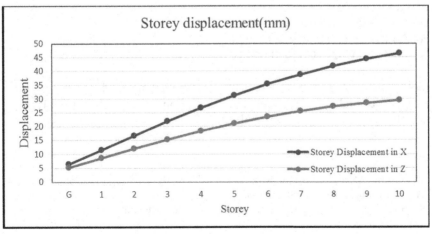

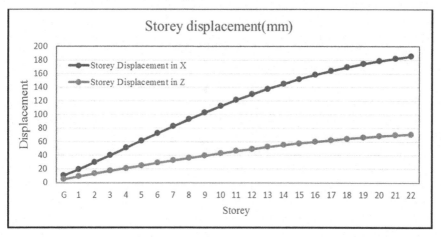

Figure 15.2 Storey displacement for G+4, G+10 and G+22 model (with P-Delta)
Source: Author

Since this study focused on P-Delta effect, a step towards were taken to determine the correlation between the values obtained for saving time of the analysis, with one condition of same structural dimensions. The correlation obtained is discussed in the conclusion section.

Conclusion

In this study, three different models of existing structures are considered under the area of high seismic zone, analyzed under StaadPro while considering the P - Δ analysis. The following data is recorded as:

1) For maximum displacement, P-Delta analysis had given higher maximum storey displacement by 1.22%, 3.16%, and 19.29% for G+4, G+10, and G+22 model, respectively.
2) For maximum storey drift, P-Delta analysis had given higher maximum storey displacement by 1.34%, 4.37%, and 24.42% for G+4, G+10, and G+22 model, respectively.
3) Correlation factor (C:
   - For G+4 and G+10; $C(y) = 0.9976x$.
   - For G+10 and G+22; $C(y) = 0.992x$.
   - For G+4 and G+22; $C(y) = 0.9943x$.

P-Delta analysis gives a significant increment in maximum storey displacement, thus must be analyzed for high-rise structures. The correlation factor tends toward 1, highlighting the efficiency and real project applicability.

References

Bhikshma, V, Hamraj, M, and Tejaswini, R (2019). Model Analysis of High-Rise Structures for Wind And Earthquake Forces. ISEC Press.

Dheeb, A S, and Abbas, R M (2019). Deterministic wind load dynamic analysis of high-rise steel. *Association of Arab Universities Journal of Engineering Sciences*. 26(1), 129–135.

Dinar, Y, Karim, S, Barua, A, and Uddin, A (2013). P- delta effect in reinforced concrete structures of rigid joint concrete structures of rigid joint. *IOSR Journal of Mechanical and Civil Engineering.* 10(4), (Nov. - Dec. 2013), 42–49.

Firoj, M, and Singh, S K (2018). Response spectrum analysis for irregular multi-storey structure in seismic zone V. 16th Symposium on Earthquake Engineering, India Paper No. 300.

Hooda Y, Kuhar P, Sharma N and Verma NK. (2021). Emerging applications of artificial intelligence in structural engineering and construction industry. *Journal of Physics: Conference Series.* 1950, 012062.

Hooda, Y, and Goyal, P K (2021). Seismic assessment of a hospital building: a case study. *IOP Conference Series: Earth and Environmental Science.* 796, 012006.

Hooda, Y, and Goyal, P K (2023). Comparative analysis of bracing configuration for retrofitting of existing structures on hill slopes. In: Marano, G C, Rahul, A V, Antony, J, Unni Kartha, G, Kavitha, P E, Preethi, M (eds), Proceedings of SECON'22. SECON 2022. Lecture Notes in Civil Engineering. (vol 284). Cham: Springer. https://doi.org/10.1007/978-3-031-12011-4_15.

IS 875 (Part 1): 1981. Code of Practice for Design Loads (Other Than Earthquake) for Buildings and Structures [Dead Loads].

IS 875 (Part 2): 1981. Code of Practice for Design Loads (Other Than Earthquake) for Buildings and Structures [Imposed Loads].

IS 1893 (Part 1): 2016. Criteria for Earthquake Resistant Design of Structures.

IS 16700: 2017. Criteria for Structural Safety of Tall Concrete Buildings

Jadav, M S, and Desai, D (2020). Study of P-delta effects on tall building. *International Journal of Engineering Development and Research.* 8(1), 42–44.

Mallikarjuna, B N (2014). Stability analysis of steel frame structures: p-delta analysis. *International Journal of Research in Engineering and Technology.* 3(8), 36–40.

Patil, S S, Ghadge, S A, Konapure, C G, and Ghadge, C A (2013). Seismic analysis of high-rise building by response spectrum method. *International Journal of Computational Engineering Research.* 3.

Verma, A, and Verma, S (2019). Seismic analysis of building frame using p-delta analysis and static & dynamic analysis: a comparative study. *I-manager's Journal on Structural Engineering.* 8(2).

16 Innovative cement-free concrete utilizing 100% recycled aggregates for sustainable construction

Gopalakrishna Banoth[a] and Dinakar Pasla[b]

School of Infrastructure, Indian Institute of Technology Bhubaneswar, Bhubaneswar, India

Abstract

In the current situation, fly ash (FA) is utilized instead of cement, which contributes to the preservation of the environment. FA alumina (Al) and rich in silica (Si) interacted with an alkaline solution such as sodium hydroxide (NaOH) and sodium silicate (Na_2SiO_3) to form a strong binding property that substituted the function of conventional ordinary Portland cement (OPC). This research studies the effect of concentration of NaOH (12, 14, & 16M), the Na_2SiO_3/ NaOH ratio maintained at 1.5, and 60°C temperature curing at one day for various molarity of mixes. In geopolymer concrete (GPC), by using 100% construction and demolish (C&D) waste recycled aggregate (RA), the workability of GPC was observed to be low because of attached mortar observed more solution. Overcome this effect in concrete by using a naphthalene-based superplasticizer. Compressive strength (CS) was used to evaluate the mechanical characteristics of GPC. The outcomes were established using specimens ages 3, 7, 14, 28, and 56 days after the curing procedure. It was concluded that RA and the molarity of NaOH affect the mechanical properties of RAGPC. The maximum compressive strength was observed from a 16M concentration of NaOH and recycled aggregate concrete.

Keywords: Compressive strength, fly ash, geopolymer concrete, molarity, recycled aggregate, workability

Introduction

Concrete is popularly used worldwide due to its low cost, ease of shaping, and availability of raw materials. However, cement production for concrete significantly contributes to CO_2 emissions, which harm the environment. One ton of cement production emits one ton of CO_2 and consumes 1.6 tons of raw materials, making it an energy-intensive process. Such CO_2 emissions led to global warming (Zhang et al., 2023). To find alternatives to cement, researchers have investigated technologies such as geopolymer (GP) technology. By utilizing cement-free binders like fly ash from thermal power plants in concrete production, the adverse effects of cement can be reduced (Biswal and Dinakar, 2021; Davidovits, 1991). Recycled aggregate (RA), which is composed of the primary components of old concrete, should be recycled for various reasons rather than being discarded. Waste aggregates should be utilized as RA in structural concrete, making it more preferable (Gopalakrishna and Dinakar, 2023). The use of RA in building engineering has become increasingly popular due to its environmental benefits, including reducing pollution and saving natural resources. Geopolymer technology is another way to make concrete more eco-friendly (Sata and Chindaprasirt, 2020). By using RA, not only can environmental issues be mitigated, but natural aggregate resources can also be conserved while significant landfill space

[a]bg13@iitbbs.ac.in, [b]pdinakar@iitbbs.ac.in

DOI: 10.1201/9781003450924-16

can be saved. Combining advances in geopolymer (GP) technology with the use of recycled aggregate (RA) instead of natural aggregate is a crucial measure in making concrete a more sustainable building material. By merging these two technologies, concrete's sustainability can be significantly improved (Guo et al., 2018). The study aims to assess the feasibility of creating geopolymer concrete from recycled materials, which can lead to an eco-friendlier construction industry.

The study focuses on examining the properties of geopolymer concrete (GPC) made with recycled aggregate (RA) in both fresh and hardened states. The objective is to analyze how altering the molarity or concentration of NaOH affects the properties of RA-based GPC. The workability of FA-based recycled aggregate geopolymer concrete remained consistent at 100 ± 10 mm, and the study also evaluated its compressive strength (CS).

Experimental investigation of materials and methodology

Materials used

This study utilized low calcium fly ash (FA) from the NALCO plant in Odisha, India, as the source material for the geopolymeric binder. X-Ray Fluorescence (XRF) analysis determined that the FA had a chemical content breakdown shown in Table 16.1, with 77.41% silicon and aluminum oxides and a Si to Al molar ratio of 1.83. Davidovits (Davidovitsm, 1999) suggested a Si/Al molar ratio of around 2 for cement and concrete production. The presence of iron oxide in the FA contributes to its greyish-white color (Fe_2O_3). The average particle diameter of the ash ranged from 1.714µm to 18.96µm, with a specific gravity (SG) of 2.1 and a specific surface area (Blaine) of 390 m²/kg. Improper disposal of fly ash can cause significant environmental damage by polluting the air, soil, and groundwater. Figure 16.1 illustrates that the two

Table 16.1: XRF analysis of fly ash.

Oxide	Al_2O_3	SiO_2	Fe_2O_3	Na_2O	CaO	MgO	K_2O	oxides	LOI
Dry weight (%)	27.28	50.13	9.28	1.12	3.25	1.52	2.82	1.76	2.9

Source: Author

Figure 16.1 Fraction of recycled aggregate
Source: Author

coarse aggregate sizes (6-12 & 12-20 mm) used in the recycled aggregate (RA) were provided by Ramky Enviro Engineers Ltd of Hyderabad, India. Fine aggregate with a size less than 4.75 mm and a specific gravity (SG) of 2.63 was obtained from the riverbed of the Mahanadi and used in the concrete mix. The experimental SG of the 12- & 20-mm size fractions of the RA were found to be 2.36 and 2.38, respectively.

The study examined the impact of Na_2SiO_3/NaOH ratio and NaOH molarity on the synthesis of a GP mortar using FA and GGBS. The highest compressive strength was observed in samples with Na_2SiO_3/NaOH ratios of 1.5 and 16M NaOH concentration. Sodium-based soluble alkalis are commonly used to compensate for AAC. NaOH and Na_2SiO_3 are the most commonly used alkaline activators for GP production. Table 16.2 provides information on the chemical composition and specific gravity of the sodium-based AAC used in the study. Table 16.3 shows the mix design proportions for recycled aggregate-based GP concrete at various NaOH concentrations.

Results and discussion

Fresh properties

The workability of GPC was analyzed using a slump cone as per IS Code 1199-1959, and its viscosity and homogeneity depended on the NaOH activator concentration. The slump of fresh GPC ranged from 75 to 110 mm, suggesting its potential use in various structures. The presence of pores in saturated surface dry RA facilitated water absorption, thereby enhancing GPC workability. However, using high NaOH concentration reduced slump due to increased alumina and silica leaching, which resulted in the rise of viscosity and stiffness as shown in Figure 16.2.

Compressive strength of RAGPC

Various percentages of NaOH in the AAC were used to produce GPC with 100% FA and RA, resulting in a CS range of 13.50 to 40.12 MPa. The concentration of NaOH in the AAC, ranging from 12 to 18M, influenced the CS of the GPC, which

Table 16.2: Chemical composition of NaOH and Na_2SiO_3.

Properties	pH	Specific gravity	Molecular weight (g/ mol)	Color	SiO_2 (%)	Na_2O (%)	H_2O (%)
NaOH	13.5	2.2	40	White	–	45.71	54.29
Na_2SiO_3	11.42	1.36	122.06	Colourless	29.9	10	60.8

Source: Author

Table 16.3: Various NaOH concentration-based mix design proportion.

NaOH molarity	Alkaline/ Fly ash	Fly ash	Total alkaline liquid	Na_2SiO_3/ NaOH ratio	Mass of NaOH	Mass of sodium silicate	20 mm	10 mm	Sand
12	0.5	400	200	1.5	80	120	489	454	655
14	0.5	400	200	1.5	80	120	489	454	655
16	0.5	400	200	1.5	80	120	489	454	655
18	0.5	400	200	1.5	80	120	489	454	655

Source: Author

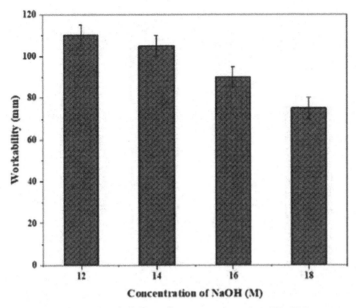

Figure 16.2 Workability of concrete at various NaOH concentration
Source: Author

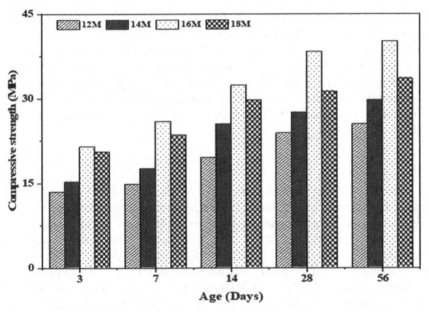

Figure 16.3 The compressive strength of RAGPC at various concentration of NaOH
Source: Author

was oven-cured at 60°C for 24 hours. Increasing the NaOH concentration up to 16M raised the GPC's CS, but further increases caused a decrease, as shown in Figure 16.3. GPC with RA had a maximum CS of 40.12 MPa at 56 days with a NaOH concentration of 16M and Na_2SiO_3/NaOH ratio of 1.5. CS values were measured in GPC experiments with varying NaOH molar concentrations and AAC/B ratios. Data from

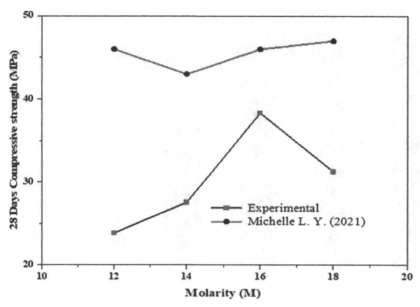

Figure 16.4 Different molarity based geopolymer concrete with 100% RA 28 days CS
Source: Author

previous studies were merged and plotted in Figure 16.4, showing a 0.5 fluctuation in NaOH molar concentration. The highest CS was achieved at 16 M NaOH for 3, 7, 14, 28 and 56 days of CS, with the lowest at 12 M. Strategic parameters such as AAC/B ratio, Na_2SiO_3/NaOH ratio, and binder type can increase CS as shown in Figure 16.4. The presence of Si and Al in the matrix enhances CS, with NASH gel formation contributing to a superior CS. The matrix composition and molar concentration both significantly impact CS. A maximum CS value for 16 M was achieved with a 40 MPa compression for 56 days.

Conclusion

In this study, the effects of varying molarities (1218 M) and a fixed AAC/B ratio (0.5) on the mechanical properties of GPC made from RA were investigated. Slump values decreased as molar concentration increased due to higher AAC solution viscosity. Factors like activator molarity, AAC/B ratio, Na_2SiO_3/NaOH ratio, binder material, size/shape, and curing conditions affected GPC performance. Increasing NaOH concentration led to higher CS values at different intervals, with the maximum observed at 16 M concentration, 1.5 Na_2SiO_3/NaOH ratio, and 60°C oven cure. A higher-molarity activator increased polycondenzation and chemical reaction rates in the polymer matrix.

Reference

Biswal, U S, and Dinakar, P (2021). A mix design procedure for fly ash and ground granulated blast furnace slag based treated recycled aggregate concrete. *Cleaner Engineering and Technology* . 5, 100314. doi:10.1016/j.clet.2021.100314.
Davidovits, J (1991). Geopolymers - inorganic polymeric new materials. *Journal of Thermal Analysis*. 37(8), 1633–1656. doi:10.1007/BF01912193.

Davidovits, J (1999). Chemistry of geopolymeric systems, terminology. *Geopolymer.* 99(292), 9–39.

Gopalakrishna, B, and Dinakar, P (2023). Mix design development of fly ash-GGBS based recycled aggregate geopolymer concrete. *Journal of Building Engineering.* 63(January), 105551. doi:10.1016/j.jobe.2022.105551.

Guo, H, Shi, C, Guan, X, Zhu, J, Ding, Y, Ling, T C, Zhang, H, and Wang, Y (2018). Durability of recycled aggregate concrete – a review. *Cement and Concrete Composites.* 89(May), 251–259. doi:10.1016/j.cemconcomp.2018.03.008.

Sata, V, and Chindaprasirt, P (2020). Use of construction and demolition waste (CDW) for alkali-activated or geopolymer concrete. *Advances in Construction and Demolition Waste Recycling.* 385–403. doi:10.1016/B978-0-12-819055-5.00019-X.

Zhang, D, Wang, Y, and Peng, X (2023). Carbon emissions and clean energy investment: global evidence. *Emerging Markets Finance and Trade.* 59(2), 312–323. doi:10.1080/15404 96X.2022.2099270.

17 Study on effect of damage location and severity on dynamic behavior of multi-storey steel framed structure

Mahalaxmi S. Sunagar[a], Jayanth K[b] and Naveen B.O.[c]

Department of Civil Engineering, The National Institute of Engineering, Mysuru, India

Abstract

Damage is defined as modifications to a system's physical property that have a negative impact on the structural performance now and in the future. The various damages are caused due to natural or man-made actions. In the present work, the dynamic behavior of undamaged and damaged multi-storey steel frame structures is studied with different damage locations in beams and columns at different levels. The damage is introduced by reducing the cross-sectional area and capacity of moments, shear, and torsion of structural steel members in beams and columns. From the results, it has been observed that there is a change in the natural frequency of the structural system with the introduction of damage of certain severity at a particular location. Further equivalent static analysis as per IS 1893:2016 has been carried out.

Keywords: Damage severity, equivalent static analysis, modal analysis, steel frame multi-storey structure, structural damage

Introduction

From the past few decades, structural steel has been extensively used of as a construction material due to its high strength to weight ratio, ductility and other important mechanical properties. Similar to reinforced concrete (RC) structures, steel structures have to resist all kinds of forces such as its self-wight, live loads, lateral loads (wind and earthquake) and dynamic loads. Due to its poor fire resistance, steel loses its strength at elevated temperature. Also, due to cyclic loading, high compressive and flexural stresses, structural steel members experiences fatigue and buckling stresses respectively. A structural engineer has to design steel structures to bear all kinds of forces by resisting various types of possible failures due to compression, tension, bending, and shear. Figure 17.1 shows types of failures in steel structural members and connections. The installation process itself frequently results in damage to the bolted structures. Damage in these situations often has delayed effects that, if ignored for a long time, can result in catastrophic events.

The reason for all the failures can be justified by the presence of various damages, location of damage and its severity. The damage present in any structural member causes changes in the physical parameters and mechanical properties of a structure. This damage in steel structures have to be assessed and mitigated, otherwise due to the progressive increase in the damage severity, structure loses its strength and ultimately leads to total collapse.

Structural health monitoring and modal analysis

Structural health monitoring (SHM) techniques can be used to measure and detect changes in vibration response which leads to quantitative measurement and can yield

[a]2021cse_mahalaxmissunagar_a@nie.ac.in, [b]jayanthk@nie.ac.in, [c]naveenbo@nie.ac.in

DOI: 10.1201/9781003450924-17

(a)   Local buckling of beams   (b)   Connection failure

*Source:*https://911research.wtc7.net/~nin11evi/911research/mirrors/guardian2/fire/SLamont.htm

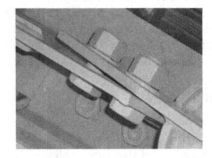

(c) Tension failure   (d)   Shear failure

*Source:*https://s3da-design.com/possible-types-of-failures-in-a-steel-
structure

Source: https://sites.google.com/site/hindustanalcoxltd/need-to-monitor-the-health-and-
safety-of-bolted-structures

Figure 17.1 Types of failures in connections

Source: https://www.passifire-sa.com/hp/how-does-structural-steel-fail-due-to-fire-exposure-how-to-protect-it/

detailed evaluation of structural condition. Numerous factors affect the structure's overall performance. The primary factor that causes the structures failure is damage or a crack. Therefore, in order to determine the health status, it is necessary to investigate the impact of cracks from the extracted natural frequencies of the structures in the early phases. The dynamic behavior of a structure is described by modal analysis in-terms of its modal characteristics, such as its natural frequency, damping ratio, and mode shape (Chinka et al., 2018).

Literature review

From the literature, various types of damage and their effect on dynamic behavior of steel structures can be found. In this section, important works from various authors have been presented and summarized. An approach to explore the failure sequence and damage mechanism of fire insulation on steel columns under monotonic and cyclic loads under ambient conditions, both experimental and numerical analyses were conducted by Wang et al. (2013). Blachowski et al. (2016) conducted an experimental and analytical examination of the relationship between a local flaw in a steel structure and its higher frequencies and higher modes. The building is made of beams that are connected together to form a flat steel frame. A specific connection's damage can be simulated by removing a few bolts. Further damage can be considered as local bucking of flanges of steel sections (Le et al., 2019). Various numerical simulation works have been carried out where in most of the cases studies and validation only modal data of modes 1-2 and 4-5 are used (Guo et al., 2019) for damage identification. The effects of damages on structures were investigated by Khan et al. (2020), which were simulated by introducing multiple cracks at different locations. It was found that, due to the presence of cracks, the modal frequencies decrease. A mathematical model called High Dimensional Model Representation (HDMR) and their applications in finite element model updating and structural damage identification has been investigated by considering, three numerical examples (i.e., beam, frame and bridge structures), where damage is introduced by reducing Young's modulus (Naveen and Balu, 2021). By considering different damage indices, a numerical case study on shear-type steel frame structure experimental prototype has been carried out by Rinaldi et al. (2022) and hence effect of damage position and intensity is evaluated to estimate the sensitivity. In the present study, dynamic behavior of multi–storey steel structures are assessed by introducing the damage at various levels in beams and columns by using structural analysis software. Dynamic response parameters such as natural frequency, time period, and displacements are recorded.

Methodology

In the present work, a single bay three storey steel moment resisting frame has been modeled with plan dimension of (5m × 3m) and with height of each storey as 3m. Damage in steel structure modeled has been introduced at level 1 in beams and columns. Seven cases have been considered including undamaged model (case-1). In the remaining are the damaged models i.e., in case-2 only in one corner column, in case-3 two long span beams at level one, in case-4 one corner column and one long span beam, case-5 one corner column, one short and long span beam at level 1, in case-6 two opposite corner columns and two long span beams and in case-7 corner columns at all levels damage has been introduced. For all the above cases the following types of damage are introduced in beams and columns i.e., by reducing 50% of the cross-sectional area and reducing moment, shear and torsion carrying capacity to 50%. Modal analysis and lateral load (Equivalent Static Analysis (ESA) as per IS 1893:2016) analysis has been carried out using ETABS software.

Results and discussions

From modal analysis, dynamic responses i.e., time period and natural frequencies and from ESA, lateral displacement, column axial loads and column moments are extracted after the analysis. Table 17.1 in this section displays the change of natural frequencies for various damage instances. From Figure 17.2(a) it has been observed that, with introduction of damage 7.6% increase in natural frequency in mode 1 is found in case-2 damage compared to case-1 i.e., undamaged case. From Figure 17.2(b), 12.7% reduction in lateral displacement is found with respect to case-1.

Similarly, due to damage in beams and columns, there is a reduction of 1.2% in axial load carrying capacity. With the reduction in moment carrying capacity, there is reduction in frequency of 21% in mode 3 for case-2. And an increase in displacement of 25% for case-5 at level 1. Similarly, an overall increase in displacements has been observed and maximum displacement has been observed in case-6 i.e., (Table 17.2) when damage is introduced in two opposite corner columns and two long span beams.

Table 17.1: Variation of natural frequency (Hz): 50% reduction in cross-sectional area.

Mode	Undamaged case	Case-1	Case-2	Case-3	Case-4	Case-5	Case-6
1	2.110	2.260	2.27	2.260	2.260	2.255	2.259
2	2.491	2.588	2.59	2.588	2.588	2.587	2.588
3	3.169	3.162	3.17	3.162	3.162	3.159	3.161

Source: Author

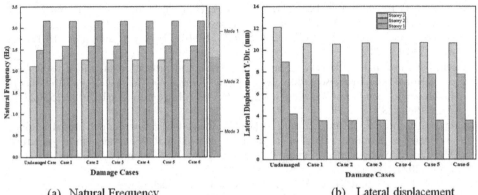

(a) Natural Frequency (b) Lateral displacement

Figure 17.2 Responses due to 50% reduction in cross sectional area
Source: Author

Table 17.2: Variation of lateral displacement (mm): 50% reduction in cross-sectional area.

Mode	Undamaged case	Case-1	Case-2	Case-3	Case-4	Case-5	Case-6
1	12.12	11.25	10.58	11.25	12.43	11.53	12.65
2	8.92	8.40	7.73	8.41	9.52	8.69	9.26
3	4.17	4.18	3.55	4.18	4.71	4.49	4.26

Source: Author

Conclusions

From the present study it can be concluded that, introduction of damage will change the dynamic behavior of the steel structure and further reduction in the moment, shear and torsional capacity is found to be critical that of cross-sectional area. Further in case of damage structure, during earthquake, lateral displacement is found to more than the undamaged case, due to which additional moments will develop in other beams and columns where there is no damage. This leads to further damage to other structural members. Hence, assessment of damage location, severity and its effects on the overall behavior of the structure is significant with respect to the safety of the structure.

References

Blachowski, B, Swiercz, A, Gutkiewicz, P, Szelążek, J, and Gutkowski, W (2016). Structural damage detectability using modal and ultrasonic approaches. *Measurement.* 85, 210-221. https:doi.org/10.1016/j.measurement.2016.02.033.

Chinka, S S B, Adavi, B, and Putti, S R (2018). Influence of crack on modal parameters of cantilever beam using experimental modal analysis. *Journal of Modeling and Simulation of Materials*. 1(1), 16–23. https://doi.org/10.21467/jmsm.1.1.16-23

Guo, J, Jiao, J, Fujita, K, and Takewaki, I (2019). Damage identification for frame structures using vision-based measurement. *Engineering Structures*. 199, 109634. https://doi.org/10.1016/j.engstruct.2019.109634.

Khan, M A, Akhtar, K, Ahmad, N, Shah, F, and Khattak, N (2020). Vibration analysis of damaged and undamaged steel structure systems: cantilever column and frame. *Earthquake Engineering and Engineering Vibration*. 19, 725–737. https://doi.org/10.1007/s11803-020-0591-9

Le, L M, Ly, H B, Pham, B T, Le, V M, Pham, T A, Nguyen, D H, Tran, X T, and Le, T T (2019). Hybrid artificial intelligence approaches for predicting buckling damage of steel columns under axial compression. *Materials*. 12(10), 1670. https://doi.org/10.1016/j.istruc.2019.04.002

Naveen, B O, and Balu, A S (2021). Structural damage identification of bridge using high dimensional model representation. *International Journal for Computational Methods in Engineering Science and Mechanics*. 22(4), 265–277. https://doi.org/10.1080/15502287.2020.1861128

Rinaldi, C, Ciambella, J, and Gattulli, V (2022). Image-based operational modal analysis and damage detection validated in an instrumented small-scale steel frame structure. *Mechanical Systems and Signal Processing*. 168, 108640. https://doi.org/10.1016/j.ymssp.2021.108640

Wang, W-Y, Li, G Q, and Kodur, V (2013). Approach for modeling fire insulation damage in steel columns. *Journal of Structural Engineering*. 139(4), 491–503. https://doi.org/10.1061/(ASCE)ST.1943-541X.0000688

18 Performance analysis of internal fin concentric pipe heat exchangers

Kailash Mohapatra[1,a] and Ashutosh Mohapatra[2,b]

[1]Professor, Mechanical Engineering Department, Oxford College of Engineering & Management, Bhubaneswar, India

[2]Student, Computer Science & Engineering Department, Institute of Technical Education & Research, Bhubaneswar, India

Abstract

Numerical simulations are carried out to study the internal fin configurations on heat transfer of an annulus tube heat exchanger by solving 3D, incompressible equations of Navier-Stokes (2-D axis-symmetric design for plain pipe) in addition to energy equations. 1500 mm length, 70 mm outer diameter, and 50 mm inner diameter of aluminum tube with four numbers of fins fixed in the axial direction on the inner wall of the outer tube between the annulus have been selected for simulation. It is found from the simulation that the Nusselt number was maximum for trapezoidal fin in comparison with other shapes. It was also found that the axial wall surface temperature distribution was minimum for maximum numbers of troughs (six in the present case). The temperature at the outlet of the test pipe was the maximum for the saw fin.

Keywords: Annulus tube, friction factor, Nusselt number, saw fin

1. Introduction

The annulus pipes with internal fins heat transfer equipment are widely seen in most of the industries and commercial utilities like power plants, chemical and petroleum industries, where the main attention was to minimize the weight, size and volume of the heat exchangers without reducing the thermal performance. This all can be possible by utilizing one of the methods like attaching extra high conducting materials onto the surface where the thermal coefficient is least. This can be achieved by fixing fins with different fin numbers, fin shapes or fin sizes. These extra materials can be applied on either side of the flow passages. These fins are used widely in different engineering and scientific applications to augment heat transfer from heated surfaces, whether heat transfer mode occurs due to forced or free convection. A great deal of research work has been conducted earlier to improve the heat transfer rate by reducing the weight and volume of heat transfer equipment with fins fixed internally or externally of the flow passage for economic gain. In this present context internally finned tube equipment received considerable attention by reducing the volume as well as weight to improve the effectiveness of the heat transition equipment. After extensive literature reviews the present study is initiated by comparing the fluid flow and thermal behavior of the internal finned annulus passage with plain tube under forced convection where the fluid flow nature is considered as developing and turbulent.

Rest of the study is arranged in the subsequent orders Section-2 contains existent literature reviews, Section-3 mathematical formulation, Section-4 numerical solution methods, Section-5 results and discussion, and Section-6 summarization of the findings work. from the research

[a]kailash72@ymail.com, [b]ashutoshmohapatra2020@gmail.com

DOI: 10.1201/9781003450924-18

2. Literature review

Kuvvet and Yavuz (2011) conducted an experiment on plain and finned concentric passages pipe and found that the average Nusselt number is increased from 3.6 times to 18 times in the finned passages compared to unfinned passages. Li et al. (2001) performed numerical simulations on annular ducts with four radius ratios and four apex angles under fully developed flow and found results well matched with the available correlations with maximum variations up to 8.3%. An experimental study conducted by Alijani and Hamid (2013) on a tube with a circular cross-section with a core rod as an insert with honeycomb construction at a very high temperature and found that the coefficient of heat transition of the center rod was enhanced by 227369% again for the honeycomb web insert, it was 409679% in comparison with an unfinned pipe at uniform wall temperature. Braga and Saboya (1999) from their experimental study with an isothermal inner wall and outer wall with insulation on which 20 numbers fins were attached observed that the Nusselt number and Reynolds number function with each other and are not dependent on the thermal properties of the materials of the fin. Yu et al carried out an experiment to find out the friction factor and thermal behavior of double pipe structure to correlate the Reynolds number with Nusselt number and pressure drop for unblocked and blocked inner pipes. Dirker and Meyer (2005) formulated correlations to predict Nusselt number for different Reynold's number based on hydraulic diameter and found that the Nusselt number was within 3% of the experimental value for a variety range of annular diameter ratios. A numerical study conducted by Qiuwang et al. (2008) found that when subjected to both constant heat flux and wall temperature for an annular wavy finned tube will have an optimum inner core outer diameter to outer pipe inner diameter.

3. Mathematical formulation

The numerical simulation is performed for a concentric tube with an external diameter, of D_1, and an internal diameter of D_2, with a length of L as viewed in Figure 18.1.

Air is admitted into the annulus passage of the pipe at the inlet whereas the outlet is opened to the surrounding atmosphere. Fins are arranged regularly between the annular passage in the longitudinal direction around the internal surface of the outer pipe. The study began with the spines having rectangular cross sections with length(L), height(H) and thickness(t) and the shapes of the fin are changed subsequently for constant mass flow rate of air. The flow field is calculated in the domain by using 3-D, energy equations and Navier-Stokes equations for incompressible fluid flow (2-D axisymmetric model for smooth tube). Air is used as working fluid for the simulation at an inlet temperature, 300K and considered incompressible as the velocity at the inlet is kept less than 10 m/s. The external surface of the exterior pipe is solid and

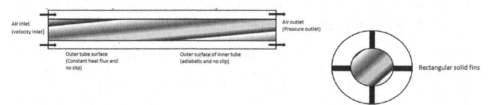

Figure 18.1 Illustrative sketch of annulus tube with internal fins
Source: Author

considered no-slip and applied to a uniform heat flux of 3000W/m² whereas the outer surface of the inner pipe is adiabatic. Appropriately the pressure outlet and velocity inlet boundary conditions were chosen for the outlet and inlet of the tube. Fins are fixed on the interior face of the tube and considered solid. Boundary conditions for each side of the fins available inner side of the pipe were considered as no-slip and coupled. The airflow through the annulus passage of the test pipe at a constant rate of 0.0196 kg/s.

Governing equations

Continuity $\rightarrow \dfrac{\partial}{\partial x_i}(\rho U_i) = 0$

Momentum $\rightarrow \dfrac{D(\rho U_i)}{Dt} = -\dfrac{\partial P}{\partial x_i} + \dfrac{\partial}{\partial x_j}\left[\mu\left(\dfrac{\partial U_i}{\partial x_j} + \dfrac{\partial U_j}{\partial x_i}\right) - \rho\,\overline{u_i u_j}\right] \& \dfrac{P}{\rho} = MT$

Energy $\rightarrow \dfrac{D(\rho T)}{Dt} = \dfrac{\partial}{\partial x_i}\left[\left(\dfrac{\mu}{Pr} + \dfrac{\mu_t}{Pr_t}\right)\dfrac{\partial T}{\partial x_t}\right]$, Conduction $\rightarrow \dfrac{\partial}{\partial x_i}\left(k\dfrac{\partial T}{\partial x_i}\right) = 0$

4. Numerical solution procedure

At proper boundary conditions and by using Fluent 14 for solving steady conservation equations of momentum, mass, energy and turbulence kinetic energy. First order upwind scheme initially used for discretizing the equations of turbulent and momentum and then second order upwind scheme is used after getting the converged solutions from first order upwind scheme to get better results (the surface temperature profile was hiked by 0.5% and approaching towards the solutions of experiment). Since density is the function of temperature both SIMPLE algorithm and pressure staggered option (PRESTO) scheme was utilized for finer convergence. The relaxation factor for all variables like momentum, pressure, k and ε are 0.8, 0.3, and 0.7 respec-tively used for convergence. When the field residual of energy and all the variables dripped down to and 10⁻⁶ and 10⁻³ respectively then all the discretized equations are converged

5. Results and discussions

Grid sensitivity is conducted as shown in Figure 18.2. Initially, coarse grids were chosen and thereafter cells are further refined from 16,800 to 2, 37,500, thereby decreasing the outlet by 1.5%. It was found that refining beyond 5, 59,500 did not have significant variation. So, 5, 59,500 cells were considered for all calculations. For the same 0.0196 kg/s mass flow rate at the inlet and 10 mm fin height, the distribution of wall surface temperature for different shapes of fin is described in Figure 18.3. There are four numbers of rectangular, trapezoidal, triangular and + (plus) shapes of fins chosen for the present study. From the above figure, it is clearly visible that the distribution of wall surface temperature for the trapezoidal fin is lowest in comparison with other designs. Hence transfer of heat is highest for trapezoidal fins because of better mixing due to turbulence. It was also found from calculation that the Nusselt number for trapezoidal fins was maximum compared with others as shown in Figure 18.4. The relation between of fin numbers and wall surface temperature shown in the Figure 18.5. Longitudinal rectangular cross-section fins are being utilized at present simulation and other geometrical configurations are shown in the figure.

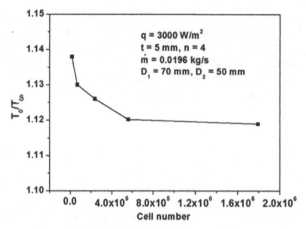

Figure 18.2 Cell number v/s tube outlet temperature
Source: Author

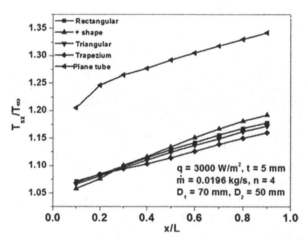

Figure 18.3 Relation between wall temperature and fin shape
Source: Author

It is clearly found from the figure that the temperature of the air at the outlet increases with fin numbers, but the wall surface temperature drops with the fin numbers.

Again, from the analysis, the Nusselt number was increased along with the number of fins as seen from the above figure. It was also observed that the rate of heat transfer to the working fluid was increased with fin numbers. From Figure 18.6, it was also seen that, at the inlet of the test tube Nusselt number was high and decreased continuously toward the exit.

This happened because of heat transferred by convection at the entrance of the test pipe was initially dominated but later on towards the exit due to the growth of the boundary layer the conductive heat transfer was more prominent. So, it is clear from the above discussions that by inserting more numbers of fins though the surface temperature is decreased considerably but on other sides the centerline friction factor is increased to a greater extent as shown in Figure 18.7. This implies that to deliver an equal quantity of fluid more energy will be consumed.

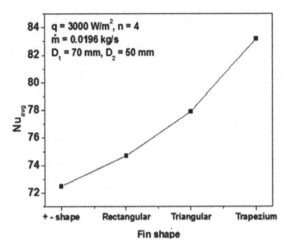

Figure 18.4 Average Nusselt number vs fin shape
Source: Author

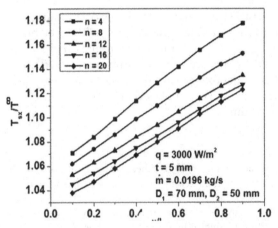

Figure 18.5 Wall temperature vs pipe length
Source: Author

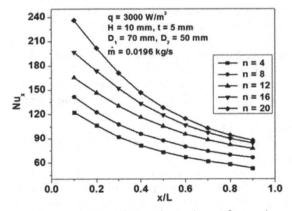

Figure 18.6 Local Nusselt number vs fin number
Source: Author

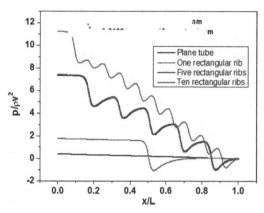

Figure 18.7 Friction factor vs fin number

Source: Author

6. Conclusion

The conservation equations for energy, momentum and mass are being solved to find out the surface temperature, exit temperature and pumping power of annulus tube with internal fins. The transfer of heat to the working fluid increases with more numbers of fin. In comparison with various shapes, the transfer of heat is highest for trapezoidal fins. It is also concluded that by inserting more numbers of fins though the surface temperature is decreased considerably but on other sides the centerline friction factor is increased to a greater extent. This implies that to deliver an equal quantity of fluid more energy will be consumed.

References

Alijani, M R, and Hamidi, A A (2013). Convection and radiation heat transfer in a tube with core rod and honeycomb network inserts at high temperature. *Journal of Mechanical Science and Technology*. 27(11), 3487–3493.

Braga, C V M, and Saboya, F E M (1999). Turbulent heat transfer, pressure drop and fin efficiency in annular regions with continuous longitudinal rectangular fins. *Experimental Thermal and Fluid Science*. 20, 55–65.

Dirker, J, and Meyer, J P (2005). Convective heat transfer coefficients in concentric annuli. *Heat Transfer. Heat Transfer Engineering*, 26(2), 38–44.

Kuvvet, K and Yavuz, T (2011). Effect of fin pitches on heat transfer and fluid flow characteristics in the entrance Z region of a finned concentric passage. *Journal of Thermal Science and Technology*. 31(2), 109–122.

Li, Y, Hung, T C, and Tao, W Q (2001). Numerical simulation of fully developed turbulent flow and heat transfer in annular-sector duct. *Heat and Mass Transfer*. 38, 369–377.

Noori, N R, and Saeed, A F (2023). The experimental impact of convective heat transfer improvement from numerous perforated shape fin array. *Al-Rafidain Engineering Journal (AREJ)*, 28(1), 280–292.

Pichler, M, Haddadi, B, Jordan, C, Norouzi, H, and Harasek, M (2021). Effect of particle contact point treatment on the CFD simulation of the heat transfer in packed beds. *Chemical Engineering and Research*, 165, 242–253.

Qiuwang, W, Mei, L, and Min, Z (2008). Effect of blocked core-tube diameter on heat transfer performance of internally longitudinal finned tubes. *Heat Transfer Engineering*. 29(1), 107–115.

Yu, B, Nie, J H, Wang, Q W, and Tao, W Q (1999). Experimental study on the pressure drop and heat transfer characteristics of tubes with internal wave-like longitudinal fins. *Heat and Mass Transfer*. 35, 65–73.

19 Fault detection of rolling bearing using artificial neural network and differential evolution algorithm

Brahma Bibhutibhusan[1,a] and Sethi Rabinarayan[2,b]

[1]Research Scholar, IGIT, Sarang, BPUT, Rourkela, Odisha, India

[2]Assisistant Professor, Department of Mechanical Engineering, IGIT, Sarang, Odisha

Abstract

This research is based on fault diagnosis of roller bearing which is the most important transmission components. Fault detection in a vibrating component is generally done by analyzing the vibration response of the component and response to loading fluctuations. At present, the condition monitoring techniques have alarmed scientists to develop new methodologies for in-line monitoring and finding of fault in the machinery. Current paper proposes a method for health monitoring of bearings using an artificial neural network (ANN) and a combination of Laplace-wavelet transform and differential evolution (DE) algorithm-based optimization technique. The method uses the ANNs to classify rolling bearing faults which is based on parameters extracted from wavelet transform in the time and frequency domains. The DE algorithm is being introduced for optimization of the parameters obtained from the Laplace-wavelet form and the ANN classifier. The results show that this approach is highly operative in diagnosing bearing fault conditions with a good success rate using only a few input characteristics which is applied and verified both for actual and simulated bearing vibration data.

Keyword: Artificial intelligence, artificial neural network , differential evolution algorithm, dynamic response, fault diagnosis, Laplace-wavelet

1. Introduction

Machines having rotating components are widespread use in various engineering applications such as power plants, marine propulsions system, airplane engine etc. To allow this relative rotational motion between components, bearings are used. So, any type of failure in bearing will lead to heavy losses for the component as well as to the industry. Therefore, it is important to find fault symptoms early using various methods, among which vibration analysis is the most widely used technique. In vibration analysis different vibration data are collected and plots are prepared showing amplitude-time, amplitude-frequency, or spectrum of time-frequency domain. If there is any fault in the bearing, there will be a change in the plots from the regular one and by measuring that change it can be concluded that whether the fault is to be addressed or not. Various methods have been developed by different scientists to acknowledge this issue and artificial neural networks (ANN) is one of the most broadly used practices in health monitoring of roller bearings (Taplak et al., 2006). ANNs are artificial intelligence systems that mimic biological neural networks. Researchers use non-linear methods for fault detection in complex and non-linear systems without reliable measurements or mathematical simulations. ANNs are beneficial for automatic recognition and classification of fault conditions and do not require in-depth knowledge of system performance. Differential evolution (DE) is a type of evolutionary algorithm

[a]bibhutibhusan6234@gmail.com, [b]rabinsethi@igitsarang.ac.in

DOI: 10.1201/9781003450924-19

that is well-suited for optimizing functions in continuous search spaces. DE is being applied to a wide range of problems, including optimization of vibration signals for predictive maintenance and fault verdict. Therefore, the aim of this paper is to unite ANN and DE algorithm to advance a competent fault diagnosis method for roller ball bearing. In future more technologies can be added along with this to make this method more robust.

2. Literature review

In the present scenario the systems trust the ANN techniques to support the toughness of investigative structures. Different types of ANN techniques have been developed in condition monitoring and fault diagnostics and have been used in modelling and resolution making in fault diagnostics as discussed. Omar and Gaouda (2012) used wavelet-based tool which is also beneficial in identifying vibrations which are working under noisy environment. Diwakar et al. (2012) represented the spectra of FFT that may yield peaks at probable error frequencies. Zou et al. (2021) advanced a system for the failure analysis of bearing of high-speed train pulling motor using discrete wavelet (DWT) and enhanced deep belief network (DBN). Haj Mohamad and Nataraj (2021) applied phase space topology (PST) and statistical functions for fault diagnosis of roller bearing of rotary machineries under varying operation circumstances. Cui et al. (2020) found that original auto encoders used for mechanical fault diagnosis have restricted feature extraction capacity because of deficiency of label information. To resolve this problem, they proposed feature distancing stack encoding method for fault detection of rollers. Elasha et al. (2015) suggested a system centered on acoustic emission (AE) for health monitoring of helicopter gearbox. Wang et al. (2020) suggested a system for monitoring bearing of wind turbine via online monitoring of bearing clearance which will indicate premature failure of bearing.

3. Methodology

3.1 Fault diagnosis

Diagnosis analyzes machine state and symptoms to detect faults early. Defects in bearings can be local or distributed and generate impulsive forces that create peaks on vibration plots of the bearing assembly as displayed in Figure 19.2. The force of impulse is given by Al-Raheem et al. (2008).

$$S(t) = Ce^{-(\zeta/\sqrt{1-\zeta^2})\omega_d t} \sin(\omega_d t) \tag{1}$$

Where ζ is the damping ratio, ω_d is the damped natural frequency of and C is a factor which scales the amplitude. As the fault travels over the load zone, the amplitude will respond accordingly.

$$x(t) = \sum i \, Ai \, S(t-T_i) + n(t) \tag{2}$$

Here, the waveform produced by the ith impact at time T_i is represented by $S(t-T_i)$. T_i is calculated as the sum of iT and a random slip parameter, r_i, where T is the meantime within consecutive impacts. The amplitude-demodulation is represented by Ai, and n(t) represents the additional circumstantial noise.

3.2 Laplace wavelet transform

The Laplace wavelet is an analytical, complex, exponentially decaying function that exists only for positive time values, and it is given by Equation 3, where β is the damping factor that limits the exponential envelope's time-domain decay rate and, consequently, the wavelet's resolution. The DE equation is then introduced which is a stochastic optimization algorithm to find the optimum result to a given problem.

$$\Psi(t) = A\, e^{-((\beta/\sqrt{1-\beta^2}+j)\omega c\, t)} \text{ when } t \geq 0 \ \& \ \ \Psi(t) = 0 \text{ when } t < 0 \tag{3}$$

3.3 Optimization using differential evolution algorithm

DE algorithm evolves a population of parameter vectors over generations using mutation and crossover operations. The objective function measures the quality of each vector and guides the evolution towards better parameters that maximize the separation between healthy and faulty vibration signals, optimizing a variety of signal processing algorithms. The DE follows the steps as follows (Saufi et al., 2018):

1. Mutation ➡ 2. Crossover ➡ 3. Selection

For every target vector $X_{i,G}$, a mutant vector V_i, where $i = 1, 2, 3,...N$, r1,r2,r3 are random numbers, V_i is the mutant vector generated. F is a scaling factor. The trial vector U_i is then created by equation (5) with the use of a crossover operator with a probability Cr. The fittest vector can be selected in an optimization problem by equation (6).

$$V_i = X_{r1i} + F \times (X_{r2i} - X_{r3i}) \tag{4}$$

$$U_{i\,G+1} = (U_{1i,\,G+1},\ U_{2i,\,G+1},\ ...,\ U\,D_{i,\,G+1}) \tag{5}$$

$$X_{i,G+1} = U_{i,G}\, f\,(U_{i,G}) < f(X_{i,G+1}) \ \& \ X_{i,G}\, f\,(U_{I,G}) \geq f(X_{i,G+1}) \tag{6}$$

Root mean square (rms), standard deviation (SD), and kurtosis factor in the time domain, as well as the ratio of the WPS peak frequency (fmax) to the shaft rotational frequency (f rpm), are the characteristics that were retrieved.

4 Experiment and calculation

Vibration signals from both healthy (30305 Tapered Roller bearing) and faulty bearings (same bearing with roller fault) are collected which were run at 2000 to 2500 rpm with varying load in time domain and are being converted to spectral units using Laplace-wavelet transform from the trial setup as displayed in Figure 19.2 below. The spectral figures are analyzed using DE algorithm for fault analysis and diagnosis.

Results

This study utilized DE method to detect the severity and location of cracks in rolling element bearings. The DE algorithm efficiently identified changes in natural frequencies caused by the presence of cracks, enabling accurate structural damage detection.

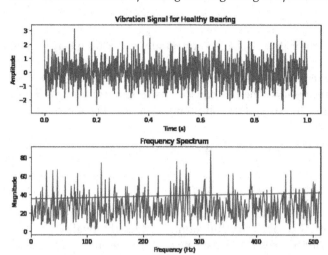

Figure 19.1 Showing vibration signal for a healthy bearing
Source: Author

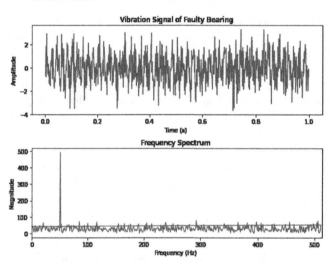

Figure 19.2 Showing vibration signal for a faulty bearing
Source: Author

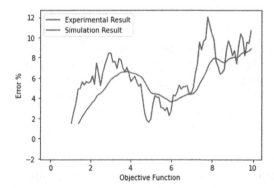

Figure 19.3 Showing error curve between simulation result and DE algorithm result
Source: Author

The study compared results obtained from simulation analysis and DE by performing error analysis as shown in Figure 19.3.

Conclusion

A new approach has been proposed for identifying and classifying issues in roller bearings which combines the Laplace-wavelet transform and an artificial neural network (ANN), and ANN parameters optimized using a differential evolution (DE) algorithm. The most significant parameters collected from the primary wavelet measures were used as input vectors for the neural network, resulting in faster and more accurate classification of simulated and real bearing vibration signals. This method achieved accuracy of 81.4% in identifying different types of rolling bearing faults.

References

Al-Raheem, K F, Roy, A, Ramachandran, K P, Harrison, D K, and Grainger, S (2008). Application of the Laplace-wavelet combined with ANN for rolling bearing fault diagnosis. *Journal of Vibration and Acoustics*. 130(5).

Cui, M, Wang, Y, Lin, X, and Zhong, M (2020). Fault diagnosis of rolling bearings based on an improved stack autoencoder and support vector machine. *IEEE Sensors Journal*. 21(4), 4927–4937.

Diwakar, G, Satyanarayana, M R S, and Kumar, P R (2012). Detection of bearing fault using vibration analysis. *International Journal of Emerging Technology and Advanced Engineering*. 2(9), 123–137.

Elasha, F, Greaves, M, Mba, D, and Addali, A (2015). Application of acoustic emission in diagnostic of bearing faults within a helicopter gearbox. *Procedia Cirp*. 38, 30–36.

Haj Mohamad, T, and Nataraj, C (2021). Fault identification and severity analysis of rolling element bearings using phase space topology. *Journal of Vibration and Control*. 27(3-4), 295–310.

Omar, F K, and Gaouda, A M (2012). Dynamic wavelet-based tool for gearbox diagnosis. *Mechanical Systems and Signal Processing*. 26, 190–204.

Saufi, S R, bin Ahmad, Z A, Leong, M S, and Lim, M H (2018). Differential evolution optimization for resilient stacked sparse autoencoder and its applications on bearing fault diagnosis. *Measurement Science and Technology*. 29(12), 125002.

Taplak, H, Uzmay, I, and Yıldırım, S (2006). An artificial neural network application to fault detection of a rotor bearing system. *Industrial Lubrication and Tribology*. 581, 32–44.

Wang, J, Xu, M, Zhang, C, Huang, B, and Gu, F (2020). Online bearing clearance monitoring based on an accurate vibration analysis. *Energies*. 13(2), 389.

Zou, Y, Zhang, Y, and Mao, H (2021). Fault diagnosis on the bearing of traction motor in high-speed trains based on deep learning. *Alexandria Engineering Journal*. 60(1), 1209–1219.

20 Application of machine learning models in Mahanadi River basin, Odisha for prediction of flood

Shradhanjalee Pradhan[1,a], Janhabi Meher[1,b] and Bibhuti Bhusan Sahoo[2,c]

[1]Department of Civil Engineering, VSSUT, Burla, Odisha, India

[2]Centurion University of Technology and Management, Bhubaneswar, India

Abstract

Machine learning (ML) models can aid in flood alerts and reduction of its disastrous effect. ML approaches have gained popularity because of their minimal computing demands and rely on observational data. The goal of this study was to develop an ML model to predict floods using a 30-year historical rainfall dataset, to be applied in three stations in the Mahanadi River system: Kesinga, Tikarpara, and Sallebhata, located in a high flood-risk area of the Odisha state. Logistic regression (LR) was found as the best model in terms of receiver operating characteristics (ROC), accuracy (ACC) and recall (R) scores as compared to the other two algorithms, which are decision tree (DT) and random forest (RF).

Keywords: Accuracy, decision tree, flood prediction, logistic regression, machine learning, recall, random forest, receiver operating characteristics

Introduction

Natural disasters and man-made calamities pose a threat to human life. Good disaster management and early warning systems can reduce their impact (Zhou, 2021). Flood prediction models play a crucial role in reducing damage caused by floods. Machine learning (ML) can help in predicting future floods based on historical data (samantaray et al. 2015). The Mahanadi River is prone to flooding due to its heavy monsoon season rainfall and the topography of the river basin. Some factors that contribute to floods in the region include deforestation, soil erosion, and the construction of dams and embankments. During heavy rainfall periods, the river can overflow its banks, causing widespread damage to crops, homes, and infrastructure. In recent past, there have been several significant floods in the Mahanadi basin, including the 2008 flood that affected over 2 million people and caused significant economic losses.

In response to the threat of floods, the Odisha government has implemented various measures to mitigate their impact, including the construction of embankments, the improvement of early warning systems, and the evacuation of at-risk populations. There are also ongoing efforts to improve the management of water resources in the region, including the implementation of sustainable agriculture and water use practices. Despite these efforts, the risk of floods remains high in the Mahanadi basin and continued efforts are needed to address the underlying causes of flooding and to prepare communities for future events.

The focus of this study is to apply an ML model to forecast floods in Mahanadi River basin, Odisha using historical rainfall data and apply it to other high-risk Indian

[a]shradhanjaleepradhan94@gmail.com, [b]jmeher_ce@vssut.ac.in, [c]bibhuti5000@gmail.com

DOI: 10.1201/9781003450924-20

basins. Despite advancements in flood prediction systems, without a precise predictive model, many trending approaches in developing nations will not be able to anticipate floods with any degree of accuracy. Choosing the right prediction model from various ML techniques is the focus of this research paper.

Material and methods

The Mahanadi River basin is a major drainage system located in eastern India. It covers an area of approximately 141,600 square km and flows through the states of Odisha, Chhattisgarh, Jharkhand, and Madhya Pradesh. Being one of the largest rivers of the nation, Mahanadi River is the principal source of water for agriculture, drinking, and industrialization in the region. It is fed by several tributaries, including the Seonath, Jonk, and Hasdeo, and its basin is characterized by fertile plains and rolling hills. The Mahanadi basin is also home to several important wildlife habitats, including the Simlipal National Park and the Saranda Forest. However, the Mahanadi basin is facing various challenges, including deforestation, soil erosion, water pollution, and over-extraction of water for irrigation and other purposes. In recent years, there has also been increased conflict between states over the sharing of water resources (Jena et al., 2014).

Selection of the study area was based on its susceptibility toward flood. Further limited study was found on application of ML approaches for prediction of flood discharge of this region compared to hydrological and statistical models, which motivated for the present study. The daily discharge data for three gauging stations of Mahanadi River basin, Kesinga, Tikarpara and Sallebhata was successfully acquired from Central Water Commission, Bhubaneswar for a period of 30 years, from 1st January 1987 to 31st December 2017.

Methodology

As part of ML, supervised learning teaches a function that converts inputs into outputs using test input-output pairs. In the considered data set table, the flood was featured with annual discharge value and flood values below average annual discharge was as leveled as NO and above average annual discharge as YES under the newly added features. This infers that the flood values were classified into two classes, YES and NO, which were labeled as 1 and 0 respectively during the data analysis (Gharakhanlou and Perez, 2023). Four statistics which include false positive, false negative, true positive and, true negative, were derived to evaluate the sample value identification and classification efficiency of a model. The number of samples that were correctly classified as flooding values is known as true positive (TP); the number of samples that were correctly classified as non-flooding values is known as true negative (TN); the number of samples that were incorrectly classified as flooding values is known as false positive (FP); and the number of samples that were incorrectly classified as non-flooding values is known as false negative (FN). Using these statistics three error indices were computed, which are accuracy, recall and ROC (Rizeei et al., 2019). To estimate the efficiency of the three considered algorithms, the accuracy, recall and ROC error indices were expressed using the expressions given below. The three scores of each flood prediction model were evaluated using 80% of the dataset for training and 20% for test of the algorithms.

$$ROC = \frac{FP}{TP} \tag{1}$$

$$ACC = \frac{TP+TN}{FP+FN+TP+TN} \tag{2}$$

$$R = \frac{TP}{FN+TP} \tag{3}$$

Three simple machine learning approaches, including LR, DT, and RF, which are based on unparalleled concepts, were implemented to accurately forecast the occurrence of the flood. The three algorithms follow entirely different approaches to classify the search space, which encourages the user to evaluate and compare their individual efficiency for classifying a given dataset. Decision trees (DTs) are a widely-used machine learning algorithm that is applied to make predictions and classify data. A DT is a tree-like graph that starts with a root node and splits into branches based on the information gain metric, which measures how much the split reduces the uncertainty in the data. The feature that provides the maximum information gain is selected as the splitting criterion for the node. The end of the branches is called leaf node and represents the final prediction. DTs are useful for both regression and classification problems and can handle both numerical and categorical data (Lawal et al., 2021). RF is a machine learning method applied for solving classification and regression problems using a supervised algorithm. It is based on the concept of decision trees and operates by constructing many decision trees (hence the "forest") and combining their results to make a final prediction. The algorithm randomly selects subsets of features and data points to train each tree in the forest, leading to high diversity among the trees and low correlation between individual trees, which improves overall accuracy. The final prediction depends on the average of predictions from individual trees in the forest, making it a highly accurate and robust algorithm (Chapi et al., 2017). LR is a statistical method used widely for classification problems of binary nature. It is used to simulate the interaction between a number of inputs features and a binary output. LR focuses on finding the best-fitting line or plane that separates the search region into positive class and negative class. The line or plane is determined by the logistic function, which maps the input features to a probability ranging from 0 to 1. The probabilities are then threshold to produce the final binary predictions (Mitchell, 1997).

The ML-inspired flood prediction model was developed using Python programming language, known for its efficient environment required for data analysis of ML. Popular Python ML tools such as Scikit-Learn and other libraries were utilized to overcome ML challenges. The data structures of pandas, including 1-D series and 2-D data frame, efficiently handle various use cases in various fields such as finance, statistics, and engineering. Pandas built on NumPy, integrates well with additional third-party libraries in a setting for scientific computing.

Results and discussion

Analysis and interpretation of the preprocessed dataset, as well as the performance assessment of the ML Models used to predict future flood situations at three stations of Mahanadi River basin are presented in this section. The accuracy, recall, and ROC scores of each flood prediction model were evaluated using 80% training and 20%

Table 20.1: Prediction results on training and test datasets.

Station	Tikarpara			Kesinga			Sallebhata		
models	ACC	R	ROC	ACC	R	ROC	ACC	R	ROC
					Training Result				
LR	52.2	70.0	80.23	45.4	53.5	75.3	78.2	60.2	68.65
DT	56.3	30.0	90.2	75.2	52.4	85.2	71.2	58.3	65.58
RF	40.2	50.0	27.26	54.3	32.3	54.3	63.3	70.0	86.66
					Test Result				
LR	50.4	66.6	71.42	54.2	35.2	57.5	71.42	70.0	86.5
DT	65.13	50.0	91.4	57.4	50.4	58.1	71.5	65.3	70.85
RF	50.82	10.4	72.6	45.1	25.00	45.2	61.42	70.0	66.66

Source: Author

test dataset. Among the three stations analyzed, Sallebhata flood could be predicted with the highest precision and was used to analyze the significance of the three models. Table 20.1 demonstrates the performance of each model, with logistic regression exhibiting scores of 71.42 accuracy, 70.0 recall, and 86.5 ROC. These results indicate that logistic regression has a high ROC score for positive predictions and a well-balanced accuracy and recall score, making it the best-recommended model for reliable flood prediction.

Conclusions

The paper explores the need for a model for predicting floods based on machine learning and compares the accuracy, recall, and receiver operating characteristics (ROC) score of logistic regression (LR), decision tree (DT) and random forest (RF) models. The best model, according to the results, was LR. However, there is a serious limitation of the applied Machine learning (ML) algorithms that were observed to work efficiently with the availability of a large sized dataset covering discharge values of continuous 30 years. Non availability of a sufficiently long and continuous time series of discharge may reduce the performance level of these models. Future work will focus on deep learning models and human-machine interaction for more accurate predictions. A flood warning system for citizens, to save lives and reduce government costs, could also be developed. Improvement of the models can be done by using data decomposition techniques, ensemble methods, and optimizers. A survey on spatial flood prediction using ML algorithms, as well as expanding flood location databases is suggested.

References

Chapi, K, Singh, V P, Shirzadi, A, Shahabi, H, Tien Bui, D, Pham, B T, and Khosravi, K (2017). A novel hybrid artificial intelligence approach for flood susceptibility assessment. *Environmental Modeling and Software*. 95, 229–245. https://doi.org/10.1016/j.envsoft.2017.06.012.

Gharakhanlou, N M and Perez, L (2023). Flood susceptible prediction through the use of geospatial variables and machine learning methods. *Journal of Hydrology*. 617, 129121 https://doi.org/10.1016/j.jhydrol.2023.129121

Jena, P P, Chatterjee, C, Pradhan, G, and Mishra, A (2014). Are recent frequent high floods in Mahanadi basin in eastern India due to increase in extreme rainfalls?. *Journal of Hydrology*. 517, 847–862. https://doi.org/10.1016/j.jhydrol.2014.06.021.

Lawal, Z K, Yassin, H, and Zakari, R Y (2021). Flood prediction using machine learning models: a case study of Kebbi state Nigeria. In IEEE Asia-Pacific Conference on Computer Science and Data Engineering (CSDE), pp. 1–6. IEEE, 2021. https://doi.org/10.1109/CSDE53843.2021.9718497.

Mitchell, T M (1997). Machine Learning. McGraw-Engineering.

Rizeei, H M, Pradhan, B, Saharkhiz, M A, and Lee, S (2019). Groundwater aquifer potential modeling using an ensemble multi-adoptive boosting logistic regression technique. *Journal of Hydrology*. 579, 124172 https://doi.org/10.1016/j.jhydrol.2019.124172

Samantaray, Dibyendu, Chandranath Chatterjee, Rajendra Singh, Praveen Kumar Gupta, and Sushma Panigrahy. (2015) Flood risk modeling for optimal rice planning for delta region of Mahanadi river basin in India. *Natural Hazards*. 76, 347–372. https://doi.org/10.1007/s11069-014-1493-9.

Zhou, Z-H (2021). Machine Learning. Springer Nature.

21 Hydraulic conveying through the slurry pipeline: environment-friendly safe disposal of Indian coal fly ash

Vighnesh Prasad[a], Anil Dubey[b] and Snehasis Behera[c]

Design and Project Engineering, CSIR - Institute of Minerals & Materials Technology, Bhubaneswar, Odisha – 751013, India

Abstract

The present investigation demonstrates the effect of coal fly ash concentration on the rheological, pipe flow, and erosion behavior during slurry pipeline transportation. Rheology data indicate that 6570 wt.% fly ash slurries exhibit shear thinning flow behavior in the shear rate range of 10300 s^{-1}. Both slurry viscosity and shear stress increase with the solid concentrations. The frictional head loss increases with solid concentration and slurry velocity; however, it decreases with an increase in pipe diameter. The frequent collisions of the fly ash particles on the surface of mild steel specimens cause erosion wear, intensifying at higher solid loading and velocity. A favorable economic flow condition is achieved when 65 wt.% of fly ash slurry is transported at a velocity of 1.5 m/s in a pipe of diameter 0.5 m.

Keywords: Coal fly ash, erosion, head loss, Rheology, slurry flow

Introduction

The disposal and handling of particulate solid wastes needs greater attention when a large throughput is transported to a longer distance. Here, hydraulic conveying through pipeline offers an economical and environmentally viable mode of transport over trucks, trains, barges, and conveyor belts. It has many advantages, including minimum environmental disruption, low air and noise pollution, low operating cost, feasibility in adverse locations, insensitivity to surface conditions, etc.

One of the major concerns related to slurry flow is obtaining favorable flow behavior in the pipeline. Here, the rheological studies play a crucial role in describing the flow characteristic of the slurries in the pipeline. Most researchers have focused on the rheological studies of dilute slurries to avoid pipe blockage during transportation. However, it does not promote handling bulk solids and optimal use of water. Rheological investigations of 46.03–46.91 wt.% lignite slurries (Wu et al., 2015), 10.3–16 wt.% corn stover slurries (Ghosh et al., 2018), and 15–50 wt.% kaolin clay (Leong et al., 2012) have been attempted previously. However, the rheological response and flow calculation for concentrated slurries (Solid > 60 wt.%) are seldom reported in past studies. In addition, the service life of an industrial pipeline is primarily controlled by erosion wear, which depends on the properties of the solid particle, suspending medium, pipe materials, slurry flow condition, operating parameters, and environmental conditions (Sadighian, 2016). Prior arts mainly focus on simulation tools to estimate erosion wear; however, the wear rate data using experimental methods for highly concentrated slurry systems are still inadequate. In the present investigation, coal fly ash (CFA), a by-product of burnt coal, is chosen as a test sample. Currently, the bulk handling and disposal of CFA is challenging for mining

[a]vprasad@immt.res.in, [b]anildubey@immt.res.in, [c]sbehera@immt.res.in

DOI: 10.1201/9781003450924-21

and thermal industries worldwide. The present study performs the rheological and erosion wear experiments with 65–70 wt.% CFA slurries to obtain a favorable flow condition in the pipeline.

Materials and methods

The CFA was collected from NALCO, Angul, Odisha, India. 65–70 wt.% CFA was added to DI water, and gently mixed using a glass rod to prepare slurries. The rheological experiments were performed using a Rheometer (HAAKE Rheostress, ThermoFisher) with a cup-bob geometry in the shear rate () range between 10-300 s^{-1}. Erosion tests were conducted using an erosion pot tester (MT-4, Micromatic Technologies) in the velocity (v_m) range of 1.0 – 3.5 m/s for 4 hrs.

Results and discussion

Sample characterization

The particle size distribution analyzed using a particle size analyzer (LA-960, HORIBA) reveals that the CFA sample contains poly-dispersed particles with median size, d_{50} = 25.32 μm (Figure 21.1a). The scanning electron microscopy (SEM), FE-SEM (SUPRA 55, Carl Zeiss) indicates that CFA particles are porous and spherical (Figure 21.1b). CFA majorly constitutes SiO_2, Al_2O_3, Fe_2O_3, TiO_2, and K_2O; however small fraction of Na_2O, MgO, P_2O_5, SO_3, and CaO are also present (Figure 21.1c). The specific gravity (Ss) of CFA estimated using a pycnometer (Borosil Ltd.) is 2.096.

Rheological measurements

Both the slurry viscosity (η) and shear stress (τ) increase with the increase in CFA concentration (C_w) from 65–70 wt.%, as shown in Figure 21.2. The increased C_w enhances the inter-particle interactions, resulting in higher η and τ. Moreover, the CFA slurries exhibit shear thinning flow behavior as η reduces with increasing $\dot\gamma$ (Figure 21.2a). It attributes to the disintegration of the slurry structure and alignment of the layers of particles in the applied $\dot\gamma$ direction. A power-law model (Equation

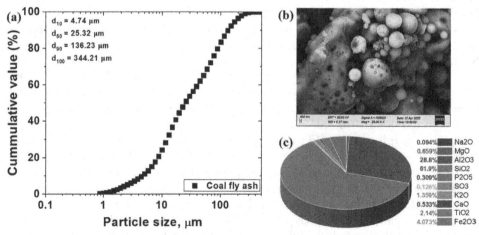

Figure 21.1 (a) Particle size distribution (b) SEM image 5000x (c) Chemical composition
Source: Author

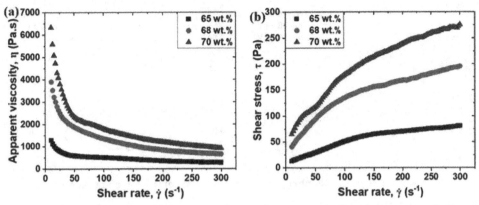

Figure 21.2 Flow curves at varying Cw = 65 - 70 wt.% (a) η versus $\dot{\gamma}$ (b) τ versus $\dot{\gamma}$ curves
Source: Author

Table 21.1: Flow coefficients during fitting of power-law model in Figure 21.2b.

C_w (wt.%)	n	k (Pa.s^n)	R^2
65	0.51	4.62	0.97
68	0.40	20.35	0.98
70	0.43	23.71	0.99

Source: Author

1) (Chhabra and Richardson, 2008) is fitted in Figure 21.2b, and the flow behavior index (n) and consistency coefficient (k) are tabulated in Table 21.1.

$$\tau = k.\dot{\gamma}^n \tag{1}$$

Slurry flow calculation

The CFA particles with specific gravity (Ss) = 2.096 shows a setting tendency in water. Thus, a critical velocity (v_c) – minimum velocity to avoid sedimentation, is estimated using a widely accepted Durand correlation (Equation 2) (Durand, 1953).

$$v_{c,durand} = F_L \sqrt{2gD(S_s-1)} \tag{2}$$

Taking Durand Froude number (F_L) as 0.4 (Durand, 1953), v_c is estimated as 0.72 - 1.31 m/s for the pipe diameters (D) = 0.15 – 0.5 m. The correlations to determine the Power-law Reynolds number (Re_{pl}), critical power-law Reynolds number (Re_{plc}), and fraction factor (f) using flow coefficients (Table 21.1) are available in our earlier publication (Prasad et al., 2020). However, the frictional pipe head loss (h_L) in the meter of water column per meter length of pipe (mWc/m) is determined using Equation (3) (Miedema and Ramsdell, 2020). Here, ρ_m and ρ_l are slurry and water densities (Kg/m³), respectively.

$$h_L = \frac{2.f.v_m^2}{gD}\left(\frac{\rho_m}{\rho_1}\right) \tag{3}$$

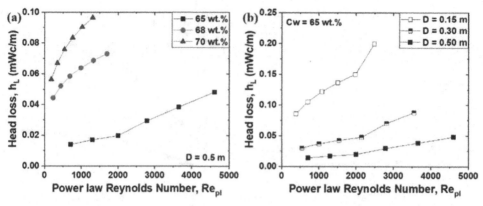

Figure 21.3 h_L calculation (a) varying $C_w = 65 - 70$ wt.% (b) varying $D = 0.15 - 0.5$ m
Source: Author

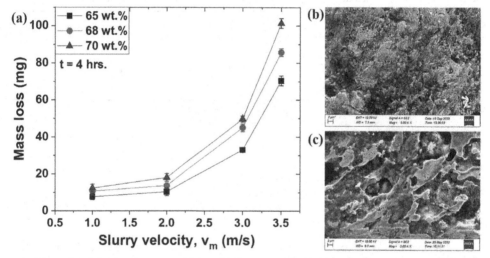

Figure 21.4 (a) C_w versus mass loss of MS specimens; (b) SEM images of MS specimens at 5000x (b) before erosion test when flow variables are absent (c) after erosion test when MS specimen is subjected to 70 wt.% CFA slurry flowing at $v_m = 3.5$ m/s
Source: Author

As shown in Figure 21.3, h_L increases with Re_{pl} or v_m regardless of C_w and D. The higher magnitude of h_L at $C_w = 70$ wt.% (Figure 21.3a) translates the additional frictional resistance between particle-particle and particle-pipe wall. Moreover, D = 0.5 m facilitates a larger area where the probability of particle-wall interaction and frictional resistance is minimal, resulting in low h_L (Figure 21.3b).

Erosion wear in pipeline

The mass loss of mild steel (MS) specimens increases with C_w and v_m due to the repetitive impingement of the CFA particles on the surface of MS. As evident in Figure 21.4c, the surface of the MS specimen contains eroded and pitting imprint structure which signifies the mass losses due to erosion wear (Figure 21.4a).

Conclusion

Results show that η and τ are the function of coal fly ash (CFA) concentration; however, the shear thinning flow behavior prevails irrespective of solid loading. h_L enhances with increasing CFA in carrier water or increasing velocity; however, it possesses a lower value in a large pipe. Erosion data suggest that the service life of the industrial pipeline and its components may degrade by transporting a concentrated slurry at a higher velocity. Although the pipeline transportation of 70 wt.% slurry promotes the bulk handling of CFA, it demands high h_L and erosion wear, which consequently enhances the cost of transportation and maintenance. Besides, $v_m = 1$ m/s shows minimum h_L and erosion rate; however, it encourages sedimentation of particles, as $v_m < v_c$. Thus, the hydraulic conveying of 65 wt.% of CFA slurry at $v_m = 1.5$ m/s in D = 0.5 m appears as a favorable and optimized flow condition in the pipeline.

Acknowledgment

This work is supported by the SERB New Delhi, under project No. EEQ/2022/000101. The authors thank the Director, CSIR – IMMT Bhubaneswar, for permitting them to publish this work.

References

Chhabra, R, and Richardson, J (2008). Non-Newtonian Flow and Applied Rheology: Engineering Applications. Oxford, UK: Butterworth-Heinemann.

Durand, R (1953). Basic relationships of the transportation of solids in pipes-experimental research. In Proceedings International Association for Hydro-Environment Engineering and Research 5th Congress, Minneapolis. 89–103.

Ghosh, S, Holwerda, E K, Worthen, R S, Lynd, L R, and Epps, B P (2018). Rheological properties of corn stover slurries during fermentation by clostridium thermocellum. *Biotechnology for Biofuels*. 11(1), 1–12.

Leong, Y K, Teo, J, Teh, Ej, Smith, J, Widjaja, J, Lee, J X, Fourie, A, Fahey, M, and Chen, R (2012). Controlling attractive interparticle forces via small anionic and cationic additives in kaolin clay slurries. *Chemical Engineering Research and Design*. 90(5), 658–666.

Miedema, S A, and Ramsdell, R C (2020). Slurry transport fundamentals, a historical overview & the delft head loss & limit deposit velocity framework (2nd edition). *World Dredging*. 52, 30–35.

Prasad, V, Thareja, P, and Mehrotra, S P (2020). Role of rheology on the hydraulic transportation of lignite coal and coal ash slurries in the pipeline. *International Journal of Coal Preparation and Utilization*. 42(4), 1263–1277.

Sadighian, A (2016). Investigating key parameters affecting slurry pipeline erosion. PhD thesis, University of Alberta, Canada.

Wu, J H, Liu, J Z, Yu, Y J, Wang, R K, Zhou, J H, and Cen, K F (2015). Improving slurryability, rheology, and stability of slurry fuel from blending petroleum coke with lignite. *Petroleum Science*. 12, 157–169.

22 Study and prediction of built-up land use for Bhubaneswar and its corresponding effects

Abhayaa Nayak[a] and Anil Kumar Kar[b]

Department of Civil Engineering, VSSUT, Burla, India

Abstract

Bhubaneswar, the capital city of Odisha, India, has been attracting the rural population to migrate towards this urban location. Socioeconomic factors like job opportunities, availability of resources, etc. might be some of the reasons for migration. Apart from such factors, this study aims to investigate topographic factors responsible for the urban expansion of this city and predict its effects on the future. From 2000 to 2020, NDBI increased by 61.2%, simultaneously reducing NDVI by 29.4%, and NDWI by 26.6%. With such figures, a correlation was developed for NDBI which indicated that by 2025 NDBI of Bhubaneswar city will increase by 41%. Being present in the average per capita availability of the Mahanadi basin, with a high population density, such a sharp increase in the built-up land usage is bound to create environmental as well as social pressure caused by a limitation of resources, especially water. Such changes call for proper strategic measures by encouraging the development of green zones to protect the local natural environment from further exploitation.

Keywords: Land usage, NDBI, urban expansion, water crisis

Introduction

Urban development or growth has added a lot of features to our life, but population growth has forced higher area expansion leading to changing of landscape scenarios from non-built-up land to impervious land. Cities with higher growth rates tend to demand higher water quantity for various domestic and non-domestic purposes like usage in hospitals and educational institutions and, most importantly in construction works and industries. Thus, one can stipulate that the amount of water that will be required in the coming future is on an increasing trend. When viewed from another angle one can hypothesize that the near future might face a crisis of usable water. This rise in water demand for a given city over a specific duration can be helpful in estimating future water consumption. In research conducted on the banks of the Hooghly River, i.e., Barrackpore subdivision, it was found that the value of NDBI increased from 0.6 to 0.65 from 1990 to 2016 (Das and Angadi, 2020). Accordingly, about 27.8% of the land from other local sources was converted into built-up land leading such urban areas to a dire loss of local vegetation. Such observance led the researchers to infer that there might be a shortage of water available for the surrounding rural areas for agricultural purposes. Later when Indian metro cities were taken into consideration for comparing the land surface temperature (LST) with NDBI and NDVI, it was found that Chennai and Mumbai had higher values of LST as compared to Delhi and Kolkata (Shahfahad et al., 2020).

The research works conducted previously show that higher built-up or population which are inter-related have an adverse effect on local water resources. Many of them

---

[a]abhayaa.nayak91@gmail.com, [b]anilkarhy@gmail.com

DOI: 10.1201/9781003450924-22

have reported and supported this theory. The reason behind taking NDBI as the main factor is that this index or parameter shows the level of built-up land used which can be directly or indirectly related to the water demand or water quantity available in a particular region. Hence, to study such effects, we have selected the capital city of Odisha, i.e., Bhubaneswar for our research.

Study area

Bhubaneswar located at 20.2640° N, 85.8184° E, provides residence to approximately 1.3 million people spread out over an area of 1,110 sq. km., which comes under the average per capita water availability indicated in Figure 22.1. Being a smart city, Bhubaneswar hosts IT companies, and many engineering, Medical, Management, and research organizations like IIT, AIIMS, ICMR, IMMT, XIMB, NISER, etc.

As observed from Figure 22.1, Bhubaneswar comes under average per capita water availability in the state. The percentage of areas with either high or low per capita water availability is much lower than the average water availability areas. So, for such areas, a high rate of unplanned urban development creates pressure on the local or neighboring water resources.

Methods

The images for conducting the research on the selected cities, i.e., Bhubaneswar and Sundergarh were downloaded from USGS (US Geological Survey, https://earthexplorer.usgs.gov/) of Landsat 7 for the year 2000, 2005 and 2010, and Landsat 8 for the year 2015 and 2020. Before 2013 Landsat 7 images were being used but after the

Figure 22.1 State water plan showing per capita availability (Department of Water Resources, Govt. of Odisha)

Source: https://dowr.odisha.gov.in/publication/plan-%26-policies/plan/state-water-plan?page=1

launch of Landsat 8 in 2013, the images were hereafter downloaded from Landsat 8. For the current study NDBI, NDWI, NDVI, Aspect, Slope, Curvature, elevation, Plan Curve and Profile curve were adopted. For the evaluation of NDBI, NDWI and NDVI, green, red, near-infrared, and short-wave infrared bands from Landsat 7 (bands 2,3,4, and 5) and Landsat 8 (bands 3, 4, 5, and 6) were used. These data were then used as input in Arc Map 10.4 for calculating the respective values for their respective years. The other parameters which generally remain constant were evaluated from digital elevation map (DEM) using Arc Map 10.4. With the help of Fishnet feature in Arc Map 10.4, the values of rest of the parameters were calculated for their respective years. The spatial resolution for both Landsat 7 and 8 was 30m. Apart from Landsat, Sentinel (launched in 2015) images are available with better spatial resolution but still Landsat was preferred because of availability of older images.

Results and discussion

With the use of linear regression, the following equation was developed with R^2 of 0.91.

$$Y_{(BM)} = 34300.12 - 0.21446 * ASPECT - 110250 * CURVATURE + 0.239609 *$$
$$DEM + 97999.7 * PLANCUR + 58403.7 * PROFILECUR + 0.026345 * SLOPE +$$
$$0.599954 * NDVI\ 2015 + 0.095773 * NDVI\ 2015 - 0.03211 * NDWI\ 2015 - 0.0033$$
$$* NDBI2010 + 0.115932 * NDVI2010 - 0.35854 * NDWI2010 + 0.41539 *$$
$$NDBI2005 + 0.127333 * NDVI\ 2005 + 0.733283 * NDWI$$
$$2005$$

(1)

Figure 22.2 showcases the predicted NDBI in comparison with the actual NDBI. The graph illustrates that the prediction equation is in good agreement with the actual data. Thus, NDBI of 2025 for Bhubaneswar was calculated using Equation 1. The

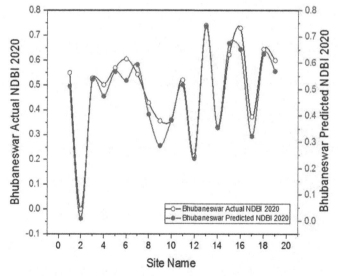

Figure 22.2 Model verification for Bhubaneswar
Source: Author

BHUBANESWAR NDBI

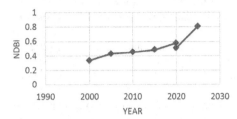

Figure 22.3 Year wise NDBI for Bhubaneswar
Source: Author

correlation for Bhubaneswar underpredicts NDBI Figure 22.3. The data clearly shows that the amount of land used in Bhubaneswar has increased by about 71%. The expansion and exploitation of land use have been more prevalent in Bhubaneswar which is having an effect on food prices, income, governance, and institutions which may slowly lead to internal violent conflicts. It has also been predicted that such rapid urbanization causing exploitation of natural reserves will lead to a 5% increase in riots and demonstrations, and a 10% increase in public-related issues. All the research works conducted both in the national or international level have suggested to use of green construction which would control local surface environmental temperature, floodings, erosions, etc., and demolishing any illegal constructions and use it for the same (Zhang et al., 2023). In accordance with the situation, it has been strongly suggested and highly recommended by (Parween et al., 2022; Chatterjee and Majumdar, 2022; Zhang et al., 2023; Rana many researches Sarkar, 2021; Purswani et al., 2022) for a systematic assessment and changes/modifications in policies for maintaining the quality of life, and surrounding water bodies which have been severely affected by urbanization.

Conclusion

The population of Bhubaneswar increased from 6,58,220 as per 2001 census to 8,43,402 as per 2011 census (as per Ministry of Home Affairs, Government of India). With the city coming under average per capita water availability, was quite higher for Bhubaneswar, indicating that with a higher population and urbanization with such rapid rates is putting the local environment as well as social life at risk. Apart from policies and ecological assessments by the local authorities to control urban expansion, measures like ecological restoration projects, ecological security patterns, and ecological barriers at required locations can be implemented.

Steps in accordance with the local landscape can be carried out enabling reasonable and sustainable urban growth with adequate ecosystem services. This study which uses NDBI as a parameter to account for urbanization tries to signify the importance of the extent of unreasonable urbanization bringing forward this scenario in front of the policymakers and city authorities to access the gravity of the situation which if unchecked may lead to future water crisis in the city.

References

Chatterjee, U, and Majumdar, S (2022). Impact of land use change and rapid urbanization on urban heat island in Kolkata city: A remote sensing-based perspective. *Journal of Urban Management.* 11(1), 59–71. https://doi.org/10.1016/j.jum.2021.09.002.

Das, S, and Angadi, D P (2020). Land use-land cover (LULC) transformation and its relation with land surface temperature changes: A case study of barrackpore subdivision, West Bengal, India. *Remote Sensing Applications: Society and Environment*. 19. https://doi.org/10.1016/j.rsase.2020.100322.

Parween, S, Siddique, N A, Mahammad Diganta, M T, Olbert, A I, and Uddin, M G (2022). Assessment of urban river water quality using modified NSF water quality index model at Siliguri city, West Bengal, India. *Environmental and Sustainability Indicators*. 16. https://doi.org/10.1016/j.indic.2022.100202

Purswani, E, Verma, S, Jayakumar, S, Khan, M L, and Pathak, B (2022). Examining and predicting land use change dynamics in Gandhinagar district, Gujarat, India. *Journal of Urban Management*. 11(1), 82–96. https://doi.org/10.1016/j.jum.2021.09.003

Rana, M S, and Sarkar, S (2021). Prediction of urban expansion by using land cover change detection approach. *Heliyon*. 7(11). https://doi.org/10.1016/j.heliyon.2021.e08437

Shahfahad, Kumari, B, Tayyab, M, Ahmed, I A, Baig, M R I, Khan, M F, and Rahman, A (2020). Longitudinal study of land surface temperature (LST) using mono- and split-window algorithms and its relationship with NDVI and NDBI over selected metro cities of India. *Arabian Journal of Geosciences*. 13(19). https://doi.org/10.1007/s12517-020-06068-1

Zhang, M, Kafy, A al, Xiao, P, Han, S, Zou, S, Saha, M, Zhang, C, and Tan, S (2023). Impact of urban expansion on land surface temperature and carbon emissions using machine learning algorithms in Wuhan, China. *Urban Climate*. 47. https://doi.org/10.1016/j.uclim.2022.101347

23 Streamflow prediction in a river using twin support vector regression and extreme learning machine

Jagdish Mallick[1,a], Safalya Mohanty[1,b] and Abinash Sahoo[2,c]

[1]Department of Civil Engineering, Parala Maharaja Engineering College, Brahmapur, India

[2]Department of Civil Engineering, National Institute of Technology, Silchar, India

Abstract

Monthly streamflow forecasting can produce significant knowledge for hydrogeological applications which includes optimization of water resources allocation, sustainable design of urban and rural water management systems, water usage, water quality and pricing assessment, and irrigation and agriculture operations. Motivation to explore and develop proficient prediction models is a continuing effort for hydrogeological applications. This study explores the potential of twin support vector regression (TSVR) method, for monthly forecasting in Rushikulya River, Odisha, India. A comparative analysis is done for evaluating prediction performance of TSVR in comparison to extreme learning machine (ELM). Forecasting metrics such as correlation coefficient (R), root mean squared error (RMSE), and Nash-Sutcliffe efficiency (ENS) are applied for assessing effectiveness TSVR model outperformed ELM model across all statistical indices. In quantifiable terms, preeminence of TSVR over ELM model was demonstrated by ENS = 0.9783 and 0.9501, R = 0.9902 and 0.9757, and RMSE = 2.356 and 6.2213, respectively.

Keywords: Extreme learning machine, rushikulya river, streamflow, TSVR

Introduction

Appropriate information on plausible variations in future streamflow is very significant for planning and managing various water resource schemes, especially water availability assessment, early drought and flood warning, hydropower generation, and planning agriculture operations. Prevailing prediction methods can be roughly classified into two categories: physical-based and data-driven (DD) approaches (Chu et al., 2017). Physical-based approaches have a probable benefit in constructing the physical understanding between the hydrological process and the variables. But, such approaches characteristically necessitate a huge amount of hydrological data, and adequate datasets are hardly accessible in developing countries, which restricts their practical usage (Peng et al., 2017). Moreover, influence and complex interaction of several arbitrary parameters result in uncertainty and poor model performances (Agnihotri et al., 2022; Ahani et al. 2018; Yaseen et al. 2018). In contrast, DD methods compared to physical-based methods, necessitate minimal data, are simple to implement, and can be quickly modelled (Sahoo and Ghose, 2022). Hence, DD approaches have turned out to be more prevalent in forecasting applications.

Boucher et al. (2020) applied multilayer perceptron and extreme learning machine (ELM) models for forecasting considering data from four catchments located in four different locations. Niu and Feng (2021) examined ANN, ANFIS, ELM, Gaussian

[a]jagadishmallick432@gmail.com, [b]mohantysafalya@gmail.com, [c]bablusahoo1992@gmail.com

DOI: 10.1201/9781003450924-23

process regression (GPR), and SVR techniques forpredicting daily series considering data from two major reservoirs in China. Their study indicated that ELM produced higher predictions in Hongjiadu reservoir whereas SVM produced better predictions in Xinfengjiang reservoir. In this work, potential efficiency of TSVR technique was compared with ELM for assessing prediction accurateness.

Study area

The River Rushikulya flows through Kandhamal and Ganjam districts of Odisha, India. Being one of the major river basins of Odisha, it originates from Eastern Ghats, South Odisha. It is located between 84⁰00'00" E to 85⁰17'17" E longitude and 19⁰30'17" N and 20⁰17'17" N latitude (Figure 23.1). The river starts from slopes of Kutrabor and Araha bity hills having a drainage area of 82,100 sq. km. and runs for roughly 165 km before meeting Bay of Bengal at Puruna Bandha of Chhatarpur area.

Methodology

ELM

Huang et al. (2006) developed ELM learning algorithm that is a kind of single-hidden layer feed-forward (SLF) neural network. This algorithm has been established as an effective one for applications in different fields. ELM's basic theory is discussed below.

Assume a dataset with random distinctive samples , an activation function and a standard SLF with hidden nodes. This can be articulated below:

$$\sum_{i=1}^{\tilde{N}} \beta_i g_i(x_j) = \sum_{i=1}^{\tilde{N}} \beta_i g_i(w_i x_j + b_i) = y_j, j = 1, \dots, N$$

here $\beta_{i1} = [\beta_{i1}, \beta_{i2}, \dots, \beta_{im}]^T$ - output weight, $w_i = [w_{i1}, w_{i2}, \dots, w_{im}]^T$ - input weight, and b_i -threshold of $it\square$ hidden node. For N samples, ELM's basic principle is to attain zero error via $\sum_{j=1}^N \|y_j - t_j\|$, here y - original output vector.

TSVR

Peng (2010) first developed TSVR which aims at discovering insensitive down-bound and up-bound aspects for training two QPPs (quadratic programming problems) in contrast to conventional SVR. Therefore, tentatively, it is swifter than SVR.

Provided a training series, $T = \{(x_1, y_1), (x_2, y_2), \dots, (x_n, y_n)\}$ where, x_i - row matrix an y_i - column matrix. Also consider, $Z = (x_1, x_2, \dots, x_n)$ and $Y = (y_1, y_2, \dots, y_n)$. Supposing, in input space the up-bound and down-bound functions are

$$f_1(x) = w_1^t x + b_1 \text{ and} f_2(x) = w_2^t x + b_2$$

Evaluating constraint

Measures such as E_{NS}, RMSE, and R are utilized to evaluate performance of models. The formulae for the iindices are

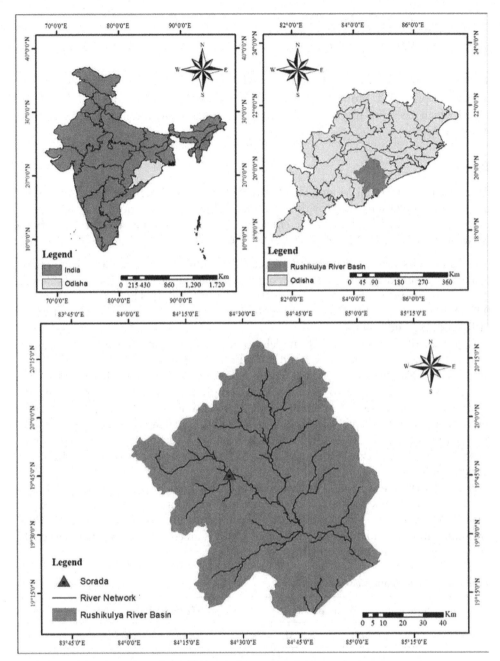

Figure 23.1 Location of specified soradagauge station
Source: Author

$$NSE = 1 - [\frac{\sum_{i=1}^{N}(M_i - N_i)^2}{\sum_{i=1}^{N}(M_i - \overline{M_1})^2}]$$ (9)

$$RMSE = \sqrt{\frac{1}{N}\sum_{i=1}^{N}(N_i - M_i)^2}$$ (10)

$$R = \frac{\sum(N_i - \overline{N_1})(M_i - \overline{M_1})}{\sqrt{\sum(N_i - \overline{N_1})^2 \sum(M_i - \overline{M_1})^2}} \tag{11}$$

where, M_i and N_i are observed and predicted $it^\square Q_f$, $\overline{M_1}$ is mean of observed Q_f and $\overline{N_1}$ is mean of predicted. Monthly average Q_f data (from 1990-2019) are collected from CWC which assists in training and testing the model. Daily data are converted into monthly data.

Result

Assessment results of ELM and TSVR models for Sorada gauge site based on different input combinations are provided in Table 23.1. Three evaluation index RMSE, R, and E_{NS} are applied for finding the efficiency of data in training and testing period. Between four models, ELM4 model $(Q_{f_{t-1}}, Q_{f_{t-2}}, Q_{f_{t-3}}, Q_{f_{t-4}}$ as input) provides excellent results with E_{NS} values0.9615 and 0.9501 during training and testing phase respectively. Similarly, for TSVR approach based on E_{NS}, when TSVR4 is considered, it provides the most excellent E_{NS} value (0.9926 and 0.9783). The best value for R is 0.9818 in training and 0.9757 in testing period for ELM4 whereas for TSVR4,value of R is 0.9982 and 0.9902.

Appraisal of ELM and TSVR models for the gauging site during both periods are demonstrated in Figures 23.2 and 23.3. From the figures, it was found that predictions by TSVR model are superior compared to ELM model. This work shows that the way in advance is to have a correct comprehension relating to performance of distinct models and use as many models as possible to gather additional proof for specified models. Time series plots of actual vs. predicted Q_f for applied models are presented in Figure 23.3. Prediction results reveal that estimated peak Q_f is 5245.83m³/S, 5409.519m³/Sfor ELM and TSVR against actual peak 5567.64m³/Sfor the station Sorada.

From observations of forecasting outcomes of this work, it can be determined that ELM does not provide satisfactory performance in monthly forecasting.Compared with ELM model, the TSVR model gives much better performances based on three evaluation indices, which demonstrates that TSVR is an effective alternative approach in enhancing forecasting accurateness of annual.

Table 23.1: Performance assessment results.

Location	Scenario	Training phase			Testing phase		
		NSE	RMSE	R	NSE	RMSE	R
Sorada	ELM I	0.9553	5.949	0.9784	0.9435	8.5502	0.9737
	ELM 2	0.9576	5.7702	0.9796	0.9452	7.7489	0.9745
	ELM 3	0.9592	5.2157	0.9799	0.948	6.914	0.9748
	ELM 4	0.9615	4.6034	0.9818	0.9501	6.2213	0.9757
	TSVR 1	0.9894	1.7381	0.9965	0.9731	3.457	0.9875
	TSVR 2	0.9904	1.0354	0.9971	0.9749	3.1865	0.9879
	TSVR 3	0.991	0.9862	0.9976	0.9764	3.1209	0.9884
	TSVR 4	0.9926	0.187	0.9982	0.9783	2.356	0.9902

Source: Author

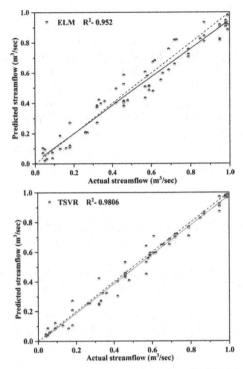

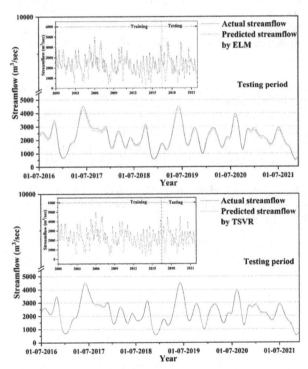

Figure 23.2 Scatter plot of applied ELM and TSVR models
Source: Author

Figure 23.3 Actual vs. predicted streamflow based on ELM and TSVR models
Source: Author

Conclusion

The aim is to analyze predictive abilities of TSVR and ELM models for monthly streamflow forecasts in Rushikulya River, Odisha, India. Based on statistical indices (R, RMSE, and E_{NS}) that were considered to carry out the evaluation of the prediction models, authors concluded that developed TSVR method outperformed the ELM method in forecasting. Even though TSVR model attains satisfactory prediction performances, this work only emphasizes univariate forecasting, i.e. not considering exogenous variables affecting , such as precipitation. In future research, other parameters affecting can be selected for developing prediction models for studying their impact on model accuracy.

References

Agnihotri, A, Sahoo, A, and Diwakar, M K (2022). Flood prediction using hybrid ANFIS-ACO model: a case study. In Inventive Computation and Information Technologies. Singapore: Springer. pp. 169–180.

Ahani, A, Shourian, M, and Rahimi Rad, P, (2018). Performance assessment of the linear, nonlinear and nonparametric data driven models in river flow forecasting. *Water Resources Management.* 32(2), 383–399.

Boucher, M A, Quilty, J, and Adamowski, J (2020). Data assimilation for streamflow forecasting using extreme learning machines and multilayer perceptrons. *Water Resources Research.* 56(6), e2019WR026226.

Chu, H, Wei, J, Li, J, Qiao, Z, and Cao, J (2017). Improved medium-and long-term runoff forecasting using a multimodel approach in the Yellow River headwaters region based on large-scale and local-scale climate information. *Water.* 9(8), 608.

Huang, G B, Zhu, Q Y, and Siew, C K (2006). Extreme learning machine: theory and applications. *Neurocomputing.* 70(13), 489–501.

Niu, W-J, and Feng, Z-K (2021). Evaluating the performances of several artificial intelligence methods in forecasting daily streamflow time series for sustainable water resources management. *Sustainable Cities and Society.* 64, 102562. https://doi.org/10.1016/j.scs.2020.102562

Peng, T, Zhou, J, Zhang, C, and Fu, W (2017). Streamflow forecasting using empirical wavelet transform and artificial neural networks. *Water.* 9(6), 406.

Peng, X (2010). TSVR: an efficient twin support vector machine for regression. *Neural Networks.* 23(3), 365–372. https://doi.org/10.1016/j.neunet.2009.07.002

Sahoo, A, and Ghose, D K (2022). Imputation of missing precipitation data using KNN, SOM, RF, and FNN. *Soft Computing.* 26(12), 5919–5936. https://doi.org/10.1007/s00500-022-07029-4

Yaseen, Z M, Allawi, M F, Yousif, A A, Jaafar, O, Hamzah, F M, and El-Shafie, A (2018). Non-tuned machine learning approach for hydrological time series forecasting. *Neural Computing and Applications.* 30(5), 1479–1491.

24 Seismic stability analysis of a small scaled hill slope reinforced with geogrid

Rasmiranjan Samal[a] and Smrutirekha Sahoo[b]

National Institute of Technology Meghalaya, Meghalaya, India

Abstract

Geotechnical and earthquake engineering put a great deal of emphasis on slope stability studies under dynamic loads. The input ground motions, and the dynamic characteristics of the soil medium greatly influence how slopes respond to seismic waves. This study uses the 1971 San Fernado Down earthquake record to evaluate the seismic behavior of a small-scaled hill slope. A two-dimensional dynamic finite element model under the plain-strain condition has been constructed for the elastic, fully plastic Mohr-Coulomb soil model. At the bottom of the slope, the analysis replicates the highest horizontal seismic accelerations. It has been stated that the deformation caused during the earthquake shaking period is how the slope responds seismically to horizontal seismic accelerations. The analysis found that the crest's horizontal and vertical displacement is higher than other parts of the slope and lower at the toe of the slope. Hence the toe of the slope is more stable than the other parts.

Keywords: Seismic analysis, finite element method, geogrid, hill slope

Introduction

Limit equilibrium (LE) stability analysis is frequently used in slope design. The limit equilibrium equations are expanded to include extra force components proportional to gravity when doing pseudo-static stability analysis of slope, which often assumes a horizontal seismic coefficient. When analyzing the stability of slopes subject to seismic loadings, the displacement approach or the pseudo-static method is typically used (Sahoo and Shukla, 2019). However, these analytical techniques fail to capture the slope's dynamic properties or failure mechanism in a time-domain context. By using numerical modelling with various slope material qualities under seismic conditions, which include the impact of ground shaking's amplitude, frequency, and duration, the slope's dynamic behavior may be obtained (Diaz-Segura, 2016). Recently, some researchers have modeled the slope using dynamic finite elements by replicating actual earthquake records. Researchers have previously noted that the most significant vertical seismic loading during an earthquake may be equivalent to or larger than the horizontal seismic loading. This could potentially impair the performance of the slope (Lee et al., 2017).

In the current research, a numerical slope stability analysis has been carried out by modeling the San Fernando Down earthquake records. This analysis utilizes finite element software, MIDAS GTS NX 2021 v1.1. Under horizontal seismic ground motion acceleration, the response of the hill slope in terms of displacement has been observed.

Methodology

Numerical investigation of the slope

Finite element analysis (FEM) is the numerical approach considered in the current study, conducted with the help of commercially available software MIDAS GTS NX

---

[a]rasmiranjannitm@gmail.com, [b]smrutirekha.sahoo@nitm.ac.in

DOI: 10.1201/9781003450924-24

2021 v1.1 computer program, which was used to construct a two-dimensional geometric model of the slope. Under the criteria of plane strain, slope stability analyzes was carried out. Triangular elements with 15 nodes were selected for the modeling to get more precise and sensitive results. The slope is located in Meghalaya, India, home to some of the wettest places on earth (Mawsynram and Sohra), and constant rain is one of the main factors causing various landslides in various parts of the states. Slope geometry is given in Figure 24.1 with a soil slope height of 15 m and slope angle of 27°. The soil parameters employed in the FEM analysis for the 3D slope Model were determined by laboratory testing of the soil samples, as indicated in Table 24.1. PET geogrids are used in this study. The geogrids are placed at a spacing of 3 m along the slope.

In static boundary conditions, roller support for the vertical and fixed boundary at the base are applied for static analysis under construction stage1. Whenever there is any input seismic wave reaches the roller support boundary, it will be reflected. In the case of infinite ground, seismic waves won't reflect. So, to represent the infinite ground condition, the free field dynamic boundary condition is used. As a 2D case, a line-free field condition is applied as the vertical boundary to absorb seismic waves. At the model's base, the bottom fixed condition is used to represent the rigid base. So,

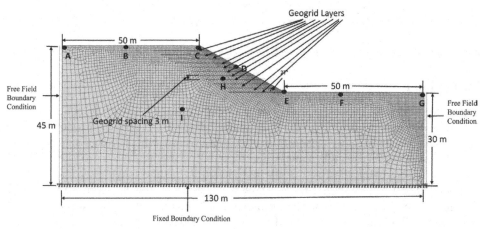

Figure 24.1 Geometry of hill slope along with meshing
Source: Author

Table 24.1: Material properties.

Parameters	Value
Frictional angle of soil (ϕ)	19.31°
Cohesion value of soil (c)	45 kN/m^2
Modulus of elasticity of the soil (E')	12717 kN/m^2
Unit weight of soil (γ)	16.775 kN/m^3
Poisson ratio of soil (ν)	0.3
Slope angle (β)	27°
Material model for soil	Mohr-Coulomb
Type of geogrid	PET
Young's modulus of geogrid	4140000 kN/m^2
Tensile strength of geogrid	85000 kN/m^2

Source: Author

two dynamic boundary conditions were applied, one is line free field, and the other is a fixed boundary condition.

The current investigation analyzed soil's elastic-plastic response using the Mohr-Coulomb model. As recommended by Ling et al. (2010), 5% Rayleigh damping is used in this study. Nine stress nodes (A, B, C, D, E, F, G, H, and I) have been assigned to distinct slope locations to study the slope's seismic behavior under earthquake conditions (Figure 24.1). For this investigation, the 1971 San Fernando Down earthquake record is used in the nonlinear time history analysis of the slope. At the bottom of the slope, the peak acceleration of 0.1563g and duration of 61.86 sec have been simulated using data from the MIDAS GTS NX earthquake database. In stage 1, gravity loading was applied to define the stress condition of the initial state. Then, a dynamic study was carried out in construction stage 2 using ground accelerations at the base of the slope.

Results and discussion

The horizontal deformation of the crest for different time steps of dynamic analysis is shown in Figure 24.2. As per the graph for node A, a peak lateral displacement of

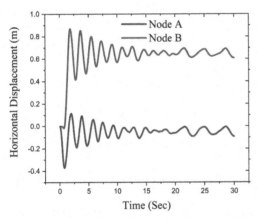

Figure 24.2 Time-varying horizontal displacement along the slope's crest
Source: Author

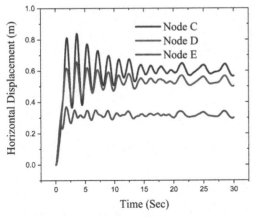

Figure 24.3 Time-dependent horizontal displacement along slope face
Source: Author

0.87 m occurs about 1.5 sec during the earthquake. Similarly, for node B, the peak displacement is 0.1 m which is very low as this is a starting node.

Figure 24.3 shows lateral deformation for the face of the slope where peak deformation of 0.85 m happened at time 4 sec and was reduced to 0.57 m at time 30 sec. The lateral deformation of node E is minimum, whereas the lateral deformation of node C is maximum in the face of the slope.

Figure 24.4 shows lateral deformation along the toe of the slope, where node leads to maximum lateral deformation, which is 0.19 m. Figure 24.5 shows lateral deformation on the slope body with time steps. The lateral deformation of node H, which is just below the slope face, is higher than the lateral displacement of node I, which is far away from the slope face. The lateral deformation at node H is 0.6 m.

Figures 24.6 and 24.7 show the vertical displacement at various nodes with respect to time. Vertical displacement at node B is maximum, whereas vertical deformation is minimum at node C. The vertical deformation at node B is 0.49 m at the slope crest. The vertical displacement at node G is 0.29 m. Das et al. (2023) observed in their study the displacement at the crest is maximum, and the displacement at the toe is minimum, which agreed with the present investigation.

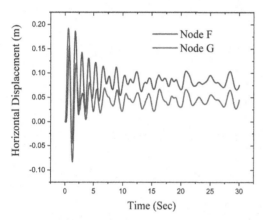

Figure 24.4 Time-varying horizontal displacement along the slope's toe
Source: Author

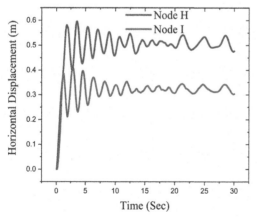

Figure 24.5 Time-dependent horizontal displacement along slope body
Source: Author

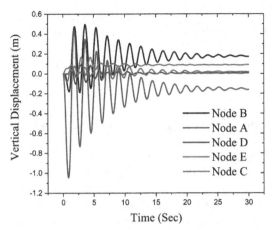

Figure 24.6 Time-dependent vertical displacement along slope crest and face
Source: Author

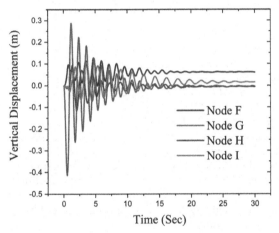

Figure 24.7 Time-dependent vertical displacement along slope toe and body
Source: Author

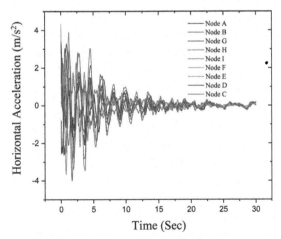

Figure 24.8 Variation of horizontal acceleration with time
Source: Author

Figure 24.8 shows the acceleration response with the different time steps with the dynamic modeling. Initially, the acceleration is higher in all nodes and reduced to zero in 30 sec. At toe node F, the maximum acceleration is 4 m/s².

Conclusion

This study examined the slope's time history under horizontal earthquake loadings using the 1971 San Fernando Down earthquake records. The following conclusions were made based above results and discussion.

In comparison to other locations of the slope, the crest and face have a higher horizontal displacement, while the toe has a lower displacement. As a result, the slope's toe is more stable than its other parts.

Similarly, the vertical displacement of the crest is higher than other parts of the slope. Vertical displacement in the toe portion experienced less effect.

As the ground acceleration is applied in the horizontal direction, horizontal displacement is more than the vertical displacement of the slope.

The maximum acceleration at the toe portion is higher among other portions, which is higher than the input acceleration.

References

Das, T, Rao, V D, and Choudhury, D (2023). Seismic stability and deformation analysis of a south India hill slope by finite elements. *Natural Hazards Review*. 24(2), 04022050.

Diaz-Segura, E G (2016). Numerical estimation and HVSR measurements of characteristic site period of sloping terrains. *Géotechnique Letters*. 6(2), 176–181.

Lee, M-G, Ha, J-G, Jo, S-B, Park, H-J, and Kim, D-S (2017). Assessment of horizontal seismic coefficient for gravity quay walls by centrifuge tests. *Géotechnique Letters*. 7(2), 211–217.

Ling, H I, Yang, S, Leshchinsky, D, Liu, H, and Burke, C (2010). Finite-element simulations of full-scale modular-block reinforced soil retaining walls under earthquake loading. *Journal of Engineering Mechanics*. 136(5), 653–661.

Sahoo, P P, and Shukla, S K (2019). Taylor's slope stability chart for combined effects of horizontal and vertical seismic coefficients. *Géotechnique*. 69(4), 344–354.

25 Role of additive and fine particle addition in the rheology of coarse bauxite slurry

N.V.K. Reddy[1,a] and J.K. Pothal[2,b]

[1]Student, Senior Research Fellow (SRF), Academy of Scientific and Innovative Research (AcSIR), Council of Scientific and Industrial Research – Human Resource Development Group (CSIR-HRDG), Ghaziabad, Uttar Pradesh, India

[2]Associate Professor, Principal Scientist, Design and Project Engineering Department, CSIR-Institute of Minerals and Materials Technology, Bhubaneswar, India

Abstract

This study examines the effect of particle size distribution and additive dosage on the rheological characteristics of bauxite slurry. The reduced and separated bauxite samples with three different particle size ranges of <45 µm, (-150+45) µm, and (-300 +150) µm were studied. The current article explore the rheological behavior through experimental investigation of varied particle size slurries at fixed mass concentration of 50%; of bimodal distribution slurries blending fine and coarse size fraction at different mass proportion; of addition of sodium hexametaphosphate (SHMP) additive at varied dosage. The bimodal slurry at fine:coarse ratio of 20:30 wt% resulted in reduced rheological properties, which further showed a shear thinning effect upon 1% w/w of SHMP dosage into the slurry well fitted to the Herschel-Bulkley model.

Keywords: Bauxite slurry, particle size distribution, bimodal distribution, packing fraction, rheology, additive dosage

Introduction

The aluminum sector among most other industries, places a much greater emphasis on a seamless integration operation by maintaining control over all resources from the raw bauxite to the completed product. With the handful of bauxite reserves worldwide, at the current production capacity, every year an average of 68 million metric tons (MMT) of aluminum is produced to cover the needs of various fields like automobile, construction, electronics, machinery, and others. The top three bauxite-producing nations in 2021 were Australia, China, and Guinea, which generated 110, 86, and 85 MMT of bauxite, respectively (Garside, 2023). The continuous supply of source bauxite is therefore extremely desirable in commercial interests, which can be ensured using pipeline slurry transportation system; this as a whole in conveying solids at high loads is uninterrupted, available all through the year, unaffected by external factors, and environmentally safe (Reddy et al., 2023).The non-Newtonian behavior of the fine colloidal particles is typically caused by the interparticle interactions, of which non-hydrodynamic forces are most significant. In the process of slurry flow these motions and interactions increase the apparent viscosity of slurry, resulting in large pressure drops demanding high energy input. The particle size distribution of the concentrate slurry will not only increase the particle mobility and maximum packing density, but also indulge many hydrodynamic interactions resulting in effective solids loading following a mixture of homogeneous and heterogeneous flow (Ahmad et al., 2014). Many studies based on the maximum packing volume fraction (PVF),

[a]nosumkiran@gmail.com, [b]jkpothal@immt.res.in

DOI: 10.1201/9781003450924-25

occupying the void spaces between the coarse particles by fines and increasing the particle mobility resulting in reduced viscosity were performed for high concentration throughput. Also, to improve the rheological behavior of slurries several findings on using various dispersants and surfactants in the form of chemical additives were reported, and some of which by Mohapatra and Kumar (2013) on coal water slurries and by Senapati et al. (2018) on iron ore slurry were dominant for the study of bauxite slurries.

The extent of study cited in the literature regarding the rheological characterization of bauxite slurry is limited, owing to which the current findings add knowledge by investigating the effect of varying particle size distribution (PSD), and dosage of additive on the flow behavior.

Experimental studies

Bauxite characterization

The current study employing bauxite ore is procured from the Samri bauxite mine through HINDALCO Industries Ltd. of Chhattisgarh state in India in less than 100 mm size. The elemental composition of this ore is interpreted by the spectra generated via, XRF spectrometer of type WDXRF-JDGIUM 4.2 Malvern PANalytical, indicating high concentration of aluminum in the ore, which is economical to refine alumina and thus the final product. The possible phase identification of the bauxite is also studied under X-ray diffractometer, upon proper interpretation of data which showed that the sample is composed of gibbsite, $Al(OH)_3$, one of the three main phases of aluminum that make up the rock bauxite. The ore after subjecting to suitable grinding media for reduction in particle size is classified using standard sieves, and three samples of particle size ranges < 45 µm, (-150+45) µm, and (-300+150) µm were prepared. The PSD of these samples measured using the laser diffraction PSD analyzer (HORIBA- LA-960) depicting the mean particle size as d_{50} = 11.61 µm, 36.83 µm, and 137.79 µm for the respective size ranges is shown in Figure 25.1. Using pycnometer (BOROSIL, 50 ml) and standard gravity technique (weighing sample and water allowing free settlement and ensuring no entrapment of air) the specific gravity of the current bauxite under study is found to be 2.72.

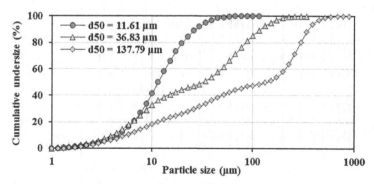

Figure 25.1 Cumulative PSD of bauxite samples of size fraction < 45 µm, (-150+45) µm, and (-300+150) µm

Source: Author

Rheological characterization

The response of bauxite slurry at varied shear rate is accessed in terms of shear stress using a coaxial cylinder rheometer of HAAKE type. The Z38 series having bob diameter 38 mm, is operated under controlled shear rate of 10200 s^{-1} at a room temperature of 25°C. It is for the reference that no chemicals were added during preparation of slurries. The fine size fraction of <45 μm used for maximum static settled concentration (Cw-max) test, showed that at the mass concentration of 55% the slurry exhibits greater Cw-max with flowability compared to other concentration of 45, 50, and 60 wt%.

Results and discussion

Effect of fine particle addition and additive in the rheology of coarse particle slurry

The (-300 + 150) μm slurry tested for change in flow behavior at a constant mass concentration of 50% was discerned to have the maximum reduction in apparent viscosity and yield stress (compared to < 45 μm slurry, and (-150 + 45) μm slurry at 50 wt%)

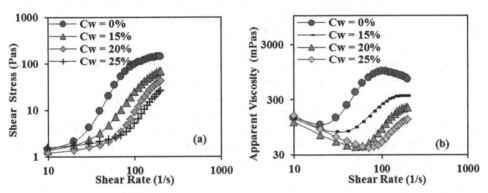

Figure 25.2 (a) Rheogram (τ vs γ˙); (b) Apparent viscosity (η) vs shear rate (γ˙) of bimodal bauxite slurry of <45 μm and +150-300 μm fraction at C$_w$ = 50%

Source: Author

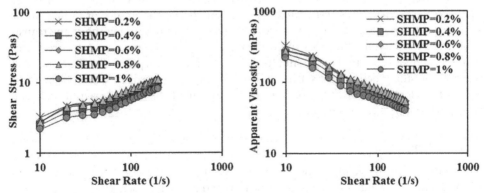

Figure 25.3 (a) Rheogram (τ Vs γ˙); (b) Apparent viscosity (η) Vs shear rate (γ˙) of bimodal bauxite slurry of <45 μm and (-300+150) μm fraction in 20:30 at total 50 wt%, with varying SHMP dosage

Source: Author

Table 25.1: List of model parameters and corresponding flow type for varying types of slurry.

Type of slurry	Model type	Goodness of fit (R²)	Yield stress (τ_o-Pas)	Consistency index (K-μ Pa.s)	Flowbehavior index (n)	Flow type
Coarse particle slurry (+150300 μm)	H-B	0.99	1.163	13.59	3.553	Dilatant
Bimodal slurry	H-B	0.99	0.9996	8366	1.068	Dilatant
Bimodal slurry with 1% w/w SHMP	**H-B**	**0.96**	**1.216**	**26240**	**0.6463**	**Pseudo-plastic**

Bold face indicates the optimum slurry composition with additive

following Non-Newtonian dilatant behavior at low shear rate. The fine size fraction of <45 μm is blended at varying concentrations of 0-25 wt% with coarse fraction of (-300+150) μm (having the minimum yield stress and viscosity), such that the slurry remains at total 50 wt%. Figure 25.2, represent a minimum yield stress and apparent viscosity with a Non-Newtonian dilatant behavior, for slurry with fine:coarse ratio of 20:30 wt%; which can be ascribed to the packing fraction, reducing surface tension and inter-particle forces. This proportion of slurry is further accessed for the changes in rheological properties upon addition of Sodium hexametaphosphate at varying concentration 0.1-1% w/w of total solids. The rheogram and viscosity variations plotted in Figure 25.3, show a decreasing stress and viscosity nature with increasing concentration of SHMP, and the minimum value is noticed at 1% w/w of SHMP. The slurry showing a non-Newtonian pseudo-plastic behavior at 1% w/w of SHMP fitted well to the Herschel-Bulkley (H-B) model with other rheological parameters can be depicted from Table 25.1.

Conclusions

The addition of fine particles of < 45 μm into (-300+150) μm fraction at optimum blending ratio of 20:30 wt%, improved the slurry mobility with minimum yield stress and apparent viscosity. At SHMP additive dosage of 1% w/w to the bimodal slurry of 50 wt% the yield stress estimated is high, nonetheless the nature of fluid is shear thinning with no slip and Taylor vortices at higher shear rates and without any transition in flow behavior. With this the transportation of high concentration ore water slurries is possible, by understanding the mechanism of adsorption of surfactants and altering the slurry behavior. As a result, the pipeline system can be made as the most efficient mode of transporting ores/minerals over long distances, where rail and road transportation is uneconomical.

Acknowledgment

The authors are thankful to the Director CSIR-IMMT for consistent support and emboldenment provided to carry out research and publish paper. The corresponding author (N.V.K. Reddy) acknowledge, AcSIR (Academy of Scientific and Innovative Research) and CSIR-HRDG for providing fellowship under CSIR-GATE-SRF

(File No. 31/GATE/09(19)/2021-EMR-1). The central characterization division of CSIR-IMMT is also acknowledged by the authors for its characterization resources.

References

Ahmad, M A, Ali, Z, and Haque, M E (2014). Fly ash slurry transportation: Indian scenario. *Fly Ash Slurry Transportation: Indian Scenario. Waste Technology*, 2(1), 1–7. doi:10.12777/wastech.2.1.2014.1-7.

Garside, M (2023). Leading countries in worldwide bauxite mine production 2022. https://www.statista.com/statistics/264964/production-of-bauxite/.

Mohapatra, S K, and Kumar, S (2013). Effect of additive on the rheological properties of coal slurry. In *The International Conference on Solid Waste Technology and Management, At Philadelphia*, 1, 1376–1381.

Reddy, N V K, Pothal, J K, Barik, R, and Senapati, P K (2023). Pipeline slurry transportation system: an overview. *Journal of Pipeline Systems Engineering and Practice*. doi:10.1061/JPSEA2/PSENG-1391.

Senapati, P, Pothal, J, Barik, R, Kumar, R, and Bhatnagar, S (2018). Effect of particle size, blend ratio and some selective bioadditives on rheological behaviour of highconcentration iron ore slurry. In *Proceedings of the 21st International Seminar on Paste and Thickened Tailings*, 227–238. Australian Centre for Geomechanics, Perth. doi:10.36487/acg_rep/1805_18_senapati.

26 Experimental research on the flexural strength of strengthened composite beams

Vijay Kumar[a]

Department of Civil Engineering, AGC Amritsar, Amritsar India

Abstract

The degradation of structures, which is primarily caused by corrosion, is one of the issues that the building and construction sector is currently dealing with. Ferrocement and fiber-reinforced polymers are becoming increasingly important to increase the strength of concrete members and so solve this issue. To check flexural behavior of traditional beams reinforced with steel bars and hybrid beams made of ferrocement and glass-fiber reinforced were casted. The 900mm-long beams had a rectangular cross-section that measured 210 mm in width and 380 mm in depth. A total of 12 beams were casted and three beams served as control examples. In order to strengthen the beams, GFP laminates, ferrocement, and GFP plus ferrocement were all used to strengthen the remaining beams. Two-point bending over an 840mm clear span was used to test the beams for flexure failure. The results of load carrying capacity and failure mode are presented. Traditional reinforcement concrete beam, ferrocement, and GFP strengthened beams as well as a mixture of GFP and ferrocement beams were compared in terms of experimental outcomes. The results of the trial demonstrated that GFP and ferrocement had a higher load-carrying capability than other beams.

Keywords: Glass fiber polymer, ferrocement, fiber reinforced polymer, reinforcement, reinforced concrete

Introduction

Strength, stability, serviceability, and safety are major factors, as per structural applications (Abdullah et al. 2003). Many structures are built in developing nations using outdated codes. However, new codes state that this construction is unsafe. These structures need to be replaced, which takes time and money. All those structures need to be maintained in order to increase their lifespan. These structures require a lot of time and money for a replacement (Xiong et al., 2011). The technique and materials chosen to improve the strength and serviceability of the construction are a major source of concern. The lightweight construction of structures must be preferred about earthquake and wind pressures (Reddy et al., 2013). The structure should have sufficient stiffness and flexibility from the perspective of earthquake design in order to withstand those forces. Ferrocement and FRP have characteristics that are connected to wind and seismic forces (Makki et al., 2014; Nordin et al., 2004). Lightweight materials that can be utilized as reinforcement for concrete buildings include ferrocement and fiber-reinforced polymers. Most structures are built to be safe, stable, and able to withstand external loads (Nassif et al., 2004).

The literature study led to the conclusion that externally bonded GFP and ferrocement can significantly boost a member's strength. The member to which externally bonded ferro cement is attached is more ductile as a result.

---

[a]er.vijaysai@gmail.com, Vijay.ce@acetedu.in

DOI: 10.1201/9781003450924-26

Objectives

1. To evaluate the strength development of composite beams strengthened with a Glass fiber cloth, ferrocement, and their combination.
2. To study cracking behavior of strengthened beams with a glass fiber cloth, Ferrocement, and their combination.
3. Correlation of the outcomes.

Methodology and experimental work

"Investigation of flexure capacity of members those are strengthened by utilizing ferrocement and GFP laminates" is the focus of the experimental effort. The dry bonding technique is used to apply ferro cement and GFP over a reinforced concrete beam (Biggs et al., 1972). The casting of controlled beams and ferro-cement beams, the beams were covered with GFP and allowed to dry for the appropriate amount of time in accordance with the code provision. A total of 12 RC beams with a 400 mm × 210 mm cross-section and a 900 mm length are cast.

After proper curing according to provision of code, strengthening of beams done. For strengthening of nine beams single layers in U- shaped ferrocement and GFP were used.

Experimental work

Examining the flexure behavior of the beams reinforced and strengthened with GFP cloth and ferro-cement the casting of reinforced concrete beams comes first. After controlled beams have been cast and reinforced beams have been properly dried in accordance with statutory requirements, GFP cloth is next applied on top of the beams (Maariappan et al., 2013). Applying GFP cloth with the aid of a stronger adhesive or any other adhesive will enable it to be joined to concrete and steel beams (ACI Committee 549-1R-88 et al. 1988).

Materials

Steel wire mesh, glass fiber polymer cloth, adhesive, reinforced bars, water, fine aggregates, coarse aggregates, cement, etc.

Results and discussion

Research work, to study the flexural capacity of RC beams and GFP and ferrocement strengthened beams. The dry bonding process was used to apply GFP cloth and ferrocement. Ferrocement and GFP cloth applied in U-shape and in single layer. Load was applied at two paints on the beams to check the flexural capacity of beams. It is

Table 26.1: Number of samples.

S. No	No of reinforced concrete beams	No of strengthened by GFP	No of strengthened by ferrocement	No of combination of GFP + ferrocement
1	03	03	03	03

Source: Author

Table 26.2: Properties of GFP obtained from tensile test.

Fiber type	Width (mm)	Length (mm)	Thickness (mm)	Break stress (MPa)	Peak stress (MPa)	Young's modulus (MPa)	Elongation at peak (%)	Elongation at beak (%)
Glass	20	200	0.40	6.33	7.03	640	13.69	31.54

Source: Author

Table 26.3: Woven wire mesh properties.

Steel wire	Length (mm)	Thickness (mm)	Break stress (MPa)	Peak stress (MPa)	Young's modulus (MPa)	Elongation at Break (%)	Elongation at peak (%)
steel	200	0.50	372.8	414.3	8.4×10^5	0.21	0.19

Source: Author

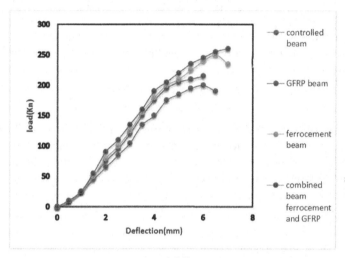

Figure 26.1 Comparison of different strengthened beams
Source: Author

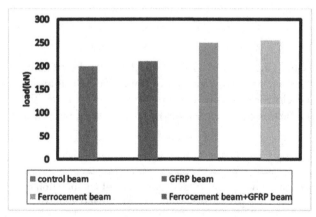

Figure 26.2 Compression of load carrying capacity of different beams
Source: Author

emphasized that ferrocement shows good tensile and ductility properties. It is noticed during experimental work, load carrying capacity of strengthened beams by ferrocement increased and it also shows minor cracks before failure. From results it is also noticed that GFP strengthened beams also show enhancement in load carrying capacity but compare to ferrocement strengthened beams it is less.

Conclusion

• The flexural resistance capacity of beams was significantly enhanced by effectively bonding ferrocement and glass fibre polymer. Ferrocement + GFP bonding produced better outcomes for boosting beams' capacity to carry loads. Ferrocement beams were able to carry higher weight than GFP-strengthened beams. Compared to control beams and GFP beams, beams strengthened with ferrocement showed a higher flexural resistance capacity. As intended, all the beams cracked in flexure. The layers of GFP and ferrocement play a role in providing resistance against the development of cracks in the beams. Flexure cracks are seen to appear all along the height of the beam.

References

Abdullah and Takiguchi, K (2003). An investigation into the behavior and strength of reinforced concrete columns strengthened with ferrrocement jackets. *Cement & Concrete Composites*, 25, 233–242.

ACI Committee 549-1R-88 and 1R-93: Guide for the Design, Construction, and Repair of Ferrocement, ACI 549-1R-88 and 1R-93, in Manual of Concrete Practice, American Concrete Institute, Farmington Hills, Michigan, 1988 and 1993.

Biggs, G W. (1972). An introduction to design for ferrocement vessels, industrial development branch. Fisheries Service, Ottawa, pp. 224.

Ragheed F M (2014).Response of Reinforced concrete beams retrofitted by ferrocement. *International Journal of Scientific & Technology Research*, 3(9).

Maariappan, G (2013). Studies on behavior of RC beam-column joint retrofitted with basalt fiber reinforced polymer sheets. *Global Journal of Research in Engineering Civil and Structural Engineering*, 13(5).

Xiong, G J (2011). Load carrying capacity and ductility of circular column confined by ferrocement including steel bars. *Construction and Building Materials*, 25, 2263–2268.

Reddy, H N J (2013). Strengthening of RC beams in flexure using natural jute fiber textile reinforced composite system and its comparative study with CFRP and GFRP strengthening systems. *International Journal of Sustainable Built Environment*, 133(1).

Nassif, H H (2004). Experimental and analytical investigation of ferrocement-concrete composite beams. *Cement & Concrete Composites*, 26, 787–796.

Nordin, H (2004). Testing of hybrid FRP composite beams in bending. Composites: Part B 35, 27–33s

27 Impact of marble powder, silica fume, and steel fiber on fresh and hardened attributes of cement concrete

Niharika Pattanayak[a], Malaya Ku. Sahu[b] and Sudhanshu Sekhar Das[c]

Department of Civil Engineering, Veer Surendra Sai University of Technology, Burla, Odisha, India

Abstract

Sustainable construction has gained significant importance due to its potential to address environmental challenges like depletion of natural resources and the harmful impact of waste disposal. Industrial wastes such as marble powder and silica fume, which have a low rate of recycling, are particularly detrimental to the environment. In this study, steel fiber (1.5) was added to a combination of varying percentages of silica fume (0%, 10%, and 15%) and marble powder (0%, 10%, and 15%) to analyze its impact on the properties of fresh and hardened cement concrete. The maximum compressive strength was attained when marble powder and silica fume substitute were both 10%, resulting in 29.43 MPa and 28.3 MPa, respectively. The optimal percentage of marble powder and silica fume to be used in a concrete mix is 10%.

Keywords: Fragile, marble powder, silica fume, steel fiber, tensile failure

Introduction

The construction industry's current growth has led to a surge in demand for concrete. However, the production of cement required for concrete is an energy-intensive process and results in significant emissions, leading to concerns regarding waste, pollution, resource depletion, and degradation. It is crucial to develop solutions to address these issues, and the concrete sector must transition toward using more eco-friendly materials to promote sustainable development. Waste products generated by companies can be utilized as a cement or aggregate replacement. To increase the compressive strength (CS) of cement concrete (CC), additional components like mineral admixtures can be added. Researchers have explored using discarded palm oil shells from the agricultural industry as a lightweight aggregate (Khan et al., 2014) and various mineral admixtures like fly ash, silica fume (SF) as cement replacements to improve both fresh and hardened concrete qualities (Khan et al., 2014).

Literature analysis

The studies that used SF, marble powder (MP), and steel fiber (STF) to make the concrete mix are reviewed in this section.

Effect of MP in concrete

Since prehistoric times, marble has been used in construction. MP is a harmful industrial byproduct that is produced during the cutting, shaping, and polishing of marble.

[a]nehapattnaik0@gmail.com, [b]malaykumar546@gmail.com, [c]ssdas8@gmail.com

DOI: 10.1201/9781003450924-27

By utilizing these wastes in concrete, not only the ecological and environmental issues will be resolved, but also it will produce more affordable and long-lasting concrete (Arel, 2016). The CS increases by 20–26%, and the splitting tensile strength (STS) increases by 1015% when MP is combined with natural sand at a ratio of 1575% (Arel, 2016). To achieve the desired results, MP-partial replacement cement requires the addition of super-plasticizers (Rodrigues et al., 2015). In the absence of super plasticizers, they found that even at replacement rates of up to 10%, the physical attributes of CC were negatively affected.

Influence of SF in concrete

SF is a waste product of silicon elemental or silicon alloy formation, which can enhances concrete's longevity, CS, abrasion resistance, permeability, and ability to shield embedded steel from corrosion when mix with CC (Holland, 2005). When SF blends with super plasticizer, the CS rises to 82.7 MPa. It is common practice to add SF to CC in order to reduce the material's abrasion susceptibility (Holland, 2005; Rashad et al., 2014). Due to its large surface area, the addition of SF to the CC mix raises water consumption (Holland, 2005) but improves its mechanical performance (Lange et al., 1997).

Influence of STF in concrete

CC is a fragile material with a low tensile strength and slow strain ability. When STF is added to CC, it boosts the material's tensile and compressive strengths. When concrete is coupled with STFs, many of its engineering characteristics are considerably improved. According to research by Behbahani et al. (2011), addition of STF increases CS from 0 to 15% and STFs have a far greater influence on concrete's flexural strength (FS). The inclusion of 2% of STF can increase the FS by 55% (OH et al., 1999).

The investigation aims to determine the ideal ratio of SF and MP to be used in concrete mix.

Materials employed

The research are conducted using Portland Pozzolana Cement (specific gravity of 2.87) from Angul, Odisha; sand (specific gravity and absorption rate found to be 2.5 and 1.2%, respectively) from the Brahmani River; crushed stone aggregate (specific gravity and water absorption were 2.65 and 0.5%, respectively); hook-end STF and SF (specific gravity -2.22) from Karanwal Infratech; MP (specific gravity -2.71) from Ganesh Marbles Shop in Angul, Odisha; and Sika Viscocrete 4021 NS as the super plasticizer.

Experimental study

The experiential plan has four phases: grading of CA and FA, mix design, casting cubes and cylinders for CS and STS tests, respectively, and curing the samples. Testing and analysis are done in the last stages.

Grading of CA and FA

20 mm (fraction I) and 10 mm (fraction II) size CA, along with sand as FA, are used for the present work. The entire grading satisfied the recommended range in

accordance with IS 383-1970, and presented in Figure 27.1. Then the mix proportion was prepared in accordance with IS 10262 (2009).

Test procedure

A total of nine mixes are casted, including the control mixes (CA, FA, and PPC). Designations for different CC mixes are also mentioned in Table 27.1. For example, mixes formed by 10% replacement of cement with SF and MP are designated as S10 and M10, respectively. Similarly, when both SF and MP replace cement by 10% each, it is designated as S10M10. After the mix was prepared, a workability test was conducted immediately. The slump test, CS (150×150×150 mm), and STS (150 mm Ø and 300 mm height) of concrete were measured at 7 and 28 days of curing age.

Results and discussion

After 28 days of curing, the entire test was performed as per the respective codes, and the results are presented in Table 27.2. The workability of control mix concrete has decreased due to the addition of finer particles, i.e., SF and MP, and plasticizer is used to maintain it (Srivastava et al., 2012; Khan et al., 2014). Mix M10, S10, and, S10M10 achieved maximum CS and STS due to the binding and pore-filling

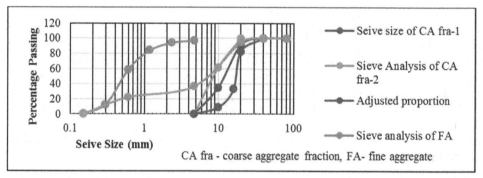

Figure 27.1 Grading curve
Source: Author

Table 27.1: Mix proportions for different designations.

Concrete mix	SF (% by wt. of cement)	MP (% by wt. of cement)	STF (% by wt. of cement)	Plasticizer (% by wt. of cement)	CA	FA
Control Mix	0	0	1.5	0		
S10	10	0	1.5	0.4		
S15	15	0	1.5	0.4	20mm and 10mm	Sand
M10	0	10	1.5	0.4		
S10M10	10	10	1.5	0.4		
S15M10	15	10	1.5	0.4		
M15	0	15	1.5	0.4		
S10M15	10	15	1.5	0.4		
S15M15	15	15	1.5	0.4		

Source: Author

Table 27.2: Comparison of properties of different concrete mixes.

	Curing period (days)	Concrete mixes								
		Control mix	S10	S15	M10	S10 M10	S15 M10	M15	S10 M15	S15 M15
Workability (mm)	-	80	120	110	118	115	112	115	110	108
CS (MPa)	7	17.00	18.89	14.39	19.61	17.94	13.59	13.44	13.22	12.86
	28	25.44	28.30	21.65	29.43	26.81	19.77	20.34	19.98	19.33
STS (Mpa)	7	1.53	1.69	1.48	1.78	1.59	1.43	1.50	1.48	1.37
	28	1.96	2.06	1.78	2.17	1.96	1.73	1.80	1.75	1.66

Source: Author

properties of fine SF and MP (Corinaldesi, et al., 2010; Shirule et al., 2012; Srivastava et al., 2012; Khan et al., 2014). The main cause of increased STS in mix M10, S10, and S10M10 is the hooking and binding properties of steel STFs and fine SF and MP (Shirule et al., 2012; Srivastava et al., 2012).

Conclusion

The experimental investigations indicate that both SF and MP have potential for use in CC as cement substitutes. In the control mixes CS and STS were 25.44 MPa and 1.96 MPa, respectively. 1.5% hook-end STFs added to the concrete resulted in a slight increase in CS and STS, but the workability of the concrete decreased as SF and MP percentage rises. The optimum percentage of MP and SF to be used in the CC mix was 10%, with maximum CS of 29.43 MPa and 28.3 MPa and highest STS 2.17 MPa and 2.06 MPa, respectively.

References

Aldred, J. M., Holland, T. C., Morgan, D. R., Roy, D. M., Bury, M. A., Hooton, R. D., ... & Jaber, T. M. (2006). Guide for the use of silica fume in concrete. ACI–American Concrete Institute–Committee: Farmington Hills, MI, USA, 234.

Ameri, M. and Nasr, D., 2017. Performance properties of devulcanized waste PET modified asphalt mixtures. *Petroleum Science and Technology*, 35(1), 99–104.

Behbahani, H, Nematollahi, B, and Farasatpour, M (2011). Steel fiber reinforced concrete: a review. http://dl.lib.mrt.ac.lk/handle/123/9505

Corinaldesi, V, Moriconi, G, and Naik, (2010). Characterization of marble powder for its use in mortar and concrete. *Construction and Building Materials*. 24(1), 113–117. https://doi.org/10.1016/j.conbuildmat.2009.08.013

Dixon, J., & Mayfield, B. (1971). Concrete reinforced with fibrous wire. Concrete (London). 5(3), 73–76.

Geyer, R. (2020). Production, use, and fate of synthetic polymers. In Plastic waste and recycling (pp. 13-32). Academic Press.

Holland, T C (2005). Silica Fume User's Manual. Federal Highway Administration.

Khan, S U, Nuruddin, , Ayub, T, and Shafiq, N (2014). Effects of different mineral admixtures on the properties of fresh concrete. *The Scientific World Journal*. 2014. https://doi.org/10.1155/2014/986567

Kumar, M., Xiong, X., He, M., Tsang, D.C., Gupta, J., Khan, E., Harrad, S., Hou, D., Ok, Y.S. and Bolan, N.S., 2020. Microplastics as pollutants in agricultural soils. *Environmental Pollution*, 265, 114980.

Lange, F, Mörtel, H, and Rudert, V (1997). Dense packing of cement pastes and resulting consequences on mortar properties. *Cement and Concrete Research*. 27(10), 1481–1488. https://doi.org/10.1016/S0008-8846(97)00189-0

OH, S-G, Noguchi, T, and Tomosawa, F (1999). Estimation of the rheological constants of high-fluidity concrete by using the thickness of excess paste. *Zairyo*. 48(10), 1193–1198. 10.2472/jsms.48.1193

Rashad, A M, Seleem, , and Shaheen, (2014). Effect of silica fume and slag on compressive strength and abrasion resistance of HVFA concrete. *International Journal of Concrete Structures and Materials*. 8, 69–81. https://doi.org/10.1007/s40069-013-0051-2

Rodrigues, R, De Brito, J and Sardinha, M (2015). Mechanical properties of structural concrete containing very fine aggregates from marble cutting sludge. *Construction and Building Materials*. 77, 349–356. https://doi.org/10.1016/j.conbuildmat.2014.12.104

Shirule, P, Rahman, A and Gupta, (2012). Partial replacement of cement with marble dust powder. *International Journal of Advanced Engineering Research and Studies*. 1(3), 2249.

Srivastava, V, Agarwal, V, and Kumar, R (2012). Effect of Silica fume on mechanical properties of concrete. *Journal of Academia and Industrial Research*. 1(4), 176–179. http://jairjp.com/SEPTEMBER%202012/07%20VIKAS%20SRIVASTAVA.pdf

Wiliamson, G (1974). The effect of steel fibers on compressive strength of concrete, fiber reinforced concrete. SP-44. ACI, pp. 195–207.

28 Waste plastic management and utilization in road construction: a review

Niketan Rana[a] and Dr. Siksha Swaroopa Kar[b]

Flexible pavement Division, CSIR-CRRI Academy of Scientific and Industrial Research New Delhi, India

Abstract

Global concern has been raised regarding global plastic waste generation and its management. At the end of its life plastic is recycled or enters the environment in the form of macro as well micro plastic. A study suggests that if waste generation trends continue then, by 2060, global waste generation would reach up to 1.1 billion tonnes yearly. Researchers are finding different ways of incorporating waste plastic in the pavement in the form of mix modifiers, bitumen modifiers and replacement to aggregates. This paper is aimed to review the background knowledge of plastic waste generation and the current status of plastic-modified asphalt around the globe.

Keywords: Asphalt, management, plastic, waste

Introduction

Our dependency on crude petroleum for its products, such as petrol, diesel, wax, naphtha, gas etc., is undeniable and one such product is bitumen, which is used as binding material for the construction of flexible pavement. Both bitumen and plastics are produced from crude petroleum as minor products. Bitumen, a primary product, is the dark-colored, highly viscous sticky residue left after the fractional distillation of crude petroleum for lighter oils and gas. Whereas, plastics, the secondary products, are obtained by the polymerization of monomers obtained after the thermal decomposition of naphtha. About 70% of bitumen is primarily used, for coating of aggregate to form bituminous concrete, in road construction.

Plastic has unquestionable advantages. The material is inexpensive, light, and simple to produce. It is extensively used in packaging, building and construction, textiles, consumer goods, transportation, electrical and electronic equipment, and industrial machinery, etc (Thiounn and Smith 2020). Over the past century, a surge in plastic production has been caused by these properties. Over the next ten to fifteen years, as global production of plastic skyrockets, this pattern will continue. We already cannot handle the amount of plastic waste we produce. Only a small percentage is recycled. Every year, approximately 13 million tonnes of plastics enter our oceans, threatening biodiversity, economies, and possibly even our own health (UNEP, 2018).

The field of research into using plastic waste as an alternative to virgin materials in infrastructure projects has gained momentum, offering a number of advantages. In order to improve the performance of asphalt pavement and prevent a variety of pavement problems like cracking, rutting, and raveling, as well as because of an increase in infrastructure, vehicular load, and traffic volume and severe climate change, asphalt modification is frequently required. One of the most effective ways to improve pavement performance has been polymer modification. Numerous studies have suggested

[a]Nikerana.rana7@gmail.com, [b]sikshaswaroopa@gmail.com

DOI: 10.1201/9781003450924-28

that polymer modification can significantly improve engineering properties at both high and low temperatures.

To layout a comprehensive representation of the importance of littered plastic in modifying binder and mixes and their importance in the production of moisture and high temperature rutting resistance bituminous mixes, this paper is aimed to review the background knowledge on plastic waste generation and current status of plastic modified asphalt around the globe. The purpose of this paper to list down the plastic waste generation and its management details along with debates regarding plastic waste and its use in asphalt modification.

Waste plastic and its management

According to OECD in 2019 total plastic used globally was 460 million tonnes (Mt) and is estimated to triple, to 1321 Mt, between 2019 to 2060, shown in Figure 28.1. All regions are expected to see an increase in the use of plastics, but Asia and Sub-Saharan Africa will see the fastest growth. Plastic use is expected to be more than six times higher in Sub-Saharan Africa in 2060 than it was in 2019. This is due to strong economic and population growth. Due to India's robust economic growth the use of plastics in India is expected to raise by more than five times. Table 28.1, compares the long-term estimates for plastic use, waste, and improper management.

It is anticipated that the total amount of plastic produced globally since 1950 will increase, rising from 9.2 billion tonnes in 2017 to 34 billion tonnes by 2050 (Geyer, 2020). Only 9% of the plastic waste that has ever been generated worldwide has been recycled. Most of it is released into the environment, landfills, or dumps. If plastic production growth continues at its current rate, the plastics industry might account for 20% of global GDP by this time.

The plastic product is recycled or incinerated. Otherwise, the plastic is disposed of in a landfill, dumped in uncontrolled locations, or littered in the environment at the

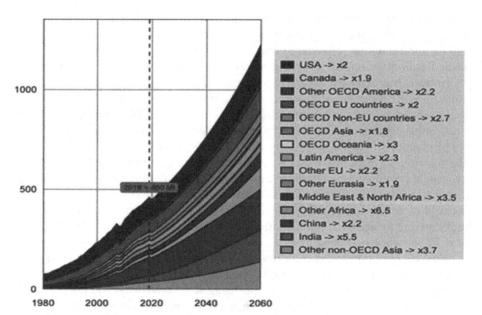

Figure 28.1 Plastics use in Mt (OECD, 2022)
Source: OECD, 2022

Table 28.1: Comparison of projections with the existing literature.

	Literature	2015/2016 (Mt)	2025 (Mt)	2040 (Mt)	2050 (Mt)	2060 (Mt)
Total plastics used	Geyer et al., 2017	380	-	-	1 100	1 371
	Ryberg et al., 2019	388				
	OECD, 2022	413	516	766	976	1 231
Total plastic waste generated	Geyer et al., 2017	302			902	
	Ryberg et al., 2019	161				
	Lebreton and Andrady, 2019	181	230	300		380
	Lau et al., 2020	220		420		
	OECD, 2022	308	409	615	799	1 014
Total uncollected plastic waste	Jambeck et al., 2015	37	70			
	Ryberg et al., 2019	41				
	Lebreton and Andrady, 2019	80	95	155		213
	Lau et al., 2020	91		240		
	OECD, 2022	74	86	111	132	153

Source: (OECD, 2022)

end of its life. Recent estimates (Geyer et al., 2017) indicate that 79% of all plastic waste ever produced is currently in landfills, dumps, or the environment, with only 9% recycled and 12% incinerated. If current use of plastics and management techniques continue, the amount of plastic waste that ends up in landfills will triple from 174 Mt in 2019 to 507 Mt in 2060, while the amount of waste that is burned will rise from 67 Mt to 179 Mt (OECD, 2022).

Waste plastic in pavement

The addition of waste plastics to bituminous mixtures seems to have a lot of potential for use in flexible pavements to either reduce the wearing course or base layer's layer thickness or extend their active service life. The use of waste plastic in bitumen modification improves the mechanistic property of pavement, as well as its capacity to withstand heavy traffic, decreases its susceptibility to deformation, and imparts improved binder ageing resistance (Mashaan et al., 2022).

LDPE and HDPE were utilized for the aggregate coating. After that, the polymer was added to the mixture, both grounded and ungrounded (Awwad and Shabeeb, 2007). Blend samples showed that the blends had better technical properties when made with milled HDPE. It has been determined that 12% of the weight of the bitumen makes the best addition of modifier to the mixture. This experiment demonstrates that the added HDPE decreases the density of the asphalt mix, decrease the number of voids as well as mineral aggregate pores (VMA) in the mix, and improve mix stability.

The basic properties of waste plastics modified asphalt in previous studies, shown in Table 28.2, indicated that using waste plastics in asphalt modification helps in reducing pollution and additional costs. Plastic-modified old bitumen leads to improved rut resistance and fatigue resistance. Waste plastic asphalt also appears to be a significant,

Table 28.2: Asphalt modification in previous studies.

Waste plastic type	Bitumen and mix type	Plastic % of bitumen	Mix conditions	Major finding	Authors
HDPE LDPE	PG 60/70	6, 8, 10, 12, 14, 16	180–190°C	Grinded 12% HDPE provide better stability results	Awwad and Shabeeb, 2007
PE	PG 40/50	2, 4, 6, 8, 10, 12	145–155°C	10% PE gave higher stability results	Ahmed, 2007
PP	SMA mix PG 50/60	1, 3, 5, 7	Mixed at 160 °C for 5 min	At 5% PP gave best results	Al-Hadidy and Yi-qiu, 2009a, 2009b
PP PE PS	Dry mix PG 80/100	5, 10, 15, 25	Not given	Using higher quantities of plastic was recommended	Vansudevan et al., 2012
HDPE	PG 80/100	1, 3, 5, 7	170°C, 2 hrs, 3000 rpm	Stability and MQ at 5% HDPE increased by about 50–55%.	Moghddam et al., 2014
HDPE LDPE	Not given	2, 3, 4, 5	Not given	4% HDPE was recommended.	Casey et al., 2008
LDPE	PG 50/60 SMA	0, 2, 4, 6, 8%	Not given	6% LDPE was recommended.	Al-Hadidy and Yi-qiu, 2009a, 2009b
PET	Pn 80/100	0, 3, 5, 10	Dry mix 150°C, Wet Mix150°C and PET 250-300°C, at 600 rpm for 30 min	Improvement in penetration and softening point were dependent on PET content and size, improved the shear stiffness and decreased phase angle at low frequency and high temperature compared to high frequency.	Kuman and khan, 2020
waste PET	PG 60/70	0, 2.5, 5, 7.5, 10, 12.5 and 15%	Mixed at 4000 rpm, for 60 min at 160 °C.	Stability increased ideally at 7.5–10% PET, improved rutting resistance and decreased moisture susceptibility	Ameri and Danial, 2017

Source: Author

useful, and cost-effective alternative to other commercial polymers and can satisfy design, coating, and building requirements.

Conclusion

The global warming has created a scenario were going for sustainability in every aspect of construction industry is need of the day. Plastic waste generation is also one of the major concerns to reduce global warming. Plastic being polymer by nature forms a homogeneous mixture with bitumen. Literature speaks of improvement in bitumen properties in terms of penetration, softening point by incorporating segregated waste plastic. This paper gives a layout of types of plastic, quantity of plastic waste that can be incorporated to enhance the binder properties.

In future, it is required to study the recyclability potential of plastic modified binder and its impact on underground water and soil contamination.

References

Ahmed, L A (2007). Improvement of marshall properties of the asphalt concrete mixture using the polyethylene as additive. *Engineering Technology*. 25, 383–394.

Al-Hadidy, A, and Yi-Qiu, T (2009a). Effect of polyethylene on life of flexible pavements. *Construction and Building Materials*. 23, 1456–1464.

Al-Hadidy, A, and Yi-Qiu, T (2009b). Mechanistic approach for polypropylene-modified flexible pavements. *Material Design*. 30, 1133–1140.

Awwad, M, and Shbeeb, L (2007). The use of polyethylene in hot asphalt mixtures. *American Journal of Applied Sciences*. 4, 390–396.

Casey, D, McNally, C, Gibney, A, and Gilchrist, M D (2008). Development of a recycled polymer modified binder for use in stone mastic asphalt. *Resources, Conservation & Recycling*. 52, 1167–1174.

Geyer, R, Jambeck, J R, and Law, K L (2017). Production, use, and fate of all plastics ever made. *Science Advances*. 3(7), e1700782.

Jambeck, J R, Geyer, R, Wilcox, C, Siegler, T R, Perryman, M, Andrady, A, Narayan, R, and Law, K L (2015). Plastic waste inputs from land into the ocean. *Science*. 347(6223), 768–771.

Lau, W.W., Shiran, Y., Bailey, R.M., Cook, E., Stuchtey, M.R., Koskella, J., Velis, C.A., Godfrey, L., Boucher, J., Murphy, M.B. and Thompson, R.C. (2020). Evaluating scenarios toward zero plastic pollution. Science, 369(6510), 1455–1461.

Lebreton, L, and Andrady, A (2019). Future scenarios of global plastic waste generation and disposal. *Palgrave Communications*. 5(1), 1–11.

Mashaan, N S, Chegenizadeh, A, and Nikraz, H (2022). A comparison on physical and rheological properties of three different waste plastic-modified bitumen. *Recycling*. 7, 18.

Moghddam, T B, Soltani, M, and Karim, M R (2014). Experimental characterization of rutting performance of polyethylene terephthalate modified asphalt mixtures under static and dynamic loads. *Construction and Building Materials*. 65, 487–494.

Organization for Economic Co-operation and Development (2022). Global Plastics Outlook: Policy Scenarios to 2060. OECD Publishing.

Ryberg, M W, Hauschild, M Z, Wang, F, Averous-Monnery, S, and Laurent, A (2019). Global environmental losses of plastics across their value chains. *Resources, Conservation and Recycling*. 151, 104459.

Thiounn, T, and Smith, R C (2020). Advances and approaches for chemical recycling of plastic waste. *Journal of Polymer Science*. 58(10), 1347–1364.

United Nations Environment Programme (2018). Single-use plastics: a roadmap for sustainability (rev. 2). https://wedocs.unep.org/20.500.11822/25496

Vansudevan, R, Sekar, A R C, Sundarakannan, B, and Velkennedy, R (2012). A technique to dispose waste plastics in an ecofriendly way-application in construction of flexible pavement. *Construction and Building Materials*. 28, 311–320.

29 Seismic response analysis and evaluation of vibration control of high-rise structure

Ushnish Roy[a] and S. Pandey[b]

Department of Civil Engineering, Techno India University, WB, India

Abstract

If high-rise structures are subjected to extreme vibrations during earthquake excitations, they may sustain serious damage or possibly collapse. To keep these high-rise constructions safe, it is crucial to reduce seismically produced vibrations. Numerous vibration control strategies have been put out in recent years, but they all have some inherent drawbacks. High rise structures require vibration control systems to withstand lateral loads, wind loads, and seismic effects. While developing high-rise buildings, we had to take into account a variety of factors, including growing wind loads, which would reduce the structures' usability and durability and eventually lead to their collapse or other damage. To build a safe, rigid structure that is impervious to lethal seismic effects, structural reaction control systems are required. This study aims to compare the evaluation of G+10 story vibration control under seismic reaction to additional linear and non-linear fluid viscous damping and has been examined in ETABS (18.0.2).

Keywords: ETABS (18.0.2), fluid viscous damping, G+10 story, high-rise structures, lateral loads, non-linear, vibration control system, wind loads

Introduction

This document covers seismic load calculation for multi-story structures under the guidelines of IS: 1893-2002 and IS: 1893-2016. The analysis and design approach for a multi-story (G+4) residential structure in zones III and IV. Buildings with an asymmetrical (or uneven) floor layout sustain greater damage, according to observations of how they performed during previous earthquakes around the world. There is a considerable stock of such exceedingly unstable buildings in seismically active places across the world (Chopra, 2020). Many studies have previously focused on the seismic behavior of such systems because asymmetric-plan structures are particularly susceptible to earthquakes (Goel, 1998). Measures for the seismic protection of structures with asymmetrical plans must be developed that are affordable. Even though seismic standards and guidelines recognize the seismic susceptibility of structures with an asymmetrical plan and make an effort to provide some side load-resisting components with more strength, they are ineffective at preventing excessive deformations that might cause damage to the building. A recent study examined the seismic response of asymmetric systems using viscoelastic and non-linear viscous dampers. The effect of damper non-linearity on structural response was shown to be rather minor, and non-linear dampers effectively reduce response to that of linear dampers with a significantly lower damp force (Sunagar et al. 2021). A more efficient analytical technique for asymmetric-plan systems with non-linear dampers has also been developed.

The main objectives of this research endeavor are, therefore, to

---

[a]uray9272@gmail.com, [b]suman.p@technoindiaeducation.com

DOI: 10.1201/9781003450924-29

- Analyze the impacts of plan asymmetry and how damper non-linearity affects them. Additionally, offered for comparison are responses from linear systems.
- Determine how damper non-linearity affects the seismic response of non-linear asymmetric systems.

Objective

In addition to meeting the aforementioned goals, the study described in this paper aims to independently validate the conclusions of past investigations. The impacts of damper non-linearity on the seismic response of linear and non-linear asymmetric systems, as well as how the consequences of plan asymmetry are impacted by the damper non-linearity, were the subjects of a detailed investigation with this objective in mind.

Methodology

Friction dampers are employed to lessen the high-rise buildings' dynamic response (Chopra, 2017). It is demonstrated that damper non-linearity generally results in a slight to substantial drop in system responses, with the exception of non-linear systems with extremely brief periods, for which the base torque and edge deformations can be greatly diminished. Moreover, the damper non-linearity-related decrease in damper forces is not severe (Gioncu and Mazzaloni, 2014) Table 29.1 and Table 29.2.

The following steps are included in the structural design process:

Planning a structure, calculating loads, analyzing a structure, designing members, drawing, and detailing.

Table 29.1: Elements of the building's geometry.

Elements	Specifications
Number of story	(G+10)
Story height	3 m
Steel Specification	Fe 500
Building height	30 m
Column size	500 × 500 mm
No. of bays facing X direction	6
No. of bays facing Y direction	4
space between bays in the X direction	4 m
space between bays in the Y direction	3 m
Beam size	200 × 500 mm
Concrete Specification	M25
Live road (kN/m2)	2
Floor finish	1
Roof road	1.5
Damping ratio	5%
Slab thickness	200 mm

Source: Author

- One building model is taken into consideration to complete the required job.
- The analysis is conducted with the fixed column supports in mind.
- The dampers are used as rigid links to provide the necessary rigidity.

Results and analysis

This technique for determining design lateral forces is also termed the static method, seismic coefficient technique, or comparable lateral force technique. Although this approach does not involve dynamic analysis, it roughly accounts for the dynamics of the building.

Building

- The building's displacement concerning height for the G+10 structure with alternate-story dampers is shown in the graph above Figure 29.2.
- According to the IS 1893 (Part 4) (2005) Law of Practice for Earthquake Resistance Industrial Structure, the allowed displacement is (H/500), or (3000/500) = 60 mm.

Table 29.2: Damper's properties.

Model	K (kN/m)	Mass (kN)
G+10	72841.580	62

Source: Author

Table 29.3: Displacement of G+10 building.

Story	Elevation (m)	Displacement (mm) without dampers		Displacement (mm) with dampers	
		X-Dir	Y-Dir	X-Dir	Y-Dir
10	30	93.05	76.04	52.26	42.98
9	27	92.61	76.51	51.97	42.79
8	24	89.79	73.74	50.33	41.35
7	21	85.36	70.14	47.86	39.02
6	18	78.69	64.41	44.16	36.18
5	15	70.83	66.30	39.71	40.08
4	12	63.54	51.62	36.14	29.03
3	9	53.03	42.89	30.10	24.48
2	6	42.23	33.12	24.15	18.76
1	3	25.22	23.87	14.49	14.11
0	0	0	0	0	0

Case	Modes	Time period(secs)		Frequency (cycles/sec)	
		WOD	WD	WOD	WD
Modal	1	1.626	1.247	0.706	0.821
Modal	2	1.54	1.235	0.754	0.842
Modal	3	1.42	1.112	0.795	0.924

Source: Author

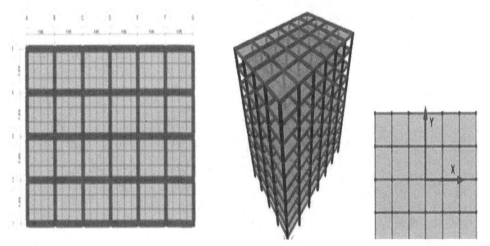

Figure 29.1 G+10 building (plan view) and G+10 building (3D view)
Source: Author

- The highest displacement measured in the X-direction is 52.72 mm.
- Without dampers, the displacement measured 94.14mm; with dampers in place, the displacement is 52.72 mm.
- By using viscous dampers, the displacement was decreased by 42.63% .

1. Static analysis method

This technique for determining design lateral forces is also termed the static method, seismic coefficient technique, or comparable lateral force technique. Although this approach does not involve dynamic analysis, it roughly accounts for the dynamics of the building. Following the model analysis, we pick the narrative response plot of the display menu, and a dialogue box where we may choose the display time for EQX and EQY appears Table 29.3.

2. Response analysis method

The linear dynamic analysis technique is known as the response spectrum approach. With this technique, the peak earthquake responses of a building are directly determined from the seismic responses. The maximum response, which may be described in terms of maximum relative velocity or maximum relative displacement, is displayed versus the un-damped natural time and for different damping levels. Following the analysis of the model, we pick the narrative response plot of the display menu, and a dialogue box where we may choose a different display time for SPECX and SPECY appears Figure 29.1 and Figure 29.2.

Discussion

The two forms of analysis linear static analysis and linear dynamic analysis also referred to as response spectrum analysis, were carried out using the ETABS program, and the results are comparable, but the lateral load distribution method used in response spectrum analysis is effective and acceptable Figure 29.3.

While static analysis only generates in the direction of loading and grows along the depth, response spectrum analysis provides story shear in both directions.

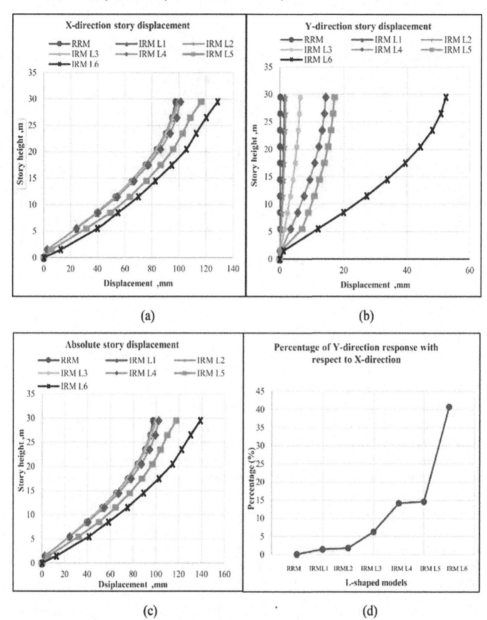

Figure 29.2 Story height vs. displacement height
Source: Author

The displacement values steadily rise with increasing story height; the top story has the greatest displacement. Story displacement for R.S. analysis is 14% less than for Static analysis, which results in an economical design.

The height of buildings increases with height, peaking at mid-levels. The narrative drift is 11% lower than similar static story drift numbers and values, which is acceptable for RS analysis.

According to IS 1893 (Part 1) (2016) Law of Practice for Earthquake Resistance Structures, story drift is within the permitted range (0.004H) in both static and dynamic instances.

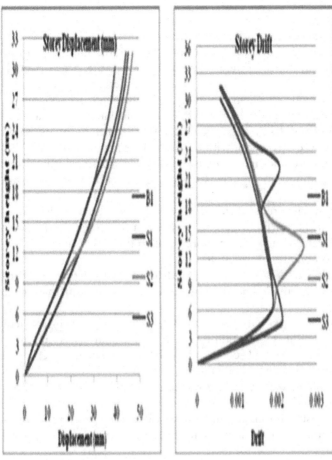

Figure 29.3 Story drift
Source: Author

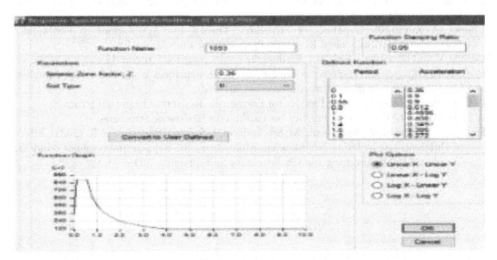

Figure 29.4 Response spectrum function
Source: Author

Conclusion

1. In the current circumstances, a structural engineer cannot afford to obtain the findings manually since it includes time-consuming procedures and difficult calculations that demand substantial time and patience. Therefore, a less complicated option is constantly needed, such as the usage of different computer-aided software that would provide excellent flexibility and efficiency. ETABS, a program that is easy to use and provides more flexible and efficient designs, was used in this project to analyze and design the structure. In this project, the analysis and design of the structure were carried out with the help of a program called ETABS, which is easy to use and provides more flexible and efficient designs. It is simple to apply various load combinations, such as seismic load and response spectrum load.

2. Vibration damping materials are used to reduce vibrations caused by the movement of machinery and the traffic of people. First, in order to reduce the noise associated with the vibration and create a more comfortable environment.

3. Changes to the load combination are simple.

4. The (G+10) Building has undergone analysis and design with details for reinforcement. The results of the modelling are more affordable and secure than those of manual procedures.

5. Due to certain soft storys (stores with less stiffness), static analysis is insufficient for high-rise structures, and it is important to give dynamic analysis. A building's stiffness of a greater value is desirable concerning earthquake damage since it can minimize the deformation demands on a building.

6. The check for deflection and shear was determined to be safe based on the study and design of the slabs and beams.

7. The dampers provide elastic movement and energy dissipation within the structure. As a result, structural elements may be improved for cost reductions, resulting in significant savings.

References

Chopra, A K (2017). Dynamics of Structures Theory and Applications to Earthquake Engineering. Hoboken, NJ: Pearson.

Chopra, A K (2020). Dynamics of Structures: Theory and Applications to Earthquake Engineering. Harlow, United Kingdom: Pearson Education Limited.

Gioncu, V, and Mazzolani, F (2014). Earthquake Engineering for Structural Design. CRC Press.

Goel, R (1998). Seismic response control of irregular structures using non-linear dampers. https://www.iitk.ac.in/nicee/wcee/article/13_3212.pdf.

IS 1893 (Part 4) (2005). Law of Practice for Earthquake Resistance Industrial Structure.

IS 1893 (Part 1) (2016) Law of Practice for Earthquake Resistance Structures.

Sunagar, P, Bhashyam, A, Shashikant, M, Sreekeshava, K S, and Chaurasiya, A K. (2021). Effect of Different Base Isolation Techniques in Multistoried RC Regular and Irregular Building, Trends in Civil Engineering and Challenges for Sustainability, 391–403, Springer.

30 Heat transfer analysis during the turning process under an air-assisted water-cooling environment

Chittibabu Gaddem[a,1], Manoj Ukamanal[b,1] and Sasmita Bal[c,2]

[1]SME, KIIT Deemed to be University, Bhubaneswar, Odisha, India

[2]Associate Professor, Mechanical engineering, Alliance University, Bangalore, India

Abstract

During the metalworking process, heat generation at the tool and workpiece interface is a major challenge for the machinist. When turning, the heat is distributed to the chip, tool, workpiece, and cutting area. This heat distribution significantly influenced the entire cutting process. In addition, heat transfer directly affects cutting attributes such as wear growth rate, tool life, surface finish, and dimensional accuracy of the workpiece achieved. In the current work, different types of machining conditions available during turning operation, the extent of heat generated on the workpiece as well as the tool, and the approaches used to minimize this heat at cutting zones were summarized. Various heat reduction methods and types of coolant applications were also discussed. During the turning of AISI 316 SS workpiece, various factors, such as cutting speed, feed rate, depth of cut, coolant pressure were taken as the input parameter for machining under an air-water atomized SIC environment. To measure the cutting zone temperature, a small diameter K-type thermocouple is placed close to the cutting edge of the tool. Maximum tool and chip-tool interface temperature of 41.8 °C and 65.1 °C with the depth of cut at 0.6 and 0.8 mm respectively noticed and minimum cutting temperatures of 32.8 °C and 33.2 °C at the lowest machining parameters was observed.

Keywords: Coolant, heat generation, machining, tool wear

Introduction

In all the machining processes, heat generation takes place which adversely affects the tool and workpiece. Manufacturers use a flood cooling technique wherein a large amount of coolant is used to remove the generated heat. As these coolants are mineral oil based and pose a serious environmental hazard. The temperature involved in the process has been predicted and measured by many researchers experimentally and theoretically. However, it was found that no concrete results were available in the literature (Sahoo et al., 2023). Machining in dry environment, atomized spray technique and sustainable machining techniques can be used as an alternate for flood cooling. Spray cooling is a process in which a liquid is sprayed onto a heated surface to remove heat. The heat transfer mechanism during spray cooling involves several stages, including droplet impingement, liquid spreading, evaporation, and convective heat transfer. Many studies have found that an increase in temperature in the cutting zone increases tool wear and dimensional inaccuracies and affects the integrity of the working surface (Abdullah et al., 2023). A gradual wear of the tool flank under dry conditions due to higher temperatures was noticed during machining operations of Ti-6Al-4V using coated cemented carbide (Subramani et al., 2023).

[a]cb.gaddemphd@gmail.com, [b]manoj.ukamanal@gmail.com, [c]sasmitabal@gmail.com

DOI: 10.1201/9781003450924-30

This paper summarizes the different types of machining conditions available during turning operation, the effects of generation of heat in the workpiece and tool, and the approaches used to minimize this heat at cutting zones. The use of various coolant applications and strategies for heat reduction were also highlighted. An experiment was conducted during the turning of AISI 316 SS workpiece considering various factors, such as cutting speed, feed rate, depth of cut, coolant pressure as the input parameters for machining under an air-water atomized SIC environment, and to measure the cutting zone temperature, a small diameter K-type thermocouple is placed close to the cutting edge of the tool.

Machining environments

Machining in dry environment

During dry machining, the cooling lubricant is completely dispensed with, which addresses the problem of mist formation from the cooling lubricants (Su et al., 2006). The literature indicates that dry machining provides a superior texture of the surface than wet machining, however, it has been found that tool wear rate and power consumption are higher compared to wet machining (Leppert, 2012).

Machining in a sustainable environment

Vegetable oil-based cutting fluids (VBCF) have emerged to address the hazardous properties of conventionally used cutting fluids as they are renewable, biodegradable, and far less toxic than petroleum-based cutting fluids (Ozcelik et al., 2011). In comparison to dry machining, MQL technique with vegetable oil-based cutting fluid produced better surface integrity with lower cutting temperatures (Subramani et al., 2023).

Machining in atomized coolant environment

The spray impingement cooling (SIC) technology, which is also ecologically benign and has a critical heat flow capacity of roughly 10 MW/m^2, has become one of the ultra-fast cooling strategies because of its high heat transfer capability. In spray impingement cooling, the bulk fluid is sheared into droplets of microscopic scale by compressed air in the spray nozzle. These microscopic droplets will be travelling at a higher velocity and impinging the target surface. Due to high temperatures, the droplets are evaporated due to radiative heat transfer as they come in the vicinity of the hot surface and form a thin vapor film on the hot surface. The impinging droplets further penetrate the vapor film and increase the heat transfer rate by forced convection. Since spray cooling yields higher heat transfer rates making it an effective ultrafast cooling technique that has a high potential to be used as an environmentally benign technique to deliver the cutting fluid during machining operations.

Experimental set-up for turning in air assisted spray environment

Air atomizing nozzle (1/4 J) was used to generate the full cone spray which was directed to the heat-affected zone during the machining of AISI 316SS. AISI 316 SS workpiece measuring 50 mm in diameter and 150 mm in length was prepared for machining. Depth of cut (t), feed rate (s), cutting speed (V), water pressure (Pw), and air pressure (Pa) were taken as the input parameter for machining under an air-water atomized SIC environment and the cutting zone temperature was determined with the help of K-type thermocouples which were attached to CHINO-data acquisition system to log the temperature data. All the machining operations were performed on HMT-NH22 high speed lathe machine. Figure 30.1 explains the total setup for the experiment.

Results and discussions

The focus was on the cooling ability of the spray impingement technique on the uncoated carbide cutting tool during the machining of AISI 316 SS. Convection is the dominant form of heat transfer phenomena in fluids which is directly proportional to the spray impingement density and the difference in temperature between the surface temperature and bulk fluid temperature. Maximum tool and chip-tool interface temperature of 41.8°C and 65.1°C with (t) at 0.6 and 0.8 mm respectively and minimum cutting temperatures of 32.8°C and 33.2°C at the lowest machining parameters were observed. The compressed air shears the water film into microscopic water droplets generating the spray which results in an increase in surface area for heat transfer. The HTC from the cutting tool can be calculated by using Equation 1 (Prinz and Bamberger, 2014). Since the spray cone covers the entire cutting zone, the radiative heat transfer is neglected in determining HTC.

$$h_{spray} = 0.69 \log \frac{(ID)_w}{0.0006} \left[\left(1.4 \sqrt{k\rho C_p} \right) exp \left(0.32 \frac{T_s - T_e}{T_f - T_e} \right) + h_v \right] \tag{1}$$

It can be observed that a rise in water pressure results in an increase in the average spray impingement density resulting in an increase in the average heat transfer coefficient. Higher HTC depicts the increased rate of heat transfer from the cutting zone to the spraying fluid resulting in reduced cutting temperatures. As the cutting temperature increases with machine time, the average HTC is seen to be decreasing as the bulk fluid temperature is constant which is evident from Figure 30.2. Figure 30.3 clearly reveals higher cutting temperatures are observed for higher feed rates under air-water SIC environment.

Figure 30.1 Spray assisted cooling during turning AISI 316 SS
Source: Self made (original) using Microsoft visio and Ansys.

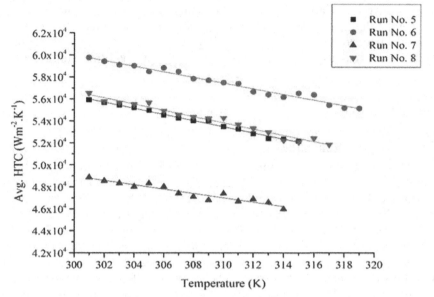

Figure 30.2 Average heat transfer coefficient in the air-water SIC environment
Source: Self made (original) using Microsoft visio and Ansys.

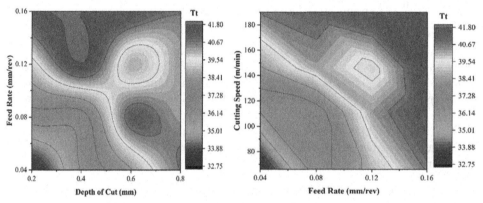

Figure 30.3 Cutting temperature contour plot under air-water SIC for varying machining parameters
Source: Self made (original) using Microsoft visio and Ansys.

The Gaussian distribution method has been employed to determine the uncertainty in the experimental data taking a confidence level of $\pm 2\sigma$ with 96% of data measured lies within the 2σ limits of mean). An uncertainty of $\pm$ 0.048 %, $\pm$ 1.4 % and ± 3.7 % was observed for the test specimen (length), location of thermocouple in insert and tool temperature measurement.

Conclusion

In this study, the heat transfer coefficient was determined using the pre-established correlation during air-water SIC technique. The heat produced during machine operation also has a substantial impact on the material's surface roughness. During machining operation, as heat generation diminishes tool life increases. Atomized air-water SIC cooling technique prevented the heat build-up in the cutting zone. Because of the evaporative heat transfer caused by the mist created by the compressed air-water spray, tool temperature, and chip temperature were sufficiently reduced for all machining circumstances under air-water SIC machining conditions. Maximum tool and chip-tool interface temperature of 41.8°C and 65.1°C with (t) at 0.6 and 0.8 mm respectively and minimum cutting temperatures of 32.8°C and 33.2°C at lowest machining parameters was observed. As the cutting temperature increases with machine time, the average HTC is seen to be decreasing as the bulk fluid temperature is constant. Maximum HTC of 5.6×10^4 $Wm^{-2}K^{-1}$ was obtained at maximum spray parameters.

References

Abdullah, M. A. A. bin, Muhammad, M. A., Ibrahim, Z., Ali, M. Y., & Purbolaksono, J. (2023). Investigation of the turning parameters on the surface finish of an aluminum bar. AIP Conference Proceedings. 2643(1), 050001.

Leppert, T (2012). Surface layer properties of AISI 316L steel when turning under dry and with minimum quantity lubrication conditions. *Proceedings of the Institution of Mechanical Engineers, Part B: Journal of Engineering Manufacture.* 226(4), 617–631.

Ozcelik, B, Kuram, E, Cetin, M H, and Demirbas, E (2011). Experimental investigations of vegetable based cutting fluids with extreme pressure during turning of AISI 304L. *Tribology International.* 44(12), 1864–1871.

Prinz, B, and Bamberger, M (2014). Determination of heat transfer coefficient of air mist sprays. *Materials Science and Technology.* 5(4), 389–393.

Sahoo, A K, Sahoo, S K, Pattanayak, S, and Moharana, M K (2023). Ultrasonic vibration assisted turning of inconel 825: an experimental analysis. *Materials and Manufacturing Processes.* 38(12), 1600–1614.

Su, Y, He, N, Li, L, and Li, X L (2006). An experimental investigation of effects of cooling/lubrication conditions on tool wear in high-speed end milling of Ti-6Al-4V. *Wear.* 261(7–8), 760–766.

Subramani, S, Sivaram, N M, and Gajbhiye, N L (2023). A study on the sustainable machining of AISI 630 stainless steel under minimum quantity lubrication. *International Journal of Materials Engineering Innovation.* 14(1), 60.

31 Performance evaluation of various concrete mixes: A new mix proportioning methodology for high-volume fly ash based concrete design mixes

Asif Ahmed Choudhury[1,a], Saurav Kar[2,b] and Anup Kumar Mondal[1,c]

[1]Department of Civil Engineering, Techno India University, Kolkata 700091, West Bengal, India

[2]Department of Civil Engineering, Heritage Institute of Technology, Kolkata West Bengal, India

Abstract

Fly ash or pulverized fuel ash is regarded as the most popular cement replacement material with cement replacement level (CRL) conventionally up to 30% or less, with higher content of fly ash not used globally. Many attempts have been made to conduct multi-dimensional research into high volume fly ash C. This paper emphasizes the determination of various properties dependent on the new mix design technique adopted for high volume fly ash concrete with varying fly ash: cement ratio at 1:1 ratio. This present study not only highlights the utilization of high-volume fly ash in high performance concrete at various percentage replacement levels but also focuses on development of a new, rational, scientific and cost-effective mix design approach. Various percentage replacement levels of fly ash (b/w) of cement are considered as 40%, 50%, 60%, and 70% with target strengths 30 MPa, 35 MPa, 40 MPa and 45MPa respectively. Various activators such as modified sulpho-napthalene formaldehyde are used. The new mix design approach provides an alternate mix proportioning technique for high volume fly ash concrete mixes unlike any national and international codes.

Keywords: Chemical activator, flow spread, high volume fly ash, new mix proportioning, slump

Introduction

Rapid changes and development of construction materials pushed the focus of engineers toward development of more sustainable, cost-effective materials. Pulverized fuel ash or fly ash used as partial cement additive with the process of activation of cement directly, with adopted replacement quantities between 15 and 25% (Berry et al., 1994). Extensive studies on high volume fly ash emerged in late 1900s (Mehta and Monteiro, 2017). Studies reveal that performance of strength values of high-volume fly ash concrete prepared with combined partnership of cement and sole pulverized fuel ash (Herrera, 2011, Ling et al., 2013). The compressive strength of high-volume fly ash concrete has been reported to be decreasing as the percentage of cementer placement kept increasing (Huang et al., 2013). Researches focused on utilization of fly ash with replacement level 55% is obtained to be very famous and hence, turning to highly market availability worldwide (Supit and Shaikh, 2014; Shaikh and Supit, 2015).

[a]asifchoudhury523@gmail.com, [b]sourav.kar@heritageit.edu, [c]anupkm1960@gmail.com

DOI: 10.1201/9781003450924-31

Experimental methodology

The new mix proportioning of HVFAC is represented in table 31.1.

The complete list of the new rational mix proportions in this new research work is presented in figure 31.1. Various hardened concrete cube and cylinder samples prepared in the research laboratory are shown in figure 31.2.

Materials used for experimental program

The class "F" fly ash required for the experimental purpose is obtained from Gujarat, India (ordered by green and nature. in website) possesses fineness modulus (FM) 2.06. The fly ash obtained stored in controlled conditions. Chemical tests conducted to determine the chemical properties of FA and the oxide composition of fly ash are as follows: SiO_2- 51.09%, Al_2O_3- 28.34%, Fe_2O_3- 12.5%, TiO_2- 1.51%, CaO-1.79%, MgO- 0.84%, SO_3- 1.09%, Na_2O- 0.42%, and loss of ignition (LOI)- 2.42%

As fine aggregate river sand has been used which fineness modulus 2.13(zone III) and specific gravity 2.55 and 20 mm nominal size coarse aggregate used, which fineness modulus 2.63 and specific gravity is 2.96. The specific gravity of the activator is 1.04. The fine and coarse aggregates taken for this experimental purpose conforms IS

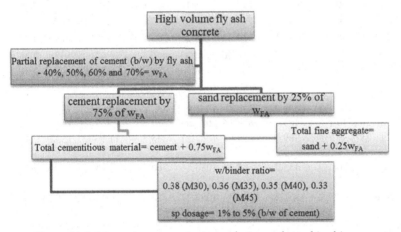

Figure 31.1 New mix proportioning technique adopted in this current research work
Source: Author

Figure 31.2 Various hardened concrete samples with varying fly ash content
Source: Author

Table 31.1: Mix proportioning stipulations: a new methodology adopted HVFAC

Sample designation	cement content (OPC) gm.	fly ash (class F)	% substitution	cementitious material (gm)	fine agg.	coarse aggregate (gm)	Total fine aggregate (gm)	Chemical activator (gm.)	water (gm.)	w/b ratio %
	gm.			C+0.75 Fly ash	gm.		sand + 0.25 FA			
Group FA I (1:1)										
CM30	442.1	0	30	442.1	580	1186.75	580.0	4.42 (1%)	168	0.38
M30_R40	249.05	166.03	40	373.6	717.57	1105.1	759.1	7.47 (2%)	142	0.38
M30_R50	202.34	202.34	50	354.1	698.9	1076.37	749.5	10.62 (3%)	134.6	0.38
M30_R60	145.28	217.914	60	308.7	740.12	1139.83	794.6	12.35 (4%)	117.31	0.38
M30_R70	101.175	236.075	70	278.2	719.62	1108.26	778.6	13.91 (5%)	105.73	0.38
Group FA II (1:1)										
CM35	420	0	30	420.0	717.6	1105.104	717.6	4.20 (1%)	151.2	0.36
M35_R40	236.6	157.73	40	354.9	717.57	1105.104	757.0	7.10 (2%)	127.8	0.36
M35_R50	184.835	184.84	50	323.5	719.62	1108.26	765.8	9.71 (3%)	116.5	0.36
M35_R60	138.01	207.02	60	293.3	615.06	947.233	666.8	14.67 (4%)	105.6	0.36
M35_R70	96.12	224.28	70	264.3	738.1	1136.7	794.2	26.4 (5%)	95.15	0.36
Group FA III (1:1)										
CM40	480	0	30	480.0	676.6	1041.95	676.6	4.8 (1%)	168	0.35
M40_R40	270.4	180.264	40	405.6	676.56	1041.95	721.6	8.1 (2%)	142.0	0.35
M40_R50	211.25	211.25	50	369.7	692.97	1067.21	745.8	11.10 (3%)	130.0	0.35
M40_R60	157.72	236.6	60	335.2	705.3	1086.16	764.5	13.41 (4%)	117.32	0.35
M40_R70	109.844	256.31	70	302.1	712.44	1097.21	776.5	15.11 (5%)	105.74	0.35
Group FA IV (1:1)										
CM45	509.1	0	30	509.1	673.5	1037.22	673.5	5.10 (1%)	168.0	0.33
M45_R40	286.8	191.2	40	430.2	676.6	1041.95	724.4	8.60 (2%)	142.0	0.33
M45_R50	224.05	224.05	50	392.1	686.82	1057.74	742.8	11.76 (3%)	129.4	0.33
M45_R60	167.3	250.932	60	355.5	669.4	1030.904	732.1	14.22 (4%)	117.32	0.33
M45_R70	116.5	271.85	70	320	692.97	1067.21	760.9	16.0 (5%)	105.6	0.33

Source: Author

standards. In this particular experimental program, the whole series of concrete mix specimens is grouped into 4 different groups with two sub groups each. The series of specimens is grouped into four groups: FAI, FAII, FA III and FA IV.

Concrete mix proportions developed: a new approach

Elaborate previous research studies show that fly ash content beyond 4050% (b/w of cement), shows non-reactive nature (Malhotra, 2002; Atis, 2003). Hence, in case of high-volume fly ash content, the extra or additional fly ash content behaves as filler materials, filling the micro pores, in concrete matrix. The additional fly ash content beyond 40% -50% b/w of cement substitutes as fine aggregate in addition to sand. This is the primary reason for adopting such a new mix proportioning methodology for high volume fly ash concrete.

Experimental observations

For concrete specimen with higher CRL (cement replacement levels with fly ash)

Observations from preparation of fresh plastic concrete with increasing percentage replacement of OPC with class F fly ash from 4070% replacement of cement in 1:1 (cement: FA) ratios, shows enhanced workability when sulphonate based and SBR latex based activator was added. The dosage of activators ranged from 1%, 2%, 3%, 4% and 5% (b/w). For various fresh concrete mixes with slump values ranged from 100 mm to 230 mm, to achieve 30 MPa, 35 MPa, 40 MPa and 45 MPa target strengths respectively. Variation in w/binder ratios with modified cementitious content and increased activator content presented satisfactory flow spread and workability results. For concrete mixes

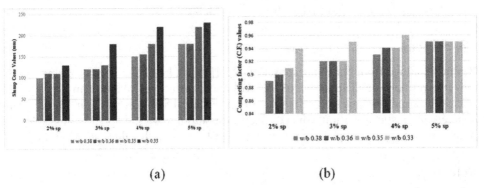

(a) (b)

Figure 31.3 Workability properties: (a) Variation of slump values at various w/b ratios. (b) Variation of compacting factor (C.F.) values for various concrete mixes at varied w/b ratios
Source: Author

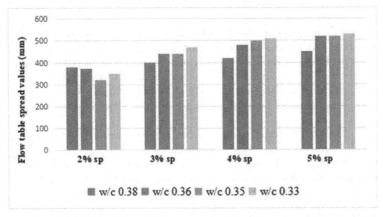

Figure 31.4 Variation of flow spread values for various concrete mixes at varied w/b ratios
Source: Author

with w/b ratio at 0.38, slump All the details concerned with slump values, compacting factor and flow spread values are shown in Figures 31.3 and 31.4 respectively.

Results and discussions

For concrete mixes with w/b ratio of 0.38, during initial period of pond curing strength values ranged from 15 MPa- 20 MPa approximately at 7-days curing, while at later days of curing showed slightly higher results of 20 MPa-30 MPa. For concrete mixes with w/b ratio of 0.36, during initial period of pond curing strength values ranged from 15 MPa- 20 MPa approximately at 7-days curing, while at later days of curing (28-days and 90-days) showed slightly higher results of 25 MPa-35 MPa. For concrete mixes with w/b ratio of 0.35, during initial period of pond curing strength values ranged from 15 MPa- 20 MPa approximately at 7-days curing, while at later days of curing (28-days and 90-days) showed slightly higher results of 40 MPa-50 MPa. For concrete mixes with w/b ratio of 0.33, during initial period of pond curing strength values ranged from 15 MPa- 25 MPa approximately at 7-days curing, while at later days of curing (28-days and 90-days) showed much higher results of 50 MPa-80 MPa. The overall trend in strength performance showed with advancement from initial or early age curing, later age strength increases manifold. With decreasing w/b ratio, later age strength gradually observed to increase.

Conclusions

Results obtained from the elaborate set of experiments, highlights the suitability of the new mix proportioning approach of high-volume fly ash concrete based on rheological, workability and crushing strength properties in practical field construction. All the fresh concrete in plastic condition achieved their best slump flow and spread values. Along with workability, all samples demonstrated their best strength values at all periods.

References

Atis, C D (2003). High-volume fly ash concrete with high strength and low drying shrinkage. *Journal of Materials in Civil Engineering*. 15(2), 153–156.

Berry, E E, Hemings, R. T, Zhang, M H, Cornelius, B J, and Golden, D M (1994). Hydration in high-volume fly ash concrete binders. *ACI Materials Journal*. 91(4), 382–389.

Herrera, A D (2011). Evaluation of sustainable high-volume fly ash concretes. *Cement and Concrete Composites*. 33(1), 39–45.

Huang, C H, et al. (2013). Mix proportions and mechanical properties of concrete containing very high-volume of class F fly ash. *Construction and Building Materials*. 46, 71–78.

Ling, X H, Setunge, S, and Patnaikuni, I (2013). Effect of different concentrations of lime water on mechanical properties of high-volume fly ash concrete. In: 22nd Australian Conference on the Mechanics of Structures and Materials, ACMSM 2012, Sydney, NSW.

Malhotra, V M (2002). High-performance high-volume fly ash concrete. *Concrete International*. 1–5.

Mehta, P. K. and Monteiro, P.J. (2017).Concrete:Microstructure, Properties and Materials. 4th Edition, McGraw Hill Education (India) Private Limited.

Macquarie Supit, S. W. and Ahmed Shaikh, F. U. (2015). Compressive strength and durability properties of high-volume fly ash (high volume fly ash) concretes containing ultrafine fly ash (UFFA). *Construction and Building Materials*. 82, 192–205.

Macquarie Supit, S. W. and Ahmed Shaikh, F. U. (2014). Durability properties of high-volume fly ash concrete containing nano-silica. *Materials and Structures*. 48(8), 2431–2445.

32 Investigation on compressive strength of ultra-high-performance concrete (UHPC) under autoclave curing

S. Revathi[a,1], Dr. D. Brindha[b,2] and R.Harshani[c,1]

[1]Assistant Professor (Sr.Gr), Department of Civil Engineering, Mepco Schlenk Engineering College, Sivakasi, India

[2]Professor Department of Civil Engineering, Thiagarajar College of Engineering, Madurai, India

Abstract

In this study, the ultimate crushing strength of a novel type of ultra-high-performance concrete (UHPC) called reactive powder concrete (RPC) is investigated by the quantity of high range water reducing (HRWR) agent used, the water-to-cement ratio, and the cement and silica fume content. The developers of the UHPC believed that eliminating coarse aggregate would greatly improve its microstructure and performance by lowering the level of variability between the cementitious materials and the aggregate. This study aimed to determine the effects of two different in-situ steam curing techniques and water curing techniques on the mechanical behavior of reactive powder concrete. Test results from more than five different mix proportions and curing techniques inform the development of a mix proportion optimized for a density of around 2300 kg/m^3 and a compressive strength of more than 150 MPa under accelerated curing conditions.

Keywords: Accelerated curing, compressive strength, SEM analysis, UHPC.

Introduction

Since the beginning of time, people have been trying to find new building materials that have high performance so that they may construct taller, stronger, and better structures. Among the several types of concrete, ultra-high-performance concrete (UHPC) is distinguished by its exceptionally high durability and strength. Also, it has a concrete strength of at least 150 MPa, an extremely high binder content, an ideal packing density to get rid of capillary holes and a very dense matrix, and an exceptionally low water-to-cementitious material ratio. To meet the needs of today's construction industry, new types of binder materials with superior qualities for high- and ultra-high-strength concrete and greatly increased durability must be researched and developed. Incredibly, technological advancements in concrete have made it possible to employ this material in massive construction projects. The scientific revolution in chemistry and physics has given researchers the tools they need to create an amazing invention in concrete technology, which has been made possible by Richard and Cheyrezy (1995). This work survey is taken to investigate the properties of micro-materials added to the concrete. For this study, a mortar cube will do in place of a cast concrete specimen because UHPC does not rely on coarse aggregate to keep the mixture homogeneous. Applying heat treatment to concrete improves its functionality. Lessly et al. (2020) examined the concrete containing low cement ultra

[a]srevathi@mepcoeng.ac.in, [b]dbciv@tce.edu, [c]harshaniramesh1050@gmail.com

DOI: 10.1201/9781003450924-32

high-performance concrete (LCUHPC). Even with conventional water curing, a sufficient range of mechanical properties was attained.

Literature review

It generally states that the ideology of the material's performance makes a clear independent review of the characteristics of UHPC. Fine and coarse aggregates make up the high volume of the components in traditional concrete and form a strong framework of connected granular pieces studied by Richard and Cheyrezy (1995). Maroliya (2012) study of the insights of reactive powder concrete (RPC) research has been on the nature of the material, its characteristics, micromechanical analysis, prospective applications, and early study into structure behavior. Flexural strength obtained using circular corrugated steel fibers is greater than that obtained using flat fibers. Although metakaolin is extremely reactive and requires more water, replacing cement with it does not increase strength. Bali and Kurnia (2018) their study showed that conducting the curing is important to maintain the concrete's wetness. As previously mentioned, adding so-called steam during the heat curing process might increase the strength of reactive powder concrete.

Material study

Design guidelines of ASTM-239R (2018), including our trial-and-error methods are shown in Table 1. Researchers have identified the following components of RPC: quartz sand, silica fumes, cement, superplasticizers, water, and quartz powder as used in this study. SEM examination is used to characterize the surface characteristics and Figure 32.1 assesses the morphological changes, including the pore size, shape, dispersion of micro-level particles or additives in the composites, and membrane texture. The process of mixing is an essential component of the research being done on these innovative cementitious materials. Abdelalim et al. (2008) and Abdelrahim et al. (2021) says that the particles in RPC are too small, users cannot use the chance

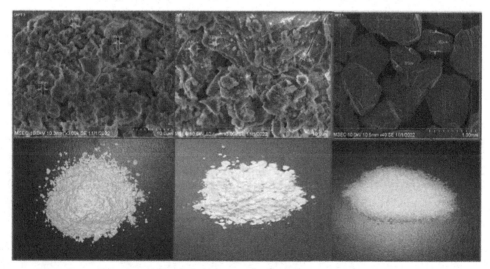

Figure 32.1 SEM image of Silica fume, quartz powder, and quartz sand
Source: Author

mixing technique. Because of very low water-cement ratio to maintain the workability of mix by using high range water reducing (HRWR) agent.

Experimental work

To complete the mixing procedure, a time commitment of between 5 and 10 minutes is necessary to attain a target compressive strength. Mixing sequence can be followed by using ASTM -C-1856 (Standard Practice for Fabricating and Testing Specimens of Ultra-High-Performance Concrete).

Figure 32.2 indicates that water curing (WC) and accelerated curing (AC) were both tested on specimens for this study. For curing, each specimen is also immersed in water curing (room temperature). Accelerated curing at a young age using makes silicate hydrates stronger and more stable. Normal curing methods (storage in water at 27°C) leave a significant quantity of unreacted silica in the concrete microstructure. The interaction with $Ca(OH)_2$ causes the silica concentration to decrease over time. Even after 28 days, the response is not quite complete. To create CSH gel, heat curing reduces the silica concentration. This is why concrete gains strength after being heated.

Table 32.1: Mix proportions of UHPC.

Ingredients	MA1	MA2	MA3	MA4	MA5
Cement(kg/m³)	860	900	950	1000	1100
Silica fume(kg/m³)	200	230	235	300	350
Quartz sand(kg/m³)	860	950	1030	1150	950
Quartz powder(kg/m³)	230	200	95	100	200
Water(kg/m³)	170	170	150	160	185
HRWR (kg/m³)	33	40	39	50	45
Water/cement	0.19	0.18	0.15	0.16	0.168
Flow table (mm)	110	114	109	118	106

Source: Author

Figure 32.2 RPC mixing techniques, autoclave curing, and testing of the specimen
Source: Author

Table 32.2: Compressive strength of RPC.

Mix name	Compressive strength (MPa)			
	Normal water curing(28days)	Autoclave curing (1MPa @ 8 hours)	Autoclave curing (2MPa @ 8 hours)	Autoclave curing (3MPa @8 hours)
MA1	90.7	112.2	122.90	116.43
MA2	95.6	106.78	117.54	112.21
MA3	119.2	138.87	152.89	148.32
MA4	85	104.6	118.78	110.99
MA5	88.8	100.8	112.43	107.67

Source: Author

Results and conclusion

A new formulation strategy based on ultra-fine materials, supported by the robust development of new admixtures, has enabled notable advancements in concrete technology over the past 20 years. The performance and properties of concrete have changed in many ways, from regular concrete to ultra-high-performance concrete to self-compacting concrete and more. The results indicate that the autoclave conditions of time, pressure, and temperature have had a major impact on the compressive strength of RPC. Although the large surface area of silica dispersions raises water requirements and diminishes workability in some ways, fresh concretes with self-flowing abilities in addition to the enhanced mechanical properties in the concrete matrix may be produced by adequately proportioning the mixture. In contrast to the normal curing method, which takes 28 days to reach the same strength, autoclave curing takes only a fraction of that time. Cement-based products are significantly altered by autoclave curing. For one thing, heating at a high temperature quickens the intensity of chemical processes; for another, adjusting the heat and pressure alters the chemical composition of hydration by-products. Conversely, high temperatures may amplify porosity and strengthen the link between aggregates and matrix. Crushing strength is maximum at 119.2 MPa for the MA3 mix proportion after 28 days of normal water curing. Autoclave curing at 2 MPa for 8 hrs is better than water curing in terms of compressive strength and also modifies the underlying microstructure of RPC. After autoclave curing, the composite is often more brittle than regular concrete. Any further elevation of the pressure level resulted in a marginal weakening of the compressive strength of the mixes. In addition to the mix MA3, the compressive strength is nearly 150 MPa, thus Silica Fume and Quartz powder must be included as shown in Table 2. UHPC exhibits extremely high autogenous shrinkage at a young age due to the use of higher cement content and a low water-to-cement ratio of approximately. This is in contrast to the autogenous shrinkage that occurs in MA4 and MA5 as a result of the high binder ratio during heat treatment. Results indicated that the accelerated curing in 2 MPa @ 8 hours specimen had high compressive strength compared to 28 days water curing, according to the test results of all mix proportions.

References

Abdelalim, A, Ramadan, M, Bahaa, T, and Halawa, W (2008). Performance of reactive powder concrete produced using local materials. *HBRC Journal.* 66.

Abdelrahim, M A A, Elthakeb, A, Mohamed, U, and Noaman, M T (2021). Effect of steel fibers and temperature on the mechanical properties of reactive powder concrete. *Civil and Environmental Engineering.* 17(1), 270–276. https://doi.org/10.2478/cee-2021-0028

ACI-239R-2018 (2018). Ultra-high-performance concrete: an emerging technology report. ISBN: 9781641950343

ASTM C1856/C1856M-17 Standard practice for fabricating and testing specimens of ultra-high performance concrete. ASTM International. doi: 10.1520/C1856_C1856M-17.

Bali, I, and Kurnia, W (2018). The curing method influence on mechanical behavior of reactive powder concrete. *International Journal of Advanced Science Engineering Information Technology.* 8(5), 1976–1983.

Lessly, S H, Kumar, S L, Jawahar, R R, and Prabhu, L (2020). Durability properties of modified ultra-high performance concrete with varying cement content and curing regime. *Materials Today: Proceedings.* 45, 6426–6432. https://doi.org/10.1016/j.matpr.2020.11.271.

Maroliya, M M K (2012). A state of art-on development of reactive powder concrete. *International Journal of Innovative Research & Development*, 1(8), 492–503.

Richard, P, and Cheyrezy, M (1995). Composition of reactive powder concrete. *Cement and Concrete Research.* 25(7),1501–1511.https://doi.org/10.1016/j.conbuildmat.2011.05.006.

33 Study on mechanical performances of polyester composites filled with pistachio shell particle

Deepak Kumar Mohapatra[a], Chitta Ranjan Deo and Punyapriya Mishra

Department of Mechanical Engineering, Veer Surendra Sai University of Technology (VSSUT), Sambalpur, India

Abstract

The study is aimed to create new polymer composites using unsaturated polyester resin and Pistachio shell particles (PSP). By using the hand lay-up technique, composite samples are fabricated with different PSP filler contents (1 wt%, 3 wt% and 5 wt%), and mechanical tests are conducted. The highest values of tensile strength, impact strength, and micro-hardness are achieved at 5 wt% PSP filler content, while the optimal flexural strength is observed at 3 wt% PSP filler content. Additionally, good adhesion between PSP filler and polyester matrix is noticed while analyzing the fracture surface micrograph by using scanning electron microscope (SEM).

Keywords: Filler, pistachio shell particles, polyester, mechanical properties, scanning electron microscope

Introduction

Today's world, a major proportion of material science study and development is directed to advancements in composite materials, owing to their desired features such as high strength and toughness, flexibility, and ease of application (Najafabadi et al., 2014; Jagadeesh et al., 2022). The aerospace and automotive industries prefer natural filler reinforced polymer composites due to their advantages over synthetic fibers such as light weight, low cost, sustainability, and eco-friendliness. Although natural fillers have potential to improve the thermal and mechanical properties of composites and reduce costs, but their hydrophilic nature limits their use as primary reinforcing agents. Numerous researchers have explored the potential of using natural fillers as reinforcing agents in polymer-based composites (Sarraj et al., 2021; Essabir et al., 2016).

Al-obaidi et al. (2020) investigated mechanical performance of epoxy-based polymer composite filled with pistachio particles. At 5 wt.% pistachio shell content and 63 μm particle size, the impact, tensile, and flexural strengths of composites increased by 75%, 56%, and 87.7%, respectively. While the amount of pistachio shell in the samples was 15 wt.% and the particle size was (63 < d < 120 μm), the hardness of epoxy composite increased by around 28%. Setty et al. (2022), in another work developed and studied the thermal properties of chemical treated *Limonia acidissima* (wood apple) shell powder (LASP). X-ray diffraction (XRD) analysis showed that treated LASPs have better water resistance due to increased crystal size and crystallinity index. Moreover, the surface modification of LASPs reduced the thermal degradation of cellulose, which was confirmed by thermo gravimetric analysis (TGA)

[a]dkmohapatra_phdme@vssut.ac.in

DOI: 10.1201/9781003450924-33

and Differential Thermal Analysis (DTA) The mechanical properties of carbonized coconut shell filler (CSF) particles containing 10, 20, 30, and 40% in polypropylene composites were investigated by Mark et al. (2020). Increasing the amount of fillers in the composite resulted in decreased flexibility and toughness, leading to lower ductility and modulus of resilience. Ikladious et al. (2019), developed an eco-friendly composite based on peanut shell powder/unsaturated polyester resin. The filler's addition to the polymer matrix enhanced all composites' mechanical capabilities, water resistance, and thermal stability. Owing to adequate wetting and excellent interfacial bonding., mechanical performances were maximum at 35 wt% of filler. Furthermore, the strength values were outstanding even after addition of 55 wt% more treated filler. Previous researchers have explored the mechanical properties of polymer composites containing agricultural wastes, but few researches on the mechanical characteristics of polyester composites filled with PSP have not yet been extensively explored. The current study investigates the potential of using PSP in creating polymer composites, and hand lay-up method was used to fabricate composite samples with PSP filler concentrations of 1%, 3%, and 5%. The mechanical performance of the samples was evaluated through several tests, and fractography study was also conducted.

Experimental procedure

Materials and method

A matrix system containing unsaturated polyester resin, methyl ethyl ketone peroxide (MEKP) acting as the initiator, and cobalt naphthenate acting as the accelerator which were collected from Abanti Enterprises, Odisha, India. The PS was procured from local sources. After that, it was subjected to grinding and sieving. The PS particles were milled to a size of 5 μm and then dried in an oven for 2 hours at 80°C before being put to use. To make PS/polyester composites, varying amounts of PS particles were added and mixed thoroughly with resin containing 0.25 wt% cobalt naphthenate and 1.5 wt% MEKP to ensure uniform distribution. The mixture was then poured into a mould and allowed to cure for 48 hours at room temperature. The resulting composites were cut into samples for mechanical testing following ASTM standards. The combinations of composite samples are presented in Table 33.1.
Mechanical characterization and morphology study

Tensile and flexural tests are conducted in accordance with standards ASTM D 638 and ASTM D790 respectively using a universal testing machine (UTM: model Instron 3382, USA). The crosshead speed is selected as 2 mm/min for testing. Izod impact test are tested with an impact tester (SCD, S-1102, India) according to ASTM D256. The hardness was measured with a micro hardness tester (Matsuzawa, MMT-X7B, Japan) according to ASTM E384. Each sample is examined five times at a load of 50gf with a 10 s indentation time. Using a scanning electron microscope (SEM) (Make: Carl Zeiss

Table 33.1: Combination of PSP/polyester composites.

Sl. No.	Designation	Wt.% of PSP	Wt. % of polyester	Thickness (mm)
1	Neat resin/ PSP-0	0	100	4.8
2	PSP-1	1	99	4.8
3	PSP-3	3	97	4.8
4	PSP-5	5	95	4.8

Source: Author

SMT, Germany, Model: EVO MA 15), the morphology of PSP/polyester composites was examined.

Results and discussion

Tensile testing

Figure 33.1 shows the tensile strength of PSP/polyester composites, which is greatly improved compared to the polyester resin (PSP-0) due to the addition of PSP filler. The PSP-5 composite has the highest strength of 36.55 MPa, which is 96.4% higher than the neat polyester matrix. The study found that increasing the PSP content from 1 wt% to 5 wt% resulted in a significant improvement in tensile strength by 51.46%, 74.07%, and 96.4%, indicating good bonding between PSP and the polyester matrix. The results suggest that PSP is a suitable filler material for enhancing the performance of composites when combined with the polyester matrix.

Flexural testing

Flexural strength of PSP/polyester composites are illustrated in Figure 33.2. The flexural strength of composite initially increases up to a filler content of 3 wt% PSP, after which it decreases. The optimal flexural value of 55.52 MPa is achieved with 3 wt% PSP filler, which is 135.61% higher than that of the neat polyester. The high stiffness value of PSP is likely responsible for this increase in flexural strength, as well as the excellent adherence and homogenous mixing of the filler with the matrix. However, the further addition of PSP filler decreases the flexural strength. This may be due to the agglomeration of PSP at higher concentration.

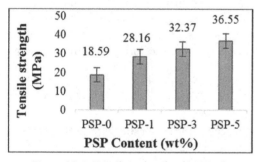

Figure 33.1 Tensile strength of PSP/polyester composites
Source: Author

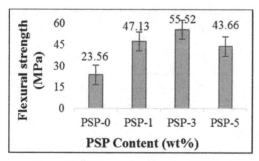

Figure 33.2 Flexural strength of PSP/polyester composites
Source: Author

Impact strength

The impact energy absorbed by the specimens was measured by using Izod impact testing. As indicated in Figure 33.3, the impact strength is improved, while the filler content of PSP is increasing from 1 wt.% to 5 wt.%. The maximum impact strength of 16.36 KJ/m² is found for PSP-5, whereas 9.61 KJ/m² for unfilled polyester, corresponding to an increment of 70.24%. This result reflects that the PSP has the potential to greatly enhance impact strength. The increase in fracture toughness due to PSP fillers reveals prevention of fracture formation in the composite.

Micro-hardness

Figure 33.4 depicts the micro-hardness of PSP/polyester composites with varying wt% of PSP contents. As the PSP content increased from 15%, there was a clear trend of increasing hardness. The maximum hardness of 13 HV is possessed by PSP-5 sample, whereas 8.5 HV is exhibited by PSP-0, which was an increase of 52.94% due to the addition of 5% PSP. This might be related to considerably greater hardness value of PSP, which leads to a stronger bond between the matrix and filler and ultimately results in increased hardness.

Fractography study

The morphology of PSP-polyester composites using SEM is presented in Figure 33.5. The fracture surface of the PSP-5 sample under tensile test is presented in Figure 5(a). A uniform dispersion of PSP in the polyester matrix with fewer voids is clearly visible. The presence voids can be attributed to the leftover moisture in the filler which

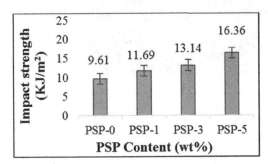

Figure 33.3 Impact strength of PSP/polyester composites
Source: Author

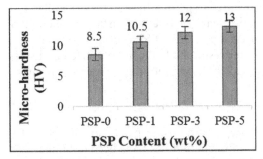

Figure 33.4 Micro-hardness of PSP/polyester composites
Source: Author

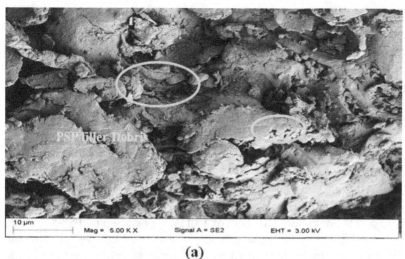

(a)

(b)

Figure 33.5 SEM image of fracture composites (PSP-5) under (a) tensile loading,
(b) flexural loading
Source: Author

absorbed into the composite throughout the manufacturing process whereas the uniform dispersion of PSP in the matrix is attributed to good adhesion between filler and matrix. Similar observation is also noticed in case of fracture surface of the PSP-5 sample under flexural test (Figure 33.5-b). As the images of SEM clearly show a good adhesion bonding between PSP and polyester resin, the PSP can be utilized as a suitable filler material to enhance the mechanical characteristics of polymer composites.

Conclusions

In this research, polymer composites are prepared by using pistachio shell particles and polyester resin. Based on studies, the following findings have been obtained.

The PSP can be utilized as suitable filler materials in the polyester matrix to improve tensile, flexural, impact and microhardness. A higher value of tensile strength is obtained at 5 wt% inclusion of PSP filler, which is 96.54% greater than that of neat resin. However, maximum value of flexural strength is evolved at 3 wt% of PSP filler, which is 135.61% higher than neat polyester.

Similarly, a higher value of impact strength and micro-hardness of value of 16.36 KJ/m^2 and 13 HV respectively are found for PSP-5 composite which are 70.24% and 52.94% higher than that of neat polyester matrix.

The SEM micrograph shows an excellent binding between polyester matrix and PSP. The homogeneous distribution of filler and strong filler-matrix bonding attributes to improve in mechanical performances of fabricated composites.

Acknowledgement

The authors are highly grateful to 'Material Testing Laboratory' of Department of mechanical Engineering, VSSUT, Burla for carrying out sample preparations and mechanical testing of samples.

References

Al-obaidi, A J, Ahmed, S J, and Abbas, A T (2020). Investigation the mechanical properties of epoxy polymer by adding natural materials. *Journal of Engineering Science and Technology*. 15, 2544–2558.

Essabir, H, Bensalah, M O, Rodrigue, D, Bouhfid, R, and Qaiss, A E K (2016). Biocomposites based on argan nut shell and a polymer matrix: effect of filler content and coupling agent. *Carbohydrate Polymers*. 143, 70–83.

Ikladious, N E, Shukry, N, El-Kalyoubi, S F, Asaad, J N, Mansour, S H, Tawfik, S Y, and Abou-Zeid, R E (2019). Eco-friendly composites based on peanut shell powder / unsaturated polyester resin. *Proceedings of the Institution of Mechanical Engineers, Part L: Journal of Materials: Design and Applications*. 233(5), 955–964.

Jagadeesh, P, Puttegowda, M, Girijappa, Y G T, Rangappa, S M, and Siengchin, S (2022). Effect of natural filler materials on fiber reinforced hybrid polymer composites: an overview. *Journal of Natural Fibers*. 19(11), 4132–4147.

Mark, U C, Madufor, I C, Obasi, H C, and Mark, U (2020). Influence of filler loading on the mechanical and morphological properties of carbonized coconut shell particles reinforced polypropylene composites. *Journal of Composite Materials*. 54(3), 397–407.

Najafabadi, M A A, Khorasani, S N, and Esfahani, J M (2014). Water absorption behaviour and mechanical properties of high density polyethylene/pistachio shell flour nanocomposites in presence of two different UV stabilizers. *Polymers and Polymer Composites*. 22(4), 409–416.

Sarraj, S, Szymiczek, M, Machoczek, T, and Mrówka, M (2021). Evaluation of the impact of organic fillers on selected properties of organosilicon polymer. *Polymers*. 13(7), 1–16.

Setty, V K S N, Goud, G, Chikkegowda, S P, Rangappa, S M, and Siengchin, S (2022). Characterization of chemically treated *Limonia acidissima* (wood apple) shell powder: physicochemical, thermal, and morphological properties. *Journal of Natural Fibers*. 19(11), 4093–4104.

34 Mechanical and tribological characterization of AA5083 reinforced with boron carbide and titanium diboride

Debasish Rout, Dikshyanta Sahoo, Chitta Ranjan Deo[a] and Rabindra Behera

Department of Mechanical Engineering, Veer Surendra Sai University of Technology (VSSUT) Sambalpur 768018, India

Abstract

As per the global spike of metal-matrix composite in R&D activity, the current investigation basically focused on the fabrication of the hybrid Al5083 matrix composite with boron-carbide (B_4C) and titanium-diboride (TiB_2) inserts and their influences on mechanical and wear characteristics. In this current research a two-step stir-casting process was followed to cast the composite samples of three distinct weight percentages of B_4C and TiB_2. Mechanical characterizations were performed as per the ASTM standard. Further, tribological characteristics were studied by carrying out dry sliding wear test. An optimum combination of reinforcement, i.e., 6 wt.% of B_4C and 9 wt.% of TiB_2, was found for both mechanical performances and tribological aspects. Again, a reduction in the coefficient of friction was also noticed at optimum reinforcement.

Keywords: Al 5083, B_4C, hybrid composite, two-step stir casting process, TiB_2, wear rate.

Introduction

Aluminum is mostly preferred due to its lightweight and corrosion resistance for making metal matrix composites (AMMCs). By incorporating foreign particles like SiC, TiB_2, B_4C, and Al_2O_3 into the aluminum matrix, AMMC properties can be altered, making it fit for aeronautical, aviation, automobile, and marine applications (Bhowmik et al., 2021). However, further enhancement in properties can also be made by supplementing hybrid reinforcement into a single matrix material. Moreover, the properties of hybrid AMMCs are largely influenced by the fabrication process as the distribution and wetting of reinforcements in the matrix play a significant role. In view of that a two-step stir casting method is preferred to conquer these challenges (Bhowmik et al., 2021; Shanmughasundaram et al., 2012). Again, this process is well accepted by several industries due to its simplicity and economic traits. In a study, Singh and Sharma (2020) investigated the physical as well as mechanical properties of Al5083 reinforced with B_4C inserts with different weight percentages, prepared by stir-casting process at 800°C. The study found that 15% reinforcement led to a notable rise in tensile strength and hardness, but the addition of B_4C resulted in a reduction in density due to increased porosity. Similarly, Raja et al. (2020) investigated the effect of different B_4C weight proportions on the mechanical and wear characteristics of AMMCs. The results of the study reflected that with increase of B_4C content led to improvements in tensile strength, hardness, and wear resistance and 10 wt% reinforcement was found to be most favorable proportion. In a different study, Gupta et al. (2020) examined the mechanical and tribological attributes of Al 1100 matrix

[a]chittadeo@gmail.com

DOI: 10.1201/9781003450924-34

composites with various weight proportion of TiB_2. They found that higher weight percentages of TiB_2 direct to improved tensile strength, hardness, and wear resistance. Likewise, Kumar et al. (2021) reported that, reinforcement of 5 wt% combination of B_4C and fly ash each is the best combination to produce a balance tensile and compression strength. However, the further increment of B4C in Al-Mg-Si-T6 matrix leads to a higher hardness value. In another study, Velavan et al. (2021) reported the wear characteristics of AA6061 matrix composite strengthened with various weight percentages of mica along with fixed weight percentage of B_4C. A sharp reduction in coefficient of friction (COF)was noticed with increase of mica content and sliding speed. Whereas an increase of friction force was also marked with increase of normal load, and sliding distance. Similar reduction of COF due to increase of granite dust content was reported by Satyanarayana et al. (2019), while examining the wear behavior of graphite/granite particles reinforced A356 matrix composite. In view of the above perspective, in the current work an effort has been taken to fabricate and study the mechanical and tribological characteristics of Al5083 composite reinforced with different weight proportion of B_4C and TiB_2.

Materials and methods

Materials used

Aluminum alloy Al5083 in billet form was procured from Suresh Metals, Maharashtra, India. The reinforcing particles, Boron Carbide (B_4C) and Titanium Diboride (TiB_2), of size 50µm were procured from Parshwamani Metals, Maharashtra, India.

Fabrication of composite

In this study, the aluminum HMMCs were developed using the two-step stir casting method. A Bottom pouring type Stir casting machine of Swam-Equip make was used. As per Table 34.1, a suitable amount of Al5083 alloy and reinforcing particles, i.e., B_4C and TiB_2 were taken for each casting to develop the desired HMMCs. The addition of reinforcement was done at 575°C temperature and stirred by mechanical at a speed of 300 rpm. Again, to ensure the proper mixing the temperature of the slurry was raised to the temperature of 750°C to achieve liquid state and stirred for 10 minutes. Once the stirring was completed, the mixture was poured into a warmed cylindrical mould cavity of 140 mm length and 13 mm diameter and allowed to solidify.

Mechanical characterization

The tensile and impact tests were carried out in accordance to the ASTM E8/E8M-09 and ASTM E23-18 standard respectively. The tensile test was conducted on a

Table 34.1: Composition of fabricated HAMMC.

SL NO.	Wt.% of (Al5083)	Wt.% of B_4C	Wt.% of TiB_2
S1	100	0	0
S2	85	4	11
S3	85	6	9
S4	85	8	7

Source: Author

Universal Testing Machine (Instron 3382) with a maximum load capacity of 100 KN at crosshead speed of 2.3 mm/min. To determine hardness, a Vickers-micro hardness tester was used by indenting the composites by a diamond-shaped indenter for 30 seconds with a dead load of 300g as per ASTM E92 standard. Three samples of each category were tested and the average value of that was recorded.

Wear test

The sliding wear test was conducted on a pin-on-disc type wear testing machine of Magnum make. As per ASTM G99-05 standards, cylindrical samples of a size of 10 mm diameter and 40 mm length were used. The experiment was conducted at ambient temperature under different normal loads (20 N, 40 N, and 60 N) and sliding velocities (1.5 m/s, 3 m/s, and 4.5 m/s) with a fixed sliding distance of 100 m against an EN-32 steel disc. The wear loss was calculated in terms of mass loss by subtracting the sample mass before and after testing. Further, the COF during the test was recorded from the computer interface.

Results and discussion

Tensile strength

Figure 34.1 (a)reflected the tensile test results of the composite samples. A sharp rise in tensile strength was marked with the hybrid reinforcement of B_4C and TiB_2 in comparison to Al5083. Again, an enhancing trend of tensile strength is also noticed with the increase of B_4C from 06 wt% and decrease of weight percentage of TiB_2 from 0% to 9%. An optimal value of tensile strength is marked at 6 wt% of B_4C and 9 wt% of TiB_2 reinforcement (154.48 MPa), which is about 25% more than that of pure Al5083. But subsequent increases in B_4C lead to a decrease in tensile value.

Impact strength

Figure 34.1 (b)represents the influence of hybridization on the impact strength of Al5083 composite. Similar to tensile strength, the impact strength of Al5083 hybrid composite is increased by 43% with inclusion of 6 wt% of B_4C and 9 wt% of TiB_2. But at the initial stage, the impact strength is found to be decreased with the inclusion of B_4C and 11 wt% TiB_2 to Al5083. It may happen due to random spreading of reinforcing particles in the matrix, which directed to energy concentration. However, a higher value of the impact strength was observed in subsequent increase of B_4C. Again, the impact strength was found to decrease with further increases in B_4C content from 6 wt% to 9 wt%. It may have happened due to the agglomeration of reinforcing B_4C at the preferred propagation sites.

Vickers micro-hardness

The variation of hardness is illustrated in Figure 34.1 (c). The pattern of the plot clearly indicates that, the hardness value of hybrid composite enhances with increase of B_4C reinforcement along with subsequent decrease of TiB_2. In accordance to the theory of plastic deformation, dislocations accumulate when the amount of reinforcement increases. Again, the inclusion of the hard reinforcements with good distribution in the matrix also increases the hardness. B_4C possesses a higher hardness value and good fusion with the Al5083 matrix in comparison to TiB_2, thus hardness of the composite is found to be enhanced with increase in B_4C content. Maximum

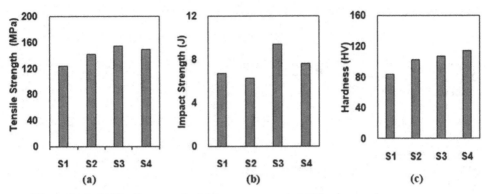

Figure 34.1 (a) Tensile strength, (b) Impact strength (c) Hardness value
Source: Author

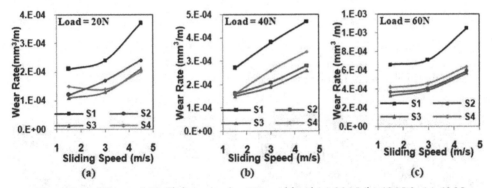

Figure 34.2 Wear rate Vs Sliding speed at Normal load (a) 20 N (b) 40 N & (c) 60 N
Source: Author

hardness value of 114.3 HV is observed in case of 8wt% of B_4C and 7wt% of TiB_2 reinforcement.

Influence of sliding speed on wear rate

Figure 34.2 (a-c) depicted the fluctuation of wear rate with respect to sliding velocity under different normal loads. Irrespective of normal load, all fabricated samples exhibit an increasing wear rate with rise in sliding velocity. It may happen due to raise of temperature at the contact area with increase in sliding velocity which promote sever plastic deformation. But inclusion of hard ceramic particles such as B_4C and TiB_2, it is substantially decreases. Sample S3 containing 6 wt% of B_4C and 9 wt% of TiB_2 exhibited a lower wear rate amongst all at different test speeds.

Influence of normal load on wear rate and coefficient of friction

The variation wear rate with applied normal load at different sliding speeds is illustrated in Figure 34.3 (a-c). It is clearly noticed that, irrespective of sliding speed the wear rate of all tested samples were increases with rise of normal load. As load increases, the frictional force also increases which resulted de-bonding and fracture. Due to increase of frictional force the asperities subjected to higher stress beyond elastic limit which leads to sever plastic deformation. Owing to sever plastic deformation the Al5083 sample shows a higher value of wear rate in comparison to B_4C and TiB_2

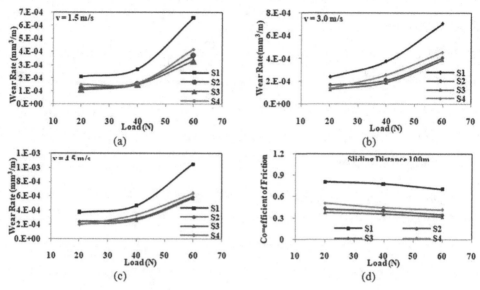

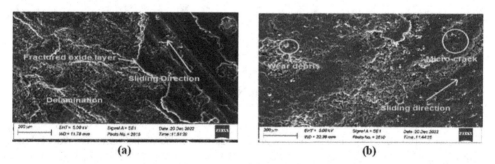

Figure 34.3 Wear rate vs normal load at sliding speed (a) 1.5 m/s (b) 3.0 m/s, (c) 4.5 m/s and (d) Co-efficient of friction vs normal load

Figure 34.4 Wear rate vs normal load at sliding speed (a) 1.5 m/s (b) 3.0 m/s, (c) 4.5 m/s and (d) Co-efficient of friction vs normal load

Source: Author

reinforced samples. A minimum wear rate was observed in case sample reinforced with 6 wt% of B4C and 9 wt% of TiB_2 but further it is increase with increase of wt % of B_4C from 6 wt% to 8 wt%. It may so happen due agglomeration of B_4C particles in higher weight percentage. Again, the decrease of wear rate with increase in wt% of B_4C and decrease in wt% of TiB_2 is also notice due to higher hardness value of B_4C than that of TiB_2. Figure 34.3 (d) depicts the variation of COF with normal load. A significant reduction of the COF with rise of normal load is noticed. The value of co-efficient of friction for S2, S3, and S4 was found to be 0.35, 0.32, and 0.42 under load of 60N, which are much lesser than the base matrix. This may so happen due to the formation of oxide layer of B_4C by successful interaction with atmosphere.

SEM analysis

Figure 34.4 shows the wear surface morphology of S3 and S4 sample under normal load of 60 N. Figure 34.4 (a) represent the wear surface morphology of S4 sample under 60N normal load at sliding velocity of 4.5 m/s. A heavy flow of material along with fractured layer due to sever delimitation in the direction of sliding velocity is observed. This could be attributed to abrasive wear mechanism under heavy normal load. Figure 34.4 (b) depicts the wear surface morphology of S3 sample under 60N normal load and 4.5 m/s sliding velocity. The presence of small micro-crack and wear debris is clearly visible on the wear surface. The presence of wear debris and the presence of small micro-crack attribute to lead to formation of protective layer due to increase of TiB_2 wt% as it has round morphology.

Conclusion

In the current study, the influence of B_4C and TiB_2 addition on mechanical and wear characteristics of Al5083 matrix composite was investigated. Based on the experimental results, following conclusions are drawn:

- The 6 wt% of B_4C and 9 wt% TiB_2 found to be the best hybrid combination for reinforcement in to Al5083 matrix to achieve highest tensile strength. But excessive inclusion of B_4C substantially reduces the strength.
- A reduction in impact strength is noticed with decrease of TiB_2 wt%. Whereas an optimum value of impact strength was observed at 6wt% of B_4C and 9wt% TiB_2. But Vickers micro-hardness increase with rising of wt% of B_4C.
- The wear resistance property of Al5083 was substantially enhanced with addition of B_4C and TiB_2 under all selected sliding speed and normal load. 6 wt% of B_4C and 9 wt% TiB_2 is found to be optimum combination of hybrid reinforcement in view of mechanical and tribological aspects.

References

Bhowmik, A, Dey, D, and Biswas, A (2021). Comparative study of microstructure, physical and mechanical characterization of SiC/TiB_2 reinforced aluminium matrix composite. *Silicon*. 13(6), 2003–2010.

Gupta, M, Gangil, B, and Ranakoti, L (2020). Mechanical and tribological characterizations of Al/TiB_2 composites. *Industrial Engineering Journal*. 13(4), 1–12.

Kumar, M S, Vasumathi, M, Begum, S R, Luminita, S M, Vlase, S, and Pruncu, C I (2021). Influence of B_4C and industrial waste fly ash reinforcement particles on the micro structural characteristics and mechanical behavior of aluminium (Al–Mg–Si-T6) hybrid metal matrix composite. *Journal of Materials Research and Technology*. 15, 1201–1216.

Raja, R, Jannet, S, Reji, S, and Paul, C S G (2020). Analysis of mechanical and wear properties of Al_2O_3 + SiC + B_4C/AA5083 hybrid metal matrix composite, d. *Materials Today: Proceedings*. 26, 1626–1630.

Satyanarayana, T, Rao, P S, and Krishna, M G (2019). Influence of wear parameters on friction performance of A356 aluminum – graphite/ granite particles reinforced metal matrix hybrid composites. *Heliyon*. 5(6), e01770.

Shanmughasundaram, P, Subramanian, R, and Prabhu, G (2012). Synthesis of Al-fly ash composites by modified two step stir casting method. *Advanced Materials Research*. 488–489, 775–781.

Singh, G, and Sharma, N (2020). Evolution of physical and mechanical properties of Al5083/B_4C composites fabricated through stir casting. *Materials Today: Proceedings*. 21, 1229–1233.

Velavan, K, Palanikumar, K, and Senthilkumar, N (2021). Experimental investigation of sliding wear behaviour of boron carbide and mica reinforced aluminium alloy hybrid metal matrix composites using box-behnken design. *Materials Today: Proceedings*. 44, 3803–3810.

35 Buckling analysis of FG plates using FEA

Abhijit Mohanty[1], Sarada Prasad Parida[2,a] and Rati Ranjan Dash[1]

[1]Odisha University of Technology and Research, Bhubaneswar, Odisha, India

[2]Templecity Institute of Technology and Engineering, Khordha, Odisha, India

Abstract

The uses of functionally-graded-material (FGM) have grown due to its precise tailoring potential. Hence, researchers are focusing on the design, manufacturing, optimization and analysis of the FGM. The finite element analysis (FEA) has its own priority where the use of the element plays a crucial factor. The accuracy of the result depends upon the degree of mobility of the element and the choice of intermediate nodes. In this work, nine-noded iso-parametric elements with intermediate nodes are introduced to increase the flexibility. Further, Navier type of solution in FEA is assumed for the buckling response of the functionally-graded-plates (FG-plates). For this, alumina-aluminum FG-plate is considered. The buckling of the FG-plate to different grading index, aspect ratio is evaluated. Further, the result is verified through the FEA of an FE software using ANSYS 16.0. The effect of mesh size in ANSYS is studied and the error percentage between the two methods is evaluated.

Keywords: ANSYS, buckling analysis, FEA, FGM, numerical analysis

Introduction

The functionally-graded-material (FGM) exhibits high hardness, thermal stability, high fatigue resistance, and wear resistance in comparison to its metallic or composite counterparts. The first example of preparation is evidenced in 1970 (Bever and Duwez, 1972). Numerous ways of preparation of the FGM are discussed in the literature (Zhang et al., 2019, Shen and Zhang 2019). The FG plates can be analyzed like the convention plates using different plate theories (Thai and Kim, 2015; Parida and Jena, 2020). The static stability analysis such as buckling analysis of the designed FG plate structure is very important from design point of view (Lee et al. 2010). The buckling strength of FG-plate is evaluated by FEA and showed that critical strength is inversely proportional to the gradient index (Rad and Panahandeh-Shahraki, 2014; Kumar et. al., 2022). Parida et al. (2022) made the static and buckling analysis of FG- laminated composite plate (FG-LCP) rested on elastic foundation. It is demonstrated that the eight noded iso-parametric elements are effective for analysis (Parida and Jena, 2022a; 2022b). In this work the intermediate nodes are introduced in the element to increase mobility. Further, Navier type of solution is assumed for the buckling analysis of the metal-ceramic functionally-graded-plates (FG-plates). The buckling response of the FG plate to different grading index, aspect ratio and mesh size is evaluated.

Theory of functionally-graded-material and method of solutions

The material property in an FGM varies along its thickness as shown in Figure 35.1(a). The material property such as; Young's modulus, density, Poison's ratio, and is defined by grading index given by;

[a]sarada800@gmail.com

DOI: 10.1201/9781003450924-35

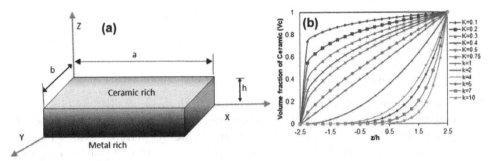

Figure 35.1 Schematic diagram of assumed FG-plate
Source: Author

$$\rho(z)=\left(\rho_c-\rho_m\right)\left(\frac{z}{h}+0.5\right)^k+\rho_m \qquad E(z)=\left(E_c-E_m\right)\left(\frac{z}{h}+0.5\right)^k+E_m$$

$$\text{(1-4)}$$

$$v(z)=\left(v_c-v_m\right)\left(\frac{z}{h}+0.5\right)^k+v_m \quad \mathrm{V_c}=\left(\frac{z}{h}+0.5\right)^k, \mathrm{V_m}+\mathrm{V_c}=1$$

Where v_c, v_m denotes volume-fraction of ceramic and metal in the FG-plate, z: layer thickness, 'h: height and 'k: power index in the FGM cross section as shown in Figure 35.1(b). Based on Navier's approach, the displacements of a simply supported FG-plate in terms of Fourier series are given by:

$$u(x,y,t)=\sum_{m=1}^{\infty}\sum_{n=1}^{\infty}U_{mn}e^{i\omega t}\cos\theta x\sin\vartheta y; \quad v(x,y,t)=\sum_{m=1}^{\infty}\sum_{n=1}^{\infty}U_{mn}e^{i\omega t}\sin\theta x\cos\vartheta y$$

$$w(x,y,t)=\sum_{m=1}^{\infty}\sum_{n=1}^{\infty}W_{mn}e^{i\omega t}\sin\theta x\,\sin\vartheta y; \quad \varphi_x(x,y,t)=\sum_{m=1}^{\infty}\sum_{n=1}^{\infty}\theta_{xmn}e^{i\omega t}\cos\theta x\,\sin\vartheta y \quad \text{(5)}$$

$$\varphi_y(x,y,t)=\sum_{m=1}^{\infty}\sum_{n=1}^{\infty}\theta_{ymn}e^{i\omega t}\sin\theta x\,\cos\vartheta y$$

Where, $\theta = m\pi/a$, $\upsilon = n\pi/b$, $\square$: Eign frequency., U_{mn}, V_{mn}, W_{mn}, θ_{xmn}, θ_{ymn} are the co-efficient of Fourier expansions. For buckling of plates, the governing equation is given by (Parida and Jena, 2022a):

$$\begin{bmatrix} K_{11} & K_{12} & K_{13} & K_{14} & K_{15} \\ K_{21} & K_{22} & K_{23} & K_{24} & K_{25} \\ K_{31} & K_{32} & K_{33} & K_{34} & K_{35} \\ K_{41} & K_{42} & K_{43} & K_{44} & K_{45} \\ K_{51} & K_{52} & K_{53} & K_{54} & K_{55} \end{bmatrix} \begin{Bmatrix} u \\ v \\ w \\ \varphi_x \\ \varphi_y \end{Bmatrix} = \begin{Bmatrix} 0 \\ 0 \\ p \\ 0 \\ 0 \end{Bmatrix} \qquad \text{(6)}$$

Applying boundary condition, i.e. $u = w = \varphi_x = N_{yy} = M_{yy} = P_{yy} = 0$ at $y = 0, b$;
Where, N_{xx}, N_{yy}, M_{xx}, M_{yy}, P_{xx}, and P_{yy} are the shear force, bending moments and external loads acting on the plate. The components of [K] can be simplified from its boundary conditions (Parida and Jena, 2022b). Upon simplification, the generalized critical buckling stress of the FG-plate in non-dimensional form is given by:

$$\sigma_{cr} = \frac{\pi^2 E}{3(1-v^2)}\left(\frac{h}{a}\right)^2 \text{ and}$$

$$P_{cr} = \frac{\pi^2 EI}{(1-v^2)a^2} \qquad (7)$$

Finite element formulation

In this, finite element analysis (FEA) is conducted both theoretically (MATLAB) and by simulation (ANSYS 16.0). The results obtained by both are found to be coherent with minimum error. For theoretical formulation, the FG-plate is divided by nine noded iso-parametric elements. The displacement field described by shape functions (N) based on Lagrange polynomials and nodal displacement (α) functions in-plane coordinates x and y as shown in Figure 35.2 is given by:

$$\{\delta\}_e = [N][\alpha]_e \qquad (8)$$

Where $[N] = [N_1 \ L \ N_9]$ *and* $[\alpha] = [\alpha_1 \ L \ \alpha_9]$

The intermediate nodes are introduced that increases the mobility of the element. The shape functions taken for the analysis are given by:

$$N_1 = \frac{1}{4}\xi(\xi\text{-}1)\eta(\eta\text{-}1) \ \ N_2 = \frac{\eta}{2}(1-\xi^2)(\eta\text{-}1) \quad N_3 = \frac{\eta\xi}{4}(\xi+1)(\eta\text{-}1) \ N_4 = \frac{\xi}{2}(\xi+1)(1-\eta^2) \ N_5 = \frac{\eta\xi}{4}(1+\xi)(1+\eta)$$

$$N_6 = \frac{\eta\xi}{2}(1-\xi)(1+\eta) \ N_7 = \frac{\eta\xi}{4}(\xi\text{-}1)(\eta+1) \quad N_8 = \frac{-\eta\xi}{2}(\xi\text{-}1)(\eta\text{-}1) \quad N_9 = (\xi^2\text{-}1)(\eta^2\text{-}1) \qquad (9)$$

Where ξ and η are the natural coordinates given by:

$$\xi = \frac{2(x-x_i)-l_x}{l_x} \ \eta = \frac{2(y-y_i)-l_y}{l_y} \qquad (10)$$

Where (x_i, y_i) denotes the co-ordinate of the ith node and the length and width of the element is expressed by l_x and l_y respectively. In this study, a FG-plate of size 200 × 200 × 5 mm is assumed. The FG-plate is made up of Aluminum (metal) and Alumina

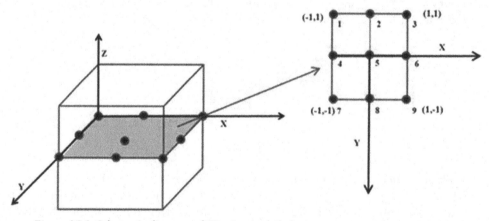

Figure 35.2 Schematic diagram of 2D nine noded element
Source: Author

(ceramic) with elastic modulus of 70 Gpa, 380 GPa, and Poison's ratio of 0.3, 0.3 respectively. The top surface is taken to be ceramic rich while the bottom of the plate is considered metal. In MATLAB program, the dimension of plate and the material properties are assigned first, and then the material property for individual layer is calculated (eq.1-4). Further the element length and shape functions are defined. The stiffness matrix for individual elements is calculated. The boundary conditions are then constrained for edge elements. The global stiffness and force matrix as stated by Equation 6 is assembled and the buckling strength is then calculated. For simulation in ANSYS, the FG-plate is assumed to compose of twenty layers to vary the material property along thickness smoothly. For simplicity of the design the FGM is assumed as layered FGM, material properties for each layer are calculated using simple rule of mixture as of composite material and saved in material library. Sold Brick 185 element is used to mesh each layer. During meshing, each layer is assumed to isotropic and assigned by its corresponding material properties. Then the degrees of freedom are constrained, and the analysis type is set for the output. the meshed model and buckling mode-shape of the FG-plate is shown by Figure 35.3(a) and (b) respectively.

Solution

In this work, the critical bucking stress for different grading index is calculated and it is found the MATLAB program and the ANSYS simulation is suitable for analysis as presented in Figure 35.4(a). The buckling strength is dependent on the constituent material and the grading index. The ceramic material has higher compressive strength compared to metals, hence FGMs with low grading value have higher buckling strength and vice versa. Figure 35.4(b) presents the comparison of buckling stress with aspect ratio by different plate theories and shows good agreement. Here, the grading index value is fixed to 1. The results obtained are coherent to each other and it can be said that, buckling strength is least affected to the change of plate dimension and remains uniform after a/h ratio of 20.

Further, it is observed that, taking the smaller elements gives better and accurate result as presented by Figure 35.5(a); however, it affects computation performance, load and time. In this regard, the element size may be chosen in between 2 to 6 mm. The error percentage while calculating buckling strength to a/h ratio between programmed FEA and ANSYS simulation is presented by Figure 35.5(b). It is observed that though the error percentage is within the permissible limit (within 67%), it decreases with increase of plate dimension in terms of length and width.

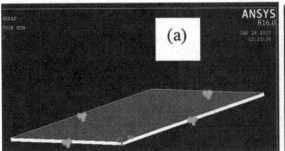 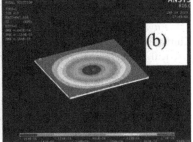

Figure 35.3 FEA models of FGM; (a) Meshed model (b) buckling mode shape
Source: Author

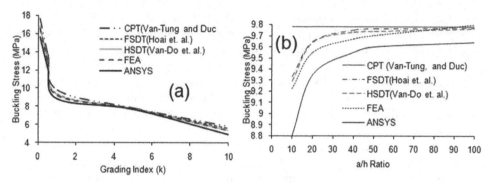

Figure 35.4 Variation of buckling stress ;(a) with grading index, (b) a/h ratio
Source: Author

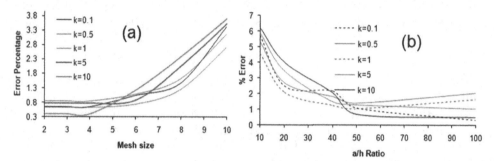

Figure 35.5 Comparison of error percentage; (a) mesh size (b) Aspect ratio
Source: Author

Conclusion

In this study, the buckling analysis of functionally-graded-material (FGM) plates are conducted using ANSYS 16.0 and by finite element analysis (FEA) modeling by MATLAB. The results obtained by both the methods show good agreement with the other theories. It is found that the ingredient, grading index, and aspect ratio affects buckling strength. However, the critical buckling stress is unchanged after an a/h ratio of 20. The mesh size less than 5mm is preferred for FE analysis. The dimension of the plate also affects the error of calculation. Smaller plates have a higher error percentage than bigger plates.

References

Bever, M B, and Duwez, P E (1972). Gradients in composite materials. *Materials Science and Engineering.* 10, 1–8.

Kumar, R, Lal, A, Singh, B. N, and Singh, J (2022). Numerical simulation of the thermomechanical buckling analysis of bidirectional porous functionally graded plate using collocation meshfree method. *Proceedings of the Institution of Mechanical Engineers, Part L: Journal of Materials: Design and Applications.* 236(4), 787–807.

Lee, Y Y, Zhao, X, and Reddy, J (2010). Postbuckling analysis of functionally graded plates subject to compressive and thermal loads. *Computer Methods in Applied Mechanics and Engineering.* 199(25–28), 1645–1653.

Parida, S P, and Jena, P C (2020). Advances of the shear deformation theory for analyzing the dynamics of laminated composite plates: an overview. *Mechanics of Composite Materials.* 56(4), 455–484.

Parida, S P, and Jena, P C (2022a). Free and forced vibration analysis of flyash/graphene filled laminated composite plates using higher order shear deformation theory. *Proceedings of the Institution of Mechanical Engineers Part C*. 236(9), 4648–4659.

Parida, S P, and Jena, P C (2022b). Selective layer-by-layer fillering and its effect on the dynamic response of laminated composite plates using higher-order theory. *Journal of Vibration and Control*. 29(11-12), 2473–2488. DOI:10775463221081180.

Rad, A A, and Panahandeh-Shahraki, D (2014). Buckling of cracked functionally graded plates under tension. *Thin-Walled Structures*. 184, 26–33.

Parida, S P, Jena, P C, and Dash, R R (2022). Dynamics of rectangular laminated composite plates with selective layer-wise fillering rested on elastic foundation using higher-order layer-wise theory. *Journal of Vibration and Control*. doi:10775463221138353.

Shen, B, and Zhang, Q L (2019). Additive manufacturing of functionally graded materials: a review. *Materials Science and Engineering: A*. 764, 138–209.Thai, H T, and Kim, S E (2015). A review of theories for the modeling and analysis of functionally graded plates and shells. *Composite Structures*. 128, 70–86.

Zhang, C, Chen, F, Huang, Z, Jia, M, Chen, G, Ye, Y, Lin, Y, Liu, W, Chen, B, Shen, Q, Zhang, L (2019) Additive manufacturing of functionally graded materials: A review. Materials Science and Engineering: A 764:138209.

36 Skin conductance based mental stress analysis using cross validation

Padmini Sethi[1,a], Ramesh K. Sahoo[2,b], Ashima Rout[3,c] and M. Mufti[4,d]

[1]Department of Psychology, RDW University, Bhubaneswar, India

[2]Department of CSEA. IGIT Sarang, Odisha, India,

[3]Department of ETCE, IGIT Sarang, Odisha, India

[4]Department of Computer Science, Nottingham University, UK

Abstract

In the modern era, mental stress is a major issue that effects the humans physically and mentally with significant psychological and physiological changes. With proper observation of stress, makes a person more productive and healthier. In the proposed work, machine learning algorithms based mental stress analyzer has been proposed. Airflow, temperature and Galvanic Skin Response (GSR) sensor has been used for collection of vital health parameters. Support vector machine (SVM), Naive Bayes, random forest and decision tree(J48) machine learning algorithms has been used in robust cross validation mode for data analysis.

Keywords: GSR sensor, machine learning, physiological data, psychological data

Introduction

Mental stress is a common problem for all in the modern era which is observed through various emotions like anger, fear, depression, anxiety, etc. Under various events/circumstances stress is generated by our body. It leads in release of sweat in the body parts such as face, sole and palm (Harker, 2013). The human body has three types of sweat glands with their respective functions. Eccrine gland is one of the types which relates to stress and emotional anxiety and predominantly present in the palm (Asahina et al., 2015). There are certain psychological and physiological activities occurring in the human body which has harmful or deadly effect on an individual mentally and physically. Thus, consistent observation of stress level of an individual and analyzing it is the need of the hour.

There is huge impact of physiological changes in a human body which can be seen through their performance, behavior, thinking ability, etc. which has a huge impact on the mental health (Glanz and Schwartz, 2008). Increase in sweating is seen due to these processes (Vijaya and Shivakumar, 2013). Physiological data should also be considered for proper analysis of an individual's stress level and its range. The skin conductance helps to get stress level of an individual which is discussed by the authors experimentally in (Navea et al., 2019). Galvanic Skin Response (GSR) has a vital role in monitoring an individual's stress with the help of its two electrodes on the two fingers. It measures the skin conductance through sweat which reflects the electrodermal response due to physiological or psychological stimulation.

GSR sensor-based technology which is measured through skin conductance (Sanchez-Comas et al., 2021; Fenz and Epstein, 1967), is in the limelight as a wearable

[a]pad.sethi@gmail.com, [b]ramesh0986@gmail.com, [c]ashimarout@gmail.com, [d]mufti.mahmud@ntu.ac.uk

DOI: 10.1201/9781003450924-36

device for the detection of stress. It is used for the early detection of mental illness due to stress levels in the human body, which will improve the quality of life. A device has been developed by Sahoo and Sethi (2015a,b) using GSR sensor, Arduino Uno microcontroller, and e-Health sensor shield to detect the mental stress of a person and is extended by in (Sahoo and Sethi, 2015a,b) by adding remote data collection by Arduino-based devices like Wi-Fi and Global System for Mobile Communication (GSM) shields. Data collection framework has been discussed in (Sethi and Sahoo, 2020). Skin conductance measured through GSR sensor can be used to estimate mental stress and also added noise due to external factors can be removed to enhance the model's accuracy (Sahoo et al., 2022). Mental stress of human body has been analyzed through different machine learning algorithm using cross validation mechanism in the proposed work.

Methodology

Data collection framework

In the proposed work, temperature sensor is used to detect body temperature in degree centigrade, and the airflow sensor is used to detect the breathing rate of a person for observation of abnormal conditions that may be high or low breathing rates. GSR sensor, placed in the middle and index finger of our plan, is used to detect the skin conductance of a person that reflects the stress level of the person. Skin conductance has been computed with the help of 5-volt constant voltage provided by GSR sensor and A minor amount of current always flows through our body that is regulated by our neural system. Whenever the stress level of a person increases, sweat gland activity of our neural system increases, that in turn increases the amount of current following through the human body. In the case of the GSR sensor, a constant 5-volt voltage has been provided; therefore, skin conductance depends on the current following through the body. Therefore, the stress level of a person is directly proportional to skin conductance.

GSR, temperature, and airflow sensors will be attached to the e-health sensor platform which is a motherboard used for collection of vital health data from the human body that will be analyzed further to detect actual health condition of human. Further, it has been attached to an Arduino Uno microcontroller and can be powered up and programmed through laptop/desktop to control various attached sensors/devices and also data can be stored for analysis. Data communication from Arduino to cloud platform for storage and analysis has been done through hybrid network platform using Wi-Fi and GSM Shield attached with Arduino uno. Initially Wi-Fi shield will look for available nearby Wi-Fi network, if available then automagically it will be connected and data will be transmitted to cloud platform otherwise, GSM shield has been used to transmit data to cloud platform using cellular network.

Working principle

e-Health sensor platform equipped with temperature, GSR and airflow sensor has been considered in the proposed for collection of vital health parameters from human body. It has been attached with Arduino Uno microcontroller along with Wi-Fi and GSM shield for data communication to cloud platform. Arduino Uno micro controller will be attached with laptop to upload Arduino sketch and also to power up the device. As per the uploaded program, various sensor will sense the body to get vital health parameters and it will be stored locally in comma separated value (CSV)

 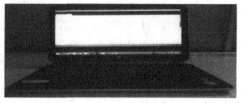

(a): Data acquisition node   (b): Observing node

Figure 36.1 Data acquisition and observing node setup [9]
Source: (Sahoo R and Sethi S, 2015b)

format in the micro SD card attached with Wi-Fi shield as reflected in Figure 36.1(a) and finally it will be uploaded to cloud platform for storage and analysis through either Wi-Fi using Wi-Fi shield or cellular network through GSM shield as per availability of network as reflected in Figure 36.1(b).

In the proposed work health data has been collected from various people in different positions like standing, sitting and laying with different moods such as normal, tension and exercise. It has been used to estimate mental stress of a person with respect to different potions and moods. Finally, it has been analyzed using various machine learning (ML) algorithms to estimate the best ML algorithm that can be used for prediction of mental stress.

Analysis through machine learning algorithms

In the proposed work, standard and well known supervised ML algorithms like Naive Bayes (Eligüzel et al., 2020), random forest (Eligüzel et al., 2020), decision tree(J48) (Eligüzel et al., 2020), and support vector machine(SVM)(Eligüzel et al., 2020) has been used to analyze the obtain dataset in Waikato Environment for Knowledge Analysis (WEKA) environment. In WEKA environment, classification can be done using ML algorithms in full training set mode, percentage split mode and cross validation mode. In case of full training set and percentage split mode, dataset will be divided into training and testing set only once to create only one model but in cross validation mode, instead of one multiples model has been created by dividing the dataset into training and testing set for multiple times in all possible ways. Therefore, cross validation mode is better and robust than other specified modes but its accuracy is less. In the proposed work, cross validation mode has been used for analysis through classification. Further, the performance of various considered ML algorithm has been evaluated through various ML evaluation parameters like true positive(TP) rate, false positive(FP) rate, precision, recall, F-measure, precision-recall curve (PRC), mean absolute error (MAE), relative absolute error(RAE), root relative squared error (RRSE), root mean squared error (RMSE), the total number of correctly classified instances, the total number of incorrectly classified instances, Kappa statistics, etc., to get the best ML algorithm for the proposed work.

Results and discussions

A comparative study through various categories on the basis of the total number of actual and predicted records has been provided in Figure 36.2. Body temp., mental stress and airflow of humans have been observed in various position like standing, sitting and laying with normal, tension and exercise mood. It is observed that predicted

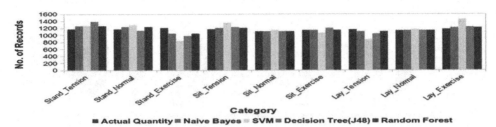

Figure 36.2 Comparative analysis of actual and predicted no. of records
Source: Author

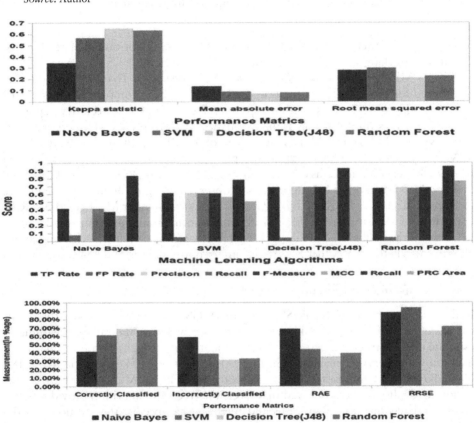

Figure 36.3 Error report of various machine learning algorithms
Source: Author

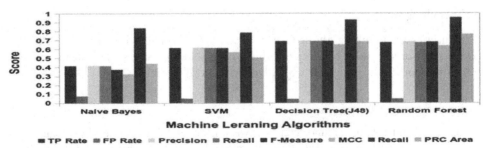

Figure 36.4 Classification report of various machine learning algorithms
Source: Author

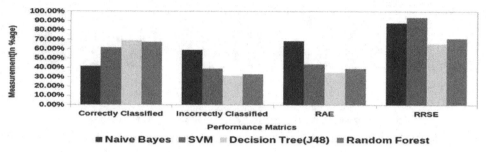

Figure 36.5 Accuracy report of various machine learning algorithms
Source: Author

record obtained after analysis through decision tree(J48) algorithm is quite closer to actual records than other ML algorithms.

Various ML algorithms like Naive Bayes, SVM, Decision Tree(J48) and random forest used for classification in the proposed work has been compared through error report in Figure 36.3, classification report in Figure 36.4 and accuracy report in Figure 36.5. It is observed that Kappa statistics of Decision Tree(J48) algorithm is better whereas MAE and RMSE is lower than other ML algorithms like Naive Bayes, SVM and random forest algorithms in Figure 3. Decision tree(J48) algorithm also provides better result in terms of precision, recall, F-measure, ROC area, PRC area, TP rate and MCC than other ML algorithms as observed in Figure 36.4. Also The FP rate of Decision Tree(J48) is lowest and TP rate is highest than other ML algorithms. RAE and RRSE of Decision Tree(J48) algorithms is also minimum as compared to other ML algorithms as observed in Figure 5. Therefore accuracy level of Decision Tree(J48 algorithm) is better than other ML algorithm used for analysis in the proposed work.

Conclusions and future scope

In the proposed work, Galvanic Skin Response (GSR), temperature and airflow sensor has been used to estimate skin conductance, body temp., and rate of breathing to estimate mental stress of a person. It has been analyzed through various standard and well-known supervised machine learning (ML) algorithms like Decision Tree(J48), Naive Bayes, random forest and support vector machine (SVM), in robust cross validation mode. It has been observed that Decision Tree(J48) algorithm provides better accuracy than other ML algorithms and further it can predict the position and the respective mental state of a person. In future, we will try to increase the accuracy of the model.

Acknowledgment

The authors are thanking about Prof. Srinivas Sethi Principal Investigator, of SERB (DST) sponsored Project to carry out this work related to Stress Analysis in wireless sensor network environment at CSEA Department, IGIT Sarang.

References

Asahina, M, Poudel, A, and Hirano, S (2015). Sweating on the palm and sole: physiological and clinical relevance. *Clinical Autonomic Research*. 25, 153–159. Https://Doi.Org/10.1007/S10286-015-0282-1.

Eligüzel, N, Çetinkaya, C, and Dereli, T (2020). Comparison of different machine learning techniques on location extraction by utilizing geo-tagged tweets: a case study. *Advanced Engineering Informatics*. 46, 101151.

Fenz, W D, and Epstein, S (1967). Gradients of a physiological arousal of experienced and novice parachutists as a function of an approaching jump. *Psychomatic Medicine*. 29, 33–51.

Glanz, K, and Schwartz, M (2008). Stress, coping, and health behavior. In Health Behavior and Health Education: Theory, Research, And Practice. pp. 211–236.

Harker, M. (2013). Psychological sweating: a systematic review focused on aetiology and cutaneous response. *Skin pharmacology and physiology*, 26(2), 92–100.

Navea, R F, Buenvenida, P J, and Cruz, C D (2019). Stress detection using Galvanic skin response: an android application. *Journal of Physics: Conference Series*. 1372, 012001. Doi:10.1088/1742-6596/1372/1/012001, Iop Publishing.

Sahoo, R.K., Prusty, A.R., Rout, A., Das, B., Sethi, P. (2022). Mental stress detection using GSR sensor data with filtering methods. In: Udgata, S.K., Sethi, S., Gao, XZ. (eds) Intelligent Systems. Lecture Notes in Networks and Systems. Singapore: Springer. 431, 537–548. https://doi.org/10.1007/978-981-19-0901-6_47

Sahoo, R., and Sethi, S. (2015a). Functional analysis of mental stress based on physiological data of GSR sensor. In Emerging ICT for Bridging the Future-Proceedings of the 49th Annual Convention of the Computer Society of India (CSI). Springer International Publishing. 1, 109–117.

Sahoo, R, and Sethi, S. (2015b). Remotely functional-analysis of mental stress based on GSR sensor physiological data in wireless environment. Information Systems Design and Intelligent Applications. New Delhi: Springer, pp. 569–577.

Sanchez-Comas, A., Synnes, K., Molina-Estren, D., Troncoso-Palacio, A., & Comas-González, Z. (2021). Correlation analysis of different measurement places of galvanic skin response in test groups facing pleasant and unpleasant stimuli. *Sensors*. 21(12), 4210.

Sethi, S, and Sahoo, R K (2020). Design of WSN in real time application of health monitoring system. Virtual and Mobile Healthcare: Breakthroughs in Research and Practice. IGI Global. pp. 643–658.

Vijaya, P. A., & Shivakumar, G. (2013). Galvanic skin response: a physiological sensor system for affective computing. *International journal of machine learning and computing*. 3(1), 31.

37 An investigation study on water quality of few lakes of Bengaluru

Amita Somya[a]

Department of Chemistry, School of Engineering, Presidency University, Bengaluru-64, India

Abstract

Current study targets to monitor the water quality of four lakes selected from east, west, north and south regions of Bengaluru viz. Ulsoor, Sankey tank, Hebbal and Bellendur lakes in Bengaluru city to assess nature of water evaluated for communal utilization, diversion and various other grounds. Various Physicochemical specifications such as pH, total dissolved solids, total suspended solids, calcium, magnesium, total hardness, chlorides, sulphates, biological oxygen demand (BOD), chemical oxygen demand (COD), sodium and potassium in addition to few heavy metal ions such as, nickel, zinc, copper, lead, chromium, iron and cadmium have been determined. It is revealed that accelerated urbanization around Bellendur lake is probably going to alter water standard which can be a danger not only to public health but also for aquatic animals and other usages.

Keywords: Bengaluru, heavy metal ions, physico-chemical parameters, lakes, water quality

Introduction

Water is one of the valuable and priceless assets gifted by nature. Lakes and reservoirs, the main source of surface water, holding approximately 90% of the world's fresh water are being utilized for industries, agriculture, fisheries, energy production, domestic and many other day to day activities. In last century, there has been an abrupt change in population without appropriate growth of their civic amenities. As a consequence, lakes and reservoirs have been utilized as sinks for the domestic and industrial discharges. Therefore, periodic assessment and monitoring of water quality (Aronoff et al., 2021; Qian et al., 2022) assists to develop management policies in order to administer surface water pollution because of increased urbanization and anthropogenic constraints on it and based on water quality statistics (Maiolo et al., 2021), it can be recommended whether the water is suitable for human and many other living species consumption. The city of Bangalore, called as "Lake City", in southern India is rapidly expanding and owing to unbridled urbanization, there is huge decrement in the number of lakes in this lake city. Moreover, most of the lakes have deteriorated because of increasing industrialization and population growth. Since lake water has been an origin of drinking and homegrown use and irrigation for country and metropolitan populace of India, it is essential to monitor it on time to time. Hence, the work under this project has been executed with a focal point to investigate the appropriateness of water for absorbing point of view by variety of living organisms and many more activities for the four lakes, Ulsoor, Sankey tank, Hebbal and Bellendur by analyzing physico-chemical parameters in addition to some heavy metals. Number of water samples were taken from these lakes and various physico-chemical parameters were analyzed in addition to some heavy metals.

[a]somya.amita@gmail.com

DOI: 10.1201/9781003450924-37

Materials and methodology

Study area

Ulsoor Lake is situated in Bangalore urban district (12.9832°N latitude and 77.6200°E longitude) of Karnataka state, India with surface elevation of 895 Meters (2936.35 Feet). Sankey tank Lake is located in Bangalore metropolitan area (13.0093°N scope and 77.5741°E longitude) with surface elevation of 929.8 Meters (3,050.5 feet). Hebbal Lake is located in Bangalore metropolitan area (13002'47.57" N scope and 77035'13.66" E longitude). The lake is arranged at 839 amsl. Bellendur Lake is located in Bangalore metropolitan area (12.9354°N scope and 77.6679°E longitude). It is a component of the Bellandur drainage system, which drains the city's southern and southern-eastern regions.

Water sampling

Three water samples from all the four lakes, Ulsoor, Sankey tank, Hebbal and Bellendur have been collected in the last week of October 2021 from surface of each lake from three different sites in 2 liter acid washed and dried cleaned plastic bottles and sent to laboratory for studies of physico-chemical parameters and few toxic heavy metals within 24 hrs, using standard methods [Bureau of Indian Standard (BIS): 10500; 2004].

Results and discussion

The results (average value) of all listed physico-chemical are depicted in Table 37.1. pH is the mathematical articulation indicating how much water is acidic or basic. pH-value for Ulsoor Lake was found to be 7.22 which is faintly alkaline, for Sankey tank 7.04 whereas for Hebbal lake 6.52 and for Bellendur lake was 6.9 which lie within the permissible limits as per BIS standard [5].

Total dissolved solids (TDS) being a significant parameter in drinking water quality norm which fosters a specific flavor to the water and at higher focus diminishes its suitability to drink that water and may cause gastro digestive aggravation with more than 500 milligram/ltr. In the current review the worth of TDS for Ulsoor lake, Sankey tank and Hebbal was found within the range of permissible limit, however, for Bellendur it was found to be exceeded the desirable limit, i.e., 601 milligram/ltr. Total suspended solids (TSS) are the waterborne molecules larger than two microns. Typically, suspended solids are the organic particles that are expelled into the water when dead animals or vegetation contaminate various water sources. Other TSS may float to the forefront or remain poised somewhere in the middle, while some silt may sink to the bottom of a body of water. The more TSS there is in a water system, the less clear it will be since TSS has an effect on water clarity. TSS were found to be in the range of 30-200 milligram/ltr which is accepted as per BIS standard.

The total hardness estimated in the order of 80 milligram/ltr to 120 milligram/ltr in all of the lakes within the permissible limit according to BIS standard. However, it is exceeded to 368 milligram/ltr in Bellendur Lake which is above the desirable limit. Bellendur Lake which is 99 milligram /ltr again, exceeded to the desirable limit. Magnesium hardness was found to exceed the permissible limits revealing 6.3 – 65 milligram/ltr.

Chloride was estimated in the range of 52 milligram/ltr to 148.6milligram/ltr, in all the lakes following the desirable limit set by BIS standard whereas sulphate was

Table 37.1: Physico-chemical parameters of collected water samples from four lakes.

Parameters	Ulsoor Lake (milligram/litr)	Sankey tank (milligram/litre)	Hebbal Lake (milligram/litre)	Bellandur Lake (milligram/litre)
pH value	7.22	7.04	6.52	6.9
Total dissolved solids	210	292	320	601
Total suspended solids	110	200	45	30
Total hardness as $CaCO_3$	76	120	80	368
Calcium	17.6	28.8	19.2	99
Magnesium	6.3	11.6	7.7	65
Chloride	52	60	130	148
Sulphate	10.9	14.6	19.2	66
BOD	20	44	16	11
COD	80	160	64	358
Sodium	15	23	23	25
Potassium	<1	<1	<1	19
Nickel	<0.01	<0.01	<0.01	<0.002
Zinc	<0.5	<0.5	<0.5	<0.1
Copper	<0.05	<0.05	<0.05	0.4
Lead	<0.01	<0.1	<0.01	0.02
Chromium	<0.05	<0.05	<0.05	<0.02
Iron	0.1	0.2	0.1	<0.3
Cadmium	<0.001	<0.001	<0.001	<0.001

Source: Author

estimated in the range of 10.9 milligram/ltr to 66 milligram/ltr, in all of the samplings point following the desirable limit set by BIS standard.

The metrics biochemical oxygen demand (BOD) and chemical oxygen demand (COD) are crucial for detecting pollution from organic wastes (Siraj et al., 2010). The amount of oxygen needed by bacteria in an aerobic environment to stabilize degradable organic matter is known as the BOD. The BOD was estimated in the range of 11 to 44 milligram/ltr. In all the lakes BOD has been found well above the permissible limits as IS standard. The higher BOD concentrations revealed the type of chemical pollution. The BOD values found in the current investigation study are higher than the CPCB limits of 5.0 milligram/ltr, which causes dissolved oxygen levels to diminish (Williamson et al., 2008).

The COD method calculates the amount of oxygen needed to chemically oxidize the majority of organic materials as well as oxidizable inorganic compounds. The COD test works well in conjunction with the BOD test to assist identify hazardous circumstances and the presence of physiologically resistant organic compounds. COD values in all these lakes have been found in the range of 64-358 milligram/ltr which is well above the permissible limits of IS standard. Higher values of COD are associated with increased anthropogenic pressure on lakes and it is evident from the results that COD values of both the lakes were very high, an indication of flooded organic matter.

Sodium aids in controlling the body's electrolyte balance. For neurons and muscles to continue to operate properly, sodium is a crucial supply. Another crucial mineral that your body needs in order to function properly is potassium. Sodium and potassium have been found within the permissible limits in all the lakes. In Ulsoor, Sankey tank and Hebbal lakes, the predominant cation trend follows the order Ca>Na>Mg>K whereas it is Ca>Mg>Na>K in Bellendur lake with calcium being dominant cation in all. It is revealed that all lakes have showed higher calcium content.

Nickel, zinc, chromium, iron and cadmium has been found well below the permissible limits in all the lakes which indicates the suitability of all lakes water for consumption by human and aquatic species as exposure of these toxic metal pollutants through water systems is the most detritus cause for decay of water animals and bad impact on other living organisms depending on water.

Copper has been found within the permissible limits in Ulsoor, Sankay tank and Hebbal lakes. However, it is found 0.4 milligram/ltr in Bellandur lake which is beyond the permissible limits. Dietary and water are the main sources of copper consumption in various living species. Recent studies have defined the threshold for the effects of copper in drinking water on the digestive system, but there is still some uncertainty regarding the long-term effects of copper on populations that are susceptible, such as carriers of the Wilson disease gene and other metabolic disorders related to copper homeostasis [World Health Organization ,2017].

And, Lead has been found within the permissible limits in Ulsoor, Sankay tank and Hebbal lakes. However, it is found 0.02 milligram/ltr in Bellandur lake which is beyond the permissible limits. In many nations around the world, lead exposure is a cause for concern since it is a cumulative metal that is exceedingly damaging to both human beings and the environment. Headache, irritability, and stomach pain are signs of Pb toxicity. Pb levels observed at greater levels have the potential to harm the brain and kidneys.

Conclusions

Physico-chemical parameters analyzed to survey the reasonableness of Ulsoor Lake, Sankey Tank, Hebbal and Bellendur lake water for human consumption and other purposes, revealed that Bellendur lake water is unfit for human consumption owing to presence of heavy metals like copper, lead and high calcium, magnesium content. Infect, this water is not appropriate for irrigation purposes as well owing to high metal contents. Moreover, higher COD and BOD values make it unfit for aquatic animals too. Hence, immediate restoration management becomes essential in order to protect this lake. Ulsoor, Sankey tank and Hebbal lakes have been found to have higher values of COD and BOD which in turn, reduce the dissolved oxygen levels leading to anaerobic conditions which is detrimental to aquatic animals and shows poor water quality.

Acknowledgements

The author is thankful to the Chancellor, Vice-Chancellor, Dean (SOE), Head (Civil Engineering) Presidency University, Bengaluru for providing research facilities. Special thanks are due to Athulya Ajaya Kumar, Sharanya Suresh, Vaishnavi S Hegde, Somu Biswas, Burla Sanjana, Lavish Bhushan Shah, Shaik Gundluru Mohammed Sayeed Hussain, Harshitha G S, Santrina Eunice J, Bandi Tejaswini, students of BBA I[st] Year, School of Management &Akarsh D V, Hemanth Kumar, and Lakshmikanth

D, students of B.Tech (final year), Civil Engineering, Presidency University for helping in this work.

References

Aronoff, R, Dussuet, A, Erismann, R, Erismann, S, Patiny, L, and Vivar-Rios, C (2021). Participatory research to monitor lake water pollution. *Ecological Solutions and Evidence.* 2, e12094.

Bureau of Indian Standard (BIS): 10500; 2004 Specification for drinking water, Indian Standard Institution, (Bureau of Indian Standard), New Delhi.

Maiolo, M, and Pantusa, D (2021). Multivariate analysis of water quality data for drinking water supply systems. *Water.* 13(13), 1766.

Qian, J, Liu, H, Qian, L, Bauer, J, Xue, X, Yu, G, He, Q, Zhou, Q, Bi, Y, and Norra, S (2022). Water quality monitoring and assessment based on cruise monitoring, remote sensing, and deep learning: a case study of qingcaosha reservoir. *Frontiers in Environmental Science.* 10, 1–13.

Siraj, S, Yousuf, A R, Bhat, F A, and Parveen, M (2010). The ecology of macrozoobenthos in shallabugh wetland of Kashmir Himalaya, India. *Journal of Ecology and the Natural Environment.* 2(5), 84–91.

Standard methods for the examination of water and waste water, American Public Health Association, American water works Association, Water Environment Federation, Washington, D.C. 1999.

Williamson, C E, Dodds, W, Kratz, T K, and Palmer, M A (2008). Lakes and streams as sentinels of environmental change in terrestrial and atmospheric processes. *Frontiers in Ecology and the Environment.* 6(5), 247–54.

World Health Organization (2017) WHO guidelines for drinking water quality: First addendum to the fourth edition. *American Water Works Association.* 109, 44–51.

38 Design and implementation of a cost-effective smart dustbin for segregating wet and dry waste

Dwarikanath Choudhury[1,a], Ranjit Kumar Behera[1,b] and Dayal Kumar Behera[2,c]

[1]Computer Science and Engineering, Silicon Institute of Technology, Bhubaneswar, India

[2]School of Computer Engineering, KIIT, Deemed to be University, Bhubaneswar, India

Abstract

On these modern days, segregation of wet and dry dust is very important. Although people understand the difference between dry and wet waste, they seem too ignorant while putting them in a bin which makes the life difficult for the garbage collector. Hence a special dustbin is conceptualized and designed which can automatically differentiate between wet and dry waste. When someone throws the waste, the system automatically segregates wet and dry dust and put them in appropriate bins. Once someone throws the waste, it come to a platform where there are three sensors, 1st one for sensing the waste, second one for measuring the quantity of dusts inside each bin and the third one for differentiate the wet and dry dust. After sensing these three parameters, the conveyor tilts on one direction to put the waste on right dustbin. These sensing data sends notification to smart phone and updates it to cloud platform as well as in a database.

Keywords: Dry waste, IR sensor, MCU, segregation, smart dustbin, wet waste

Introduction

The production of garbage likewise rises in tandem with urbanization. The garbage cans scattered throughout are overflowing with the rubbish created daily as a result of faulty waste management and negligence by human beings. The everyday process of determining which garbage is full or empty is not simple. This paper's main goals are to keep the environment tidy and smart by keeping an eye on the trash can. Although many initiatives were taken by different municipality bodies at different cities, the major problem is that the bins create a lot of bad and foul smell and makes the city dirtier to live.

Garbage has been a major problem in most cities. Although many steps are being taken by the municipal corporations to maintain the order but due to poor management system, scenes are degrading day by day. Although many articles have already published regarding the managing the waste in cities, but most of those published works are only for managing the waste in dustbins and send notifications. For example, Rajapandian et al. (2019) and Patel et al. (2019) have built a model using Arduino and ultrasonic sensors to measure the quantity of bins and send notification to the cities' municipality system, where workers and managers of municipality can see the status and act accordingly. Authors such as Chaudhary et al. (2019), Bhatt et al. (2019) and Joshi et al. (2019) have improvised these concepts by introducing GPS tracking system for vehicles which collects the waste along the cities. Even few researchers went ahead with many advanced Internet of Thing (IoT) mechanism for building concepts for overall city management as well to provide good experience to

[a]dncdwarika@gmail.com, [b]ranjit.behera@gmail.com, [c]dayalbehera@gmail.com

DOI: 10.1201/9781003450924-38

the user without their direct monitoring, by sharing the information of many stand-alone IoT applications called cross-vertical IoT application for smart cities Roy et al. (2018), Behera et al. (2019) and Behera et al. (2022). But none has considered the fact that the primary problem lies in the proper way of waste collection. As most of the time people mix the wet dust and dry dust while throwing it into a bin, it needs to be separated out for efficient management of the dust, which indeed will definitely manage the bad smell too.

Hence this paper aims to design of a dustbin that can differentiate between dry and wet waste and allows the bins to be filled by wastes according to the amount of space present so that the wastes don't overspill. This article offers a useful strategy for achieving that end.

The rest of the paper is structured as follows. Section-2 describes the proposed model. Section-3 explains the methodology in detail with the help of a flow chart. Section-4 explains the experimental setup which was done using ESP-32 microcontroller and other components. Secttion-5 discusses the results finding and Section-6 summarizes the paper.

Model description

Block diagram of the model is shown in Figure 38.1. First connect the two ultrasonic sensors to Esp32 VCC pin of both ultrasonic sensors go to the 3.3 vpin of Esp32. Then the GND pin to GND of Esp32. Trigger pin of the first ultrasonic Sensor will go to GPIO 4 of Esp32 which give ultrasonic sensor a 10-millisecond trigger pulse for creating a high frequency ultrasonic sound wave. Then the echo pin will go to GPIO 5 which gives a pulse according to the distance between the sensor and object to MCU (Esp32). After getting the signal from ultrasonic sensor the MCU calculates the distance between the sensor and object by using the ALU unit of MCU which pre-programed by the user what kind of formulas use for extracting the data from that pulse. Then the second ultrasonic sensor's trigger pin go to GPIO 19 of Esp32 for the trigger pulse. Then the echo pin of the second ultrasonic sensor go to GPIO 18 of

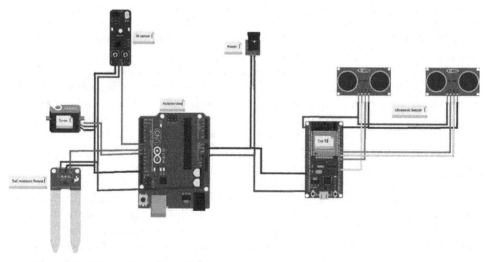

Figure 38.1 Block diagram of the model
Source: Author

Esp32 then the ALU unit of MCU which pre-programed calculate the distance just like the first ultrasonic sensor.

In our model, ultrasonic sensors are used to recognize how much percentage of dustbin is full and this raw data it sends to MCU. Then MCU converts the raw distance data to percentage data which indicates the dustbin's full or empty status. As this MCU is WIFI enabled, it can send data to cloud data base if user want to observe the same status. To differentiate between wet and dry an Arduino Uno, IR sensor, soil moisture sensor and an actuator also known as digital Servomotor(mg995r) are used. As the MCU is connected to Internet, when the dust bin is full it can send a notification to the smartphone via Telegram of the user. There is an actuator which consists of three wires, one is VCC which goes to 5vpin Arduino and GND pin to GND of Arduino, another is signal which take PWM signal, and it is connected to D9 of Arduino. In this product IR sensor uses for sensing the presence of dust and the soil moisture use for sensing the wet dust after sensing. When IR sensor Sens presence of dust it will give a digital signal to MCU. then MCU take input from soil moisture sensor and according to the program, MCU move servo on one direction and throw the dust from conveyor to the actual dustbin.

Methodology

The ultrasonic sensors' purpose is to recognize dustbin status, whether it is full or empty. There is a microcontroller Esp32 which reads data from the ultrasonics sensor by applying some logic, converts to percentage according to this data and sends a notification to the smart phone via telegram. The most important feature of this dustbin is to perform segregation between wet and dry waste and it is done by using two sensors, first is IR sensor and the second is moisture sensor. When we throw some dust into bin, first IR Senses that: and moisture sensor detects waste is dry or wet. If waste is wet, then due to water resistance some current flows. Measuring current flow through the MCU according to the calibration threshold point, then it throws the dust in wet or dry side.

Experimental setups

The setup comprises of Arduino board, two ultrasonic sensors, one infrared sensor, one moisture sensor, old screen, two playback modules, ESP-32 microcontroller and servo motor. The ultrasonic sensor helps in determining the approx. distance present inside the bins after waste are put in it. It is directly connected to ESP32. The IR sensor helps in determination of object kept on the plate. The ESP-32 is the low powered micro-controller used in the circuit.

Results and discussion

Two concurrent tests were done, and the result is shown in Table 38.1. The first one was related to distinguishing of wastes according to dryness. Ten objects were taken. Objects with odd serial numbers are wet and objects with even serial numbers are dry. The voltage readings from the IR and moisture sensors were recorded. It was observed that all dry waste was collected in the dry bin and all wet waste were collected in wet bin. It is clear from the experiment that when we have taken only dry waste, the reading of soil sensor is 0Volt whereas, when we have taken wet dust, both the soil moisture sensor and IR sensor gives a positive voltage reading. We have

Table 38.1: Data captured by different sensors.

Sl no	Arduino as MCU			ESP 32 as MCU				
		Dry dustbin		Wet dustbin				
	Dry status	Wet status	Soil moisture	IR	%full	Dist. (CM)	%full	Dist. (CM)
1	no	yes	4.8V	4.5V	0	22	0	22
2	yes	no	0V	5.1V	10	20.2	10	20.2
3	no	yes	4.9V	5.3V	20	18.4	20	18.4
4	yes	no	0V	4.8V	30	16.5	30	16.5
5	no	yes	5.1V	4.6V	40	14.3	40	14.3
6	yes	no	0V	5.4V	50	12.3	50	12.3
7	no	yes	4.5V	5.2V	60	10.1	60	10.1
8	yes	no	0V	4.7V	70	8.6	70	8.6
9	no	yes	5.5V	5.0V	80	6.4	80	6.4
10	yes	no	0V	4.9V	90	4.2	90	4.2

Source: Author

considered only 10 readings out of which five are for wet and five are for dry and the average voltage reading of soil moisture sensor and IR sensor is found to be 4.96 and 4.95 volts respectively. The second experiment was about the space management systems of both the bins. Simultaneously one dry and one wet object were put on the bins ten times. It was observed that at the beginning the space was 0% full on both the bins. It is clearly found that while distance is decreasing, the % of full is increasing, which shows the efficacy of our system.

Conclusion

As collection of garbage smartly by segregating dry and wet dust, this research work proposed a cost effective and user-friendly Smart Dustbin Monitoring system. The efficiency comes from the fact that real-time monitoring of waste management enables prompt rubbish collection and stops any hazardous waste from endangering the environment. This system will really be helpful to maintain a clean and smell free society particularly in urban areas. The effectiveness of the system increased manifold by the use of an advanced controller in the form of an Arduino combined with a GSM and GPS enabled system. The proposed model makes sure that dry and wet waste management is carried out efficiently, which can play a significant part in achieving a clean and green environment.

References

Behera, R K, Patro, A, and Roy, D S (2022). A resource-aware load balancing strategy for real-time, cross-vertical IoT applications. In Biologically Inspired Techniques in Many Criteria Decision Making: Proceedings of BITMDM 2021. Springer Nature, Singapore. pp. 15–27.

Behera, R K, Reddy, K H K, and Roy, D S (2019). A novel context migration model for fog-enabled cross-vertical IoT applications. In International Conference on Innovative Computing and Communications: Proceedings of ICICC 2019, (Vol. 2). Springer Singapore. pp. 287–295.

Bhatt, M C, Sharma, D, and Chauhan, A (2019). Smart dustbin for efficient waste management. *International Research Journal of Engineering and Technology*. 6(07), 967–969.

Chaudhary, V, Kumar, R, Rajput, A, Singh, M, and Singh, T (2019). Smart dustbin. *Research Journal of Engineering and Technology*. 6(05), 7647–7651.

Joshi, S, Singh, U K, and Yadav, S (2019). Smart dustbin using GPS tracking. *International Research Journal of Engineering and Technology (IRJET)*. 6(06), 165–170.

Patel, D, Kulkarni, A, Udar, H, and Sharma, S (2019). Smart dustbins for smart cities. *International Journal of Trend in Scientific Research and Development*. 3, 1828–1831.

Rajapandian, B, Madhanamohan, K, Tamilselvi, T, and Prithiga, R (2019). Smart dustbin. *International Journal of Engineering and Advanced Technology (IJEAT)*. 8(6), 4790–4795.

Roy, D S, Behera, R K, Reddy, K H K, and Buyya, R (2018). A context-aware fog enabled scheme for real-time cross-vertical IoT applications. *IEEE Internet of Things Journal*. 6(2), 2400–2412.

39 A study on abnormal medical image classification

Khirod Kumar Ghadai[1,a], Subrat Kumar Nayak[1,b], Biswa Ranjan Senapati[1,c] and Binaya Kumar Patra[2,d]

[1]Department of Computer Science and Engineering, Siksha 'O' Anusandhan (Deemed to be) University, Bhubaneswar India

[2]Department of Computer Science Engineering and Applications, Indira Gandhi Institute of Technology (IGIT), Sarang, Dhenkanal, India

Abstract

Image processing with artificial neural networks (ANN) has been used in several fields, including civil engineering, mechanics, industrial surveillance, the defense department, health care solution, the information technology industry, and transportation. Medical data can be categorized, and abnormalities can be identified using ANN. The learning methods have a significant impact on classification performance in ANN training. In the last few years, ANN techniques are being used widely in different medical areas in which the factor of time is very crucial to diagnose the disease in the patient at an early stage, ANN is an important machine learning tool that can handle a variety of real-world challenges. Image classification is a common approach and popular method that is used to classify aberrant medical images because of their various benefits and significant advantage artificial neural networks are frequently used in image processing. Although, ANN has several advantages, one of its most significant drawbacks is the convergence difficulty. A high rate of convergence is a factor that evaluates how practical the system is for solving computational tasks. The aim of this paper is to analyze medical image classification using artificial neural networks and to present a comparative analysis with other methodologies. While doing the comparison, we have considered different parameters and it has been observed that the Kohonen-Hopfield neural network (KHNN) model provides higher forecasts of classification accuracy and convergence rate than conventional neural networks.

Keywords: Artificial neural networks, convergence rate, Image classification, medical data classification

Introduction

Artificial neural networks (ANN) for image analysis is being used more and more in the healthcare sector. The classification of health picture records is one of its key applications. It helps divide aberrant images into several groups based on some type of similarity metric. Conventional methods rely on human observation, which has a high error rate. Due to this, automated methods that can accurately classify aberrant photos are required. The automated system's convergence rate must be realistically achievable in addition to having high accuracy. Several automatic techniques for classifying images have been developed based on these criteria. Due to their many benefits, the majority of them rely on artificial intelligence (AI) technology. In terms of performance metrics, ANN-based classification techniques outperform other techniques significantly. The extensive use of ANN approaches for

[a]khirod1990@gmail.com, [b]subratnayak@soa.ac.in, [c]biswaranjansenapati@soa.ac.in, [d]binaya.patra@gmail.com

DOI: 10.1201/9781003450924-39

applications such as brain image categorization is revealed by a literature review. This study's primary goal is to evaluate various ANN-based classifier models and does a comparative analysis for a better image classification outcome with less convergence rate and more accuracy. In the first review, we analyzed and observed that the modified counter propagation neural network (MCPN) and modified Kohonen neural network (MKNN) model is presented for the brain image classification and showcased the best in improved convergence rate and accurate results (Hemanth et al., 2014).

González and Dorronsoro (2008) describes the multilayer perceptron's gradient training approach. This study supports the artificial neural network's superiority over more traditional classifiers like linear discriminant classifiers. Brain image categorization using a neural network based on bidirectional associative memory (BAM) is applied (Sharma et al., 2008). Among aberrant brain pictures, pattern classification using feed-forward neural networks is used in (Maitra and Chatterjee, 2006). According to Zhang et al. (2010), the metaheuristic approach particle swarm optimization (PSO) has been used to optimize the magnetic resonance brain image classification. Song et al. (2007), a modified probabilistic neural network for the analysis of brain images using a well-defined maximum-margin training technique is presented and in Ludwig and Nunes (2010) supervised neural network performs better and the network model is tested. Although back propagation neural networks (BPN) are also frequently employed for classification, their applicability in real-world applications is constrained by the need for a fast convergence rate which is described in (Chen et al., 2004). As a result, the traditional neural network-based classification system has several drawbacks. Researchers have also modified traditional neural networks to improve the effectiveness of automatic classification. In Yu et al. (2006) it is described as to increase the multilayer perceptron (MLP) network's convergence rate. The author trained the hidden layer neurons using a state-of-the-art technique i.e., changes are made to the training process. Muezzinoglu and Zurada (2006) uses neural networks with radial basis functions to classify images. In this work, nearest-neighbor classification concepts are also applied. Alsultanny and Aqel (2003) have presented a hybrid method for pattern recognition using MLP and GA to enhance the neural network's performance accuracy and convergence speed, GA is used to optimize the parameters of the neural networks. According to earlier research, automated systems' performance can be improved by using stochastic algorithms like a genetic algorithm (GA) and metaheuristics like PSO. Kishore et al. (2000) studied the effectiveness of genetic programming (GP) for multi-category pattern classification is investigated. Here in review, we had gone through several traditional classifier neural network models and also gone through some of the automated hybrid classifier network models and presented the details of the best suitable classifier model.

The article is set up in a systematic sequential order as follows: the first section contains the introduction and a detail of the literature review, and the second section includes the neural network architecture and image classification procedure. Section 3 presents different classifier models w.r.t performance metrics for classification accuracy outcomes and the convergence rate of the classifier models. The outcomes conclusion and future aspects are included in Section 4 as per the review. Furthermore, emphasis can be given to the computational complexity of the system and its utilization established on different optimization algorithms and their execution time.

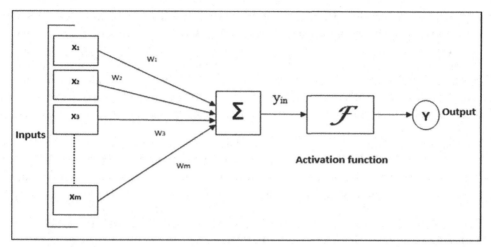

Figure 39.1 Artificial neural network model diagram
Source: Author

Methodology

ANN and its design

Figure 39.1 depicts a straightforward model of a synthetic neuron.

The artificial neuron's inputs are represented by X1, X2, X3, and Xm, and its weights are represented by W1, W2, W3, and Wm. Nodes receive all the inputs, sum them and the activation function is applied to the weighted sum and produces output. 'Y$_{in}$' is the output of the input layer and 'b' stands for bias.

$$Y_{in} = X_1W_1 + X_2W_2 + X_3W_3 \ldots X_mW_m + b \tag{1}$$

It is now necessary to send this sum over an activation function to retrieve the output, which will serve as an input for the following layer. It has been observed that many authors used the sigmoid activation function for better artificial neural network performance.

Mathematically, the sigmoid activation function is presented below:

$$f(x) = \frac{1}{1 + e^{-x}} \tag{2}$$

To train the ANN for medical data classification tasks, a variety of metaheuristics were investigated as learning methods such as artificial bee colony (ABC), ant lion optimizer, and equilibrium optimizer (EO). Backpropagation and other traditional neural network techniques are also commonly used to train ANN. However, the traditional approach has some drawbacks. These methods like Levenberg-Marquardt (LM) model tend to stay locally optimal and are so dependent on initialization that's very difficult for the prediction of the improved results of the new Input pattern (Si et al., 2022). The basic steps of image processing are briefly discussed below.

- **Image database:** It is nothing but the collection of real-time medical image data which is collected from different sources like scan centers or diagnostic centers

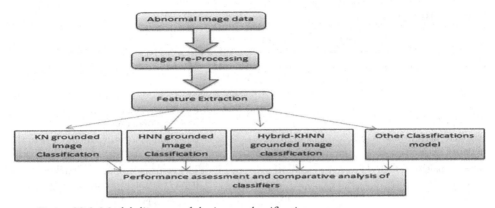

Figure 39.2 Model diagram of the image classification process
Source: (Anitha et al.,2014)

with different type of class which is available in different formats like (TIFF), GIF, and PNG.

- **Noise removal:** The method used to eradicate or reduce noise from an image is called a noise removal algorithm. The most common types of image noise are Impulse noise, additive noise, and multiplying noise.
- **Segmentation:** It involves dividing images into several pieces or objects. There are three kinds of image segmentation techniques: Manual segmentation, semi-automated segmentation and fully automatic segmentation. The fully automatic segmentation techniques do not involve human interaction and are based on supervised learning techniques that require training data.
- **Image pre-processing:** Any medical imaging application must include the steps of feature extraction and image pre-processing. These stages simply help in improving the final results' accuracy. This process reduces the noise from the raw input data or images because the raw images are not suitable for classification. Mostly Gabor filters are used for medical image filtering which is proved in (Si et al., 2022). The 2D Gabor filter's filtering equation is given by

$$g_{\lambda,\theta,\gamma,\sigma}(x,y) = exp\left(-\frac{x'^2 + \gamma^2 y'^2}{2\sigma^2}\right)\cos\left(2\pi\frac{x'}{\lambda}\right)$$

$$(x' = x\cos\theta + y\sin\theta)$$
$$(y' = -x\sin\theta + y\cos\theta)$$

Here λ represents wavelength, θ represents orientation in degrees, γ represents aspect ratio, and $\sigma = 0.56\ \lambda$

Automated image classification process

Figure 39.2 block diagram represents the automated image classification process using a different model (Alsultanny et al., 2003).

Feature extraction: Feature extraction seeks to reduce the size of the vast input data set by locating distinctive qualities, or features, that distinguish one input pattern from another. The extracted feature aims to give the classifier characteristics of the input type by taking into account the description of the relevant features of the image in a feature space (Hemanth et al., 2017).

Comparative analysis

Table 39.1: Performance metrics of different classifiers

Classifier/techniques	Accuracy percentage	Convergence rate (CPU seconds)	Model
Kohonen neural network (KNN)	73	550	Inputs are clustered together to get fired o/p neuron
Hopfield neural network (HNN)	85	920	Based on associative memory neural network
Backpropagation neural network (BPN)	93	1210	The error is propagated backward from o/p units to hidden units
Linear vector quantization network	89	990	Uses supervised learning
Hybrid Kohonen-Hopfield network (KHNN)	95	985	Comprises the mathematical measures of both KN and HNN
Support vector machine with the linear kernel (SVM)	92	--	Supervised learning models with associated learning algorithms
Radial basis function network (RBF)	92	--	The activation function used is a Gaussian function
Probabilistic neural network	75	2400	Feed-forward neural network
Multilayer perceptron	93.41	--	Fully connected class of feed-forward ANN
Feed-forward neural network with the feature search technique	80	11 hrs	Information moves in only one direction
Minimum distance classifier	64	540	Used to classify unknown image data into classes
Bayesian approach	72	19,120	Knowledge of probability models

(EI Emary and Ramakrishnan, 2008; Selvaraj et al., 2007; Georgiadis et al. 2008; Hemanth et al., 2014; Anitha et al., 2012)

Conclusion

For the early detection of an illness, it is crucial to classify medical images accurately for abnormal ones with low operating costs and computing complexity. For the classification of medical images, numerous writers have put forth various methods. A thorough analysis of articles utilizing various neural network classifier models is presented in this article. It is observed that the hybrid Kohonen-Hopfield neural network (KHNN) model provides good prediction accuracy with a short convergence time. Studies employing hybrid-KHNN models produce superior outcomes to those using classical/traditional neural network models. This is suitable for practical applications due to its many advantages over the traditional approach.

References

Alsultanny, Y A, and Aqel, M M (2003). Pattern recognition using multilayer neural-genetic algorithm. *Neurocomputing*. 51, 237–247.

Anitha, J, Vijila, K S, Selvakumar, I A, and Hemanth, J D (2012). Performance- enhanced PSO-based modified Kohonen neural network for retinal image classification. *Journal of the Chinese Institute of Engineers*. 35(8), 979–991.

Chen, Y J, Huang, T C, and Hwang, R C (2004). An effective learning of neural network by using RFBP learning algorithm. *Information Sciences*. 167(1–4), 77–86.

El Emary, I M, and Ramakrishnan, S (2008). On the application of various probabilistic neural networks in solving different pattern classification problems. *World Applied Sciences Journal*. 4(6), 772–780.

Georgiadis, P, Cavouras, D, Kalatzis, I, Daskalakis, A, Kagadis, G C, Sifaki, K, and Solomou, E (2008). Improving brain tumor characterization on MRI by probabilistic neural networks and non-linear transformation of textural features. *Computer Methods and Programs in Biomedicine*. 89(1), 24–32.

González, A, and Dorronsoro, J R (2008). Natural conjugate gradient training of multilayer perceptrons, *Neurocomputing*. 71(13–15), 2499–2506.

Hemanth, D J, Vijila, C K S, Selvakumar, A I, and Anitha, J (2014). Performance improved iteration-free artificial neural networks for abnormal magnetic resonance brain image classification. *Neurocomputing*. 130, 98–107.

Hemanth, J D, Anitha, J, and Ane, B K (2017). Fusion of artificial neural networks for learning capability enhancement: application to medical image classification. *Expert Systems*. 34(6), e12225.

Kishore, J K, Patnaik, L M, Mani, V, and Agrawal, V K (2000). Application of genetic programming for multicategory pattern classification. *IEEE Transactions on Evolutionary Computation*. 4(3), 242–258.

Ludwig, O, and Nunes, U (2010). Novel maximum-margin training algorithms for supervised neural networks. *IEEE Transactions on Neural Networks*. 21(6), 972–984.

Maitra, M, and Chatterjee, A (2006). A Slantlet transform-based intelligent system for magnetic resonance brain image classification. *Biomedical Signal Processing and Control*. 1(4), 299–306.

Muezzinoglu, M K, and Zurada, J M (2006). RBF-based neurodynamic nearest neighbor classification in real pattern space. *Pattern Recognition*. 39(5), 747–760.

Selvaraj, H, Selvi, S T, Selvathi, D, and Gewali, L (2007). Brain MRI slices classification using least squares support vector machine. *International Journal of Intelligent Computing in Medical Sciences and Image Processing*. 1(1), 21–33.

Sharma, N, Ray, A K, Sharma, S, Shukla, K K, Pradhan, S, and Aggarwal, L M (2008). Segmentation and classification of medical images using texture-primitive features: application of BAM-type artificial neural network. *Journal of Medical Physics/Association of Medical Physicists of India*. 33(3), 119.

Si, T, Bagchi, J, and Miranda, P B (2022). Artificial neural network training using metaheuristics for medical data classification: an experimental study. *Expert Systems with Applications*. 193, 116423.

Song, T, Jamshidi, M M, Lee, R R, and Huang, M (2007). A modified probabilistic neural network for partial volume segmentation in brain MR image. *IEEE Transactions on Neural Networks*. 18(5), 1424–1432.

Yu, C, Manry, M T, Li, J, and Narasimha, P L (2006). An efficient hidden layer training method for the multilayer perceptron. *Neurocomputing*. 70(1-3), 525–535.

Zhang, Y D, Wang, S, and Wu, L (2010). A novel method for magnetic resonance brain image classification based on adaptive chaotic PSO. *Progress in Electromagnetics Research*. 109, 325–343.

40 Studies on direct drop mode of atomization in a spinning cup

Brahmotri Sahoo[1,a], Kshetramohan Sahoo[2,b] and Chandradhwaj Nayak[1,c]

[1]Department of Chemical Engineering, Indira Gandhi Institute of Technology, Sarang, Dhenkanal, Odisha, India

[2]Department of Chemical Engineering Indian Institute of Science, Bangalore, India

Abstract

A rotating cylindrical cup is a simple spray forming device. Drops are formed at the lip of the cup without formation of intermediate ligament or sheet in direct drop mode, which renders it promising for use in processes in need of controlled drop formation. A spinning cup in direct drop mode can be a scalable atomizing device of high value in pharmaceutical and food processing in which a large number of suspensions and colloids which are Non-Newtonian need to be atomized. In the present work we have used a spinning cup with variable speed (100 to 450 radian per second) to atomize three base liquids, viz. water, ethyl alcohol, and 40% glycerol, which are Newtonian. Blur free images of the flying drops at high resolution (24 MP) were captured using Macro lens equipped DSLR camera and exposure controlled flashlight. The drop size distribution at specific speed, Sauter mean diameter (SMD) vs. cup speed, and dependence of mean drop size on Reynolds number (Re) and Weber number (We) are presented. The findings are consistent with basic physical understanding of drop formation under the influence of centrifugal force.

Keywords: Direct drop mode, spinning cup

Introduction

The sprays of fine droplets with high specific surface also known as atomized liquid can intensify heat/mass transport in chemical processes. Unlike pressure atomizers which need high velocity jet formation by pressurization, the rotary atomizers rely on centrifugal force to atomize bulk liquid. Low cost, simplicity of installation and operational flexibility are salient attractive features of this class of atomizers. A cup or a disc (Dombrowski and Llloyd, 1974; Ahmed and Youssef, 2014; Sahoo and Kumar, 2021) or a wheel (Bizjan et al., 2014) can be easily converted to a Rotary atomizer. Spinning cups have been studied for use in agricultural spray (Craig et al. 2014), drying (Huang and Mazumdar, 2008), encapsulation (Saifulla et al., 2019), and nano particle synthesis (Nandiyanto and Okuyama, 2011).

Liquid fed to the center of the spinning cup rises is driven towards the cup wall by the action of centrifugal force, and the centrifugal force based pressure gradient active on the film on the cup inner wall causes the film flow upto the lip region. Drop formation can take place in three modes; viz. sheet breakup mode, ligament breakup mode, and direct droplet mode from the outer end of the lip. At sufficiently high flow rates the liquid film on the cup wall extends beyond the lip and the unsupported film becomes unstable to turn into fragments or threads before eventual appearance as droplets. Kevin-Helmholtz instability caused by relative velocity between two fluids (the liquid sheet and the surrounding air) explains this instability and this mode of

[a]brahmotri.s@gmail.com, [b]kshetramohansahoo@gmail.com, [c]chandradhwaj@gmail.com

DOI: 10.1201/9781003450924-40

atomization known as sheet breakup mode is expected to produce more poly disperse drops. However, for industrial applications demanding large scale processing it can be useful. At moderate flow rates the film is limited to the exit of the lip, and a number of spiraling long threads called ligaments become drop generators. On further reduction of flow drop formation takes place directly near the disc edge and this mode is known as direct droplet mode.

We are interested to use a spinning cup for a drop forming device which will be useful in several pharmaceutical processes and segments in food processing. The liquids to be handled will be suspensions or colloids which fall in Non-Newtonian or complex fluid category. As drop formation takes place just at the edge of the lip in direct drop mode, the configuration of a cup can be utilized to control drop size distribution. However, studies on direct droplet mode of atomization using a spinning cup are limited in literature even for a Newtonian fluid. Therefore, the objective of this work is to understand the atomization characteristics of some Newtonian fluids, which are used as continuous phases in the aforesaid applications.

A brief account of earlier attempts made for control of droplet characteristics predominantly in sheet breakup and ligament mode are presented next.

Literature review

The detachment of drop at the exit of the lip in direct droplet mode shown schematically through Figure 1 can be explained as a consequence of competition between the outward centrifugal force and the inward pull due to surface tension both acting on a drop on the verge of detachment.

The properties of liquid which are of interest in atomization are; density (ρ), surface tension (σ), and viscosity (μ). The outward centrifugal force (F_C) acting on the drop of diameter d formed at the lip of the cup of diameter D rotating at angular speed ω

$$F_C = \frac{\pi}{6}d^3\rho\frac{D}{2}\omega^2 \tag{1}$$

The inward pull due to surface forces is:

$$F_S = \sigma d \tag{2}$$

Equating equations (1) and (2) we get:

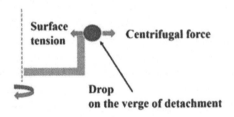

Free body diagram of a detaching drop

Figure 40.1 Free body diagram showing action of competing forces during drop detachment from a spinning cup

Source: Author

$$d = \sqrt{\frac{12\sigma}{\pi\rho\omega^2}} \qquad (3)$$

The equation can be written in terms of dimensionless drop size ($d^* = d/D$) and Weber number ($We = \frac{\rho\omega^2D^3}{\sigma}$) as:

$$d^* = cWe^{-0.5} \qquad (4)$$

Eisenklam,1972 has showed that the average inter ligament spacing coincides with the wavelength of Rayleigh-Taylor instability. Liu et al. (2012) investigated drop formation in direct and ligament mode in a spinning cup. They reported a decrease of drop size with increase in cup speed and gave a more detailed explanation for drop size in ligament mode. Ahmed and Youssef (2014) used cylindrical cups of different height and flat bottom conical cups with different cone angles to study the influence of cup configuration on drop characteristics in ligament breakup mode. Their experiments suggest that geometry related parameters such as cup height, base diameter, and cone angle have negligible influence on droplet size and droplet velocity.

Experimental

Figure 40.2 shows the schematic of the spinning cup which measures 35 mm in diameter. The cup wall is 10 mm and the liquid fed to the center of the cup would move radially outward, run through the inner wall and will finally manifest as detached droplet at the lip of the cup.

   Figure 40.3 schematically brings up the experimental facility used for the study. The liquid is supplied through a peristaltic pump to the center of the cup. The cup, rotated by a variable speed DC motor is imaged by a DSLR camera (*Nikon D7100*) fitted with a Macro lens to capture fine drops. The flashlight with controllable exposure time helps to obtain high resolution images of moving droplet without blur. The images are analyzed using *image j*.

   The viscosity of the liquids was measured using viscometer (*Brookfield, Amtec*). The density measurement was carried out using metered cuvette and weighing balance. The surface tension measurement was carried out using standard drop weight method.

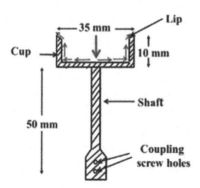

Spinning Cup Atomizer

Figure 40.2 A schematic of the spinning cup atomizer used in this work
Source: Author

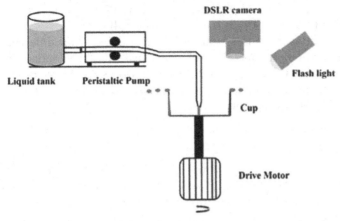

Figure 40.3 A schematic of the experimental setup (Liquid supply and Imaging). The spinning cup receives inlet flow of 100 mL/min
Source: Author

Table 40.1: Properties of the atomized liquids measured at 20°C.

Liquid	Density (kg/m³)	Viscosity (kg/m.s)	Surface tension (N/m)
Water	998	0.001	0.072
Ethyl alcohol	789	0.001	0.022
Glycerol (40%)	1099	1.5	0.063

Source: Author

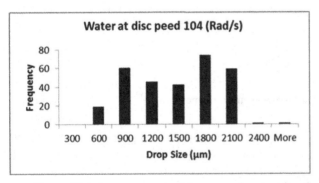

Figure 40.4 A representative histogram showing distribution of drop size (Water) produced from the cup spinning at 104 rad/s. The cup was continuously fed at 100 mL/min
Source: Author

Results and discussion

Table 40.1 lists the properties of the three liquids used in this work. Ethyl alcohol and 40% glycerol are representative of liquids with low surface tension and high viscosity respectively.

Figure 40.4 shows through histogram the size distribution of water droplets at 104 rad/s. The bimodal nature of the size distribution is reminiscent of similar distribution reported by Sahoo and Kumar (2021) for a spinning disc. The first mode with peak near 900 μm is due to secondary drops and the second one at 1700 μm stands for mean size of primary drops. This histogram is representative of all liquids.

Figure 40.5 shows the variation of Sauter mean diameter of drops at for different speeds. Sauter mean diameter of water and glycerol shows insignificant variation. It establishes the fact that viscosity doesn't influence the mean drop size. On the other hand smaller drops are obtained with ethyl alcohol as the atomizing liquid. Detailed analysis in log-log plot yields the correlation:

$$d_{32} = c.\omega^{-0.97} \tag{5}$$

Figure 40.6 presents the variation of d^* $(\frac{d}{D})$ with We. The correlation obtained is:

$$d^* = c.We^{-0.48} \tag{6}$$

The above correlation is close to theoretically expected $We^{-0.5}$ relationship.

The spinning cup atomization characteristic in direct drop mode is very similar to that of spinning disc discussed in literature. Currently we are focusing on the study of the behavior of non-Newtonian fluids with above three liquids as base or continuous phase.

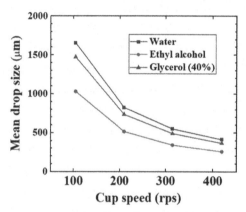

Figure 40.5 Effect of cup speed on Sauter mean diameter (d_{32})
Source: Author

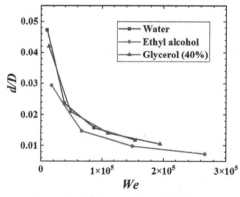

Figure 40.6 Plot showing Weber number-dimensionless drop size relationship
Source: Author

Conclusion

In this work the influence of liquid properties on the atomization characteristics of a spinning cup has been studied in direct droplet mode using high resolution static imaging. The drop size distribution of a liquid shows bi modal distribution due to presence of satellites associated with each single release of primary drop. The Sauter mean diameter of drops formed from liquids scales as ($\omega^{-0.97}$). The smallest size drops produced by ethyl alcohol at all speeds is consistent with basic understanding of drop formation dynamics. The dimension less drop size varies with Weber number as $We^{-0.48}$.

References

Ahmed, M and Youssef, M (2014). Influence of spinning cup and disk atomizer configurations on droplet size and velocity characteristics. *Chemical Engineering Science.* 107, 149-157.

Bizjan, B Širok, B Hočevar, M, and Orbanić, A (2014). Ligament-type liquid disintegration by a spinning wheel. *Chemical Engineering Science.* 116, 172-182.

Craig, I P Hewitt, A and Terry, H (2014). Rotary atomiser design requirements for optimum pesticide application efficiency. *Crop Protection.* 66, 34-39.

Dombrowski, N and Lloyd, T L (1974). Atomisation of liquids by spinning cups. *Chemical Engineering Journal.* 8(1), 63-81.

Eisenklam, P (1964). On ligament formation from spinning discs and cups. *Chemical Engineering Science.* 19(9), 693-694.

Huang, L X and Mujumdar, A S (2008). The effect of rotary disk atomizer RPM on particle size distribution in a semi-industrial spray dryer. *Dry. Technology.* 26(11), 1319-1325.

Liu, J, Yu, Q, and Guo, Q (2012). Experimental investigation of liquid disintegration by rotary cups. *Chemical Engineering Science.* 73, 44-50.

Nandiyanto, A B D and Okuyama, K (2011). Progress in developing spray-drying methods for the production of controlled morphology particles: From the nanometer to submicrometer size ranges. *Advance Powder TechnologyTechnol.* 22(1), 1-19.

Sahoo, Kand Kumar, S (2021). Atomization characteristics of a spinning disc in direct droplet mode. *Industrial & Engineering Chemistry Research.* 60(15), 5665-5673.

Saifullah, M Shishir, M R I Ferdowsi, R Rahman, M R T, and Van Vuong, Q. (2019). Micro and nano encapsulation, retention and controlled release of flavor and aroma compounds A critical review. *Trends Food Science Technology.* 86, 230-251.

41 A comparative analysis on energy efficient narrow band-IoT technology

Sunita Dhalbisoi[1,a], Ashima Rout[1,b], Srinivas Sethi[2,c] and Ramesh Kumar Sahoo[2,d]

[1]Department of ETCE, IGIT Sarang, India
[2]Department of CSE & A, IGIT Sarang, India

Abstract

The Internet of thing is a sensor-based application for larger coverage with a smaller number of resources. It can handle huge devices in terms of billions to provide smart facilities. It makes us more resourceful, whereas energy consumption is high. Low power wide area network (LPWAN), technology is used to optimize resource utilization and power levels. The third-generation partnership project known as 3GPP is presented in new Narrow Band-Internet of things mostly called NB-IoT in Release-13. NB-IoT technology will emerge in current 4G network as well as 5G access network. NB-IoT in cellular network is developed to utilize the necessities of long-range, low-power, deep penetration, low-cost, long battery life applications. NB-IoT is a global coverage solution that operates their services worldwide. Here we have worked on allocation of resource issues, including energy efficiency calculationin terms of uplink and downlink technical features of Narrow Band-IoT in different modes of operation. The performance analysis of various applications could also be reviewed under this technology.

Keywords: Internet of things, third-generation partnership project, narrow band-internet of things, low power wide area network

Introduction

The Internet of Things (IoT) is the system of devices composed of sensors, hardware, actuators, software and connectivity of network which allows objects to interface and information exchange. New technology needs to connect large IoT objects to internet and build communication infrastructure. Narrow Band-Internet of Things (NB-IoT) is a radio access technology introduced by 3GPP in Release 13 to 15 (Feltrin et al., 2019). Cellular NB-IoT is a technology for fast-expanding application and improved features like less power consumption, high system capacity and more utilization of spectrum efficiency of user devices, especially profound coverage. It is observed that, with an increase in IoT devices, the wireless sensor industry market has taken an escalation shape. At same time, IoT offers improved solutions to connect huge number of devices. It needs to reduce complexity of device. For all problems, it needs to change signals and channels a lot in physical layer. The low power wide area network (LPWAN), technologies work in two groups that is an unauthorized spectrum like Sigfox and Lora and another one authorized spectrum like NB-IoT (Liang et al., 2018; Huawei, 2015). If NB-IOT technology is compared with existing GSM/GPRS then the cost of devices will reduce rapidly even with increasing demand in near future. This technology supports large devices with low data rates and latency in reduction with the power consumption.

[a]sunita.muskan1@gmail.com, [b]ashimarout@gmail.com, [c]igitsethi@gmail.com, [d]ramesh0986@gmail.com

DOI: 10.1201/9781003450924-41

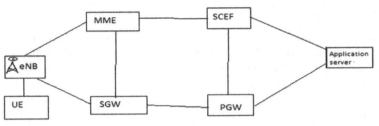

Figure 41.1 Architecture of NB-IoT
Source: Author

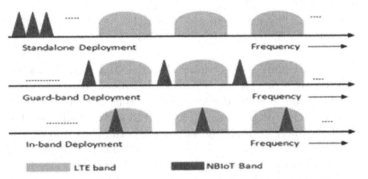

Figure 41.2 Operation modes of NB-IoT (Kanj et al., 2020)
Source: Author

NB-IoT architecture consists of different elements. User equipment (UE) transmits the data to mobile mobility entity (MME) through a random-access network, commonly known as evolved node B(e-NB). MME is connected to serving gateway (SGW) as well as service capability exposure function (SCEF). Both are connected to packet data network gateway (PGW). Finally, the user plane or control plane is connected to application server.

NB-IoT technology is derived to minimize the overhead of signaling that is overhead in radio interface and provides appropriate security to a user defined system. It improves battery life and supports SMS as deployment. It can deliver both IP and non-IP data.

Related work

NB-IoT can optimize data rate by using LTE-Advanced Pro Release 13 and more cost effective with improve power efficiency. Because it eliminates complex signaling overhead which is required for LTE based systems. NB-IoT wireless technology can exist together with current cellular technology like GSM, LTE, UMTS etc (Kanj et al., 2020). The LTE cellular network helps to produce standard NB-IoT uplink baseband waveforms.

It represents a narrow band carrier of 180kHz which is suitable for different applications. NB-IoT operates in three different modes of operation (Zhu et al., 2019). These are stand alone, In-band and guard band. In this standalone mode of operation, carrier deployed outside the LTE spectrum and used for GSM frequency with bandwidth of 200 kHz. It has guard band on both side of spectrum. In guard band mode of operation, carrier deployed in between two LTE carriers and utilize unused guard band. In In-band mode of operation, carrier arrayed in LTE carrier and utilize resource blocks.

Methodology

User equipment can either transmit or receive the data during uplink and downlink at a time. The frequency division half-duplex is used to separate frequency of uplink and downlink. During switching from UL to DL or vice-versa, there must be one subframe (SF) working as a guard in between them.

Uplink and downlink

The NB-IoT uplink consists of the following physical layer channels and signals:

(i)   Narrowband Uplink Physical Shared Channels-NPUSCH
(ii) Narrowband Random-Access Physical Channels-NPRACH
(iii) Narrowband Demodulation of Reference Signals-DM-RS

The transport layer channels are Uplink share channel Data (UL-SCH) and Random-access channel (RACH). Data are sent over the NPUSCH. In uplink, frequency division multiple access with single carrier technique (SC-FDMA) is used with subcarrier frequency spacing of 15 kHz or 3.75 kHzas reference. Evolved node B(eNB) decides which resource grid is used. Resource grids are the same e for both uplink and downlink during 15 kHz subcarrier spacing. Mostly, it has seven OFDM symbols in one time slot with slot length of 2 msec. Symbol duration in 3.75 kHz becomes four times than 15 kHz. Resource element of NB-IoT uplink consists of DM-RS signal and NPUSCH signal. Physical signal and channel are forming resource element grid which includes mapping between logical and transport channel, subframe repetition and different configuration grid. The base band signal with resource grid contains DM-RS and NPUSCH signals. NPUSCH can carry information of uplink control or uplink shared channel in two different format that is in first format uplink shared channel and in second format uplink control information. Here resource unit repeats 128 times which improves reliability and coverage. The smallest unit of mapping is the resource unit. Uplink shared channel consists of common control channel, dedicated control channel. In uplink (DMRS) is multiplexed with the data and transmitted in resource unit for data transmission using single antenna port.

The downlink of NB-IoT consists of three different physical channels. These are:

(i)   Primary and secondary synchronization signals- NPSS and NSSS
(ii) Narrowband reference signals- NRS
(iii) NarrowbandPhysical Broadcast Channel- NPBCH
(iv) NarrowbandPhysicalDownlinkControlChannel- NPDCCH
(v)   NarrowbandPhysicalDownlinkSharedChannel-NPDSCH

Results and discussion

NB-IoT is analyzed in uplink and downlink with different operation mode. It uses 180KHz of bandwidth in multitone or single tone with different value of carrier spacing.

NB-IoT uplink consists of NPUSCH and DMRS with 12 subcarriers and each can use 15KHz or 3. 75KHz.It also has slot duration of 0.5 ms or 2 ms.

Resource element of NB-IoT supports anchor for initial cell allocation and non-anchor for exchange of data in guard band, in-band and standalone operation mode.

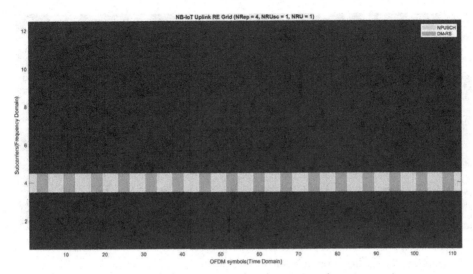

Figure 41.3 NB-IoT Uplink resource grid with 15KHz subcarrier spacing
Source: Author

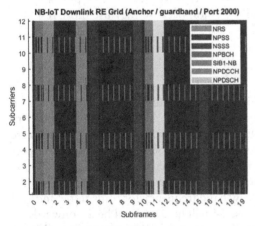

Figure 41.4 NB-IoT Downlink resource grid in operation mode of guard band
Source: Author

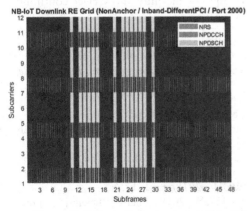

Figure 41.5 Downlink resource grid in NB-IoT operation mode of non-anchor In-band
Source: Author

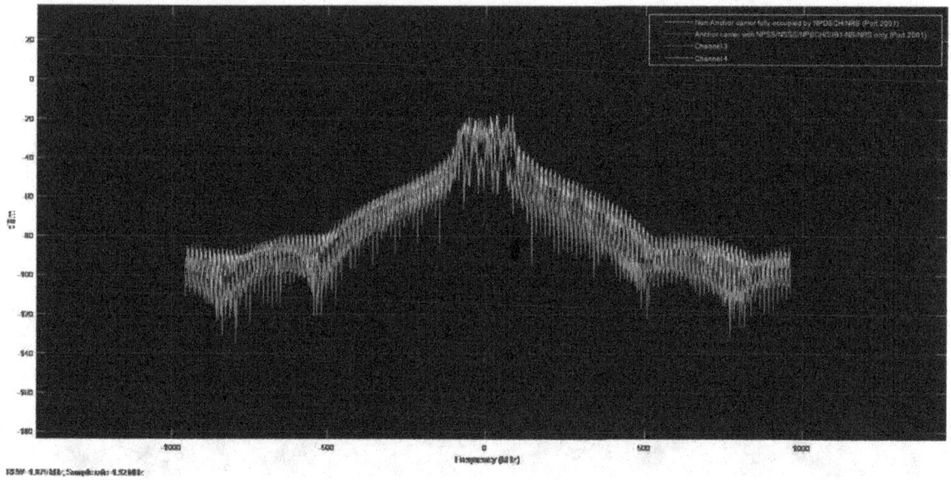

Figure 41.6 Power allocation in NB-IoT
Source: Author

Figure 41.6 shows the power required by NB-IoT. Power required by blue and yellow color has more power than green and red color. Because they use more resource element than green and red color.

Conclusion

3GPP has provided a new cellular radio air interface by using NB-IoT technology in Release 13 which is required in machine type communication. It is enhanced to small as well as rare data packets. So, user equipment can be utilized in cost efficient approach and uses less battery power. Due to the presence of multi-band structure a wider spectrum of frequencies may be used. The NB-IoT carrier in signaling part reduced to few numbers only, but the remaining is utilized for data transfer. Hence, we can utilize only significant amount of band width for control and data transfer, though the user equipment's have only narrow band frequency for transmitter and also receiver. The evolution of NB-IoT with release 14 will be continued hence onwards. In future we will try to implement NB-IoT in 5G massive MIMO with further enhancement. By using current plans, the NB-IoT can be extended to include multicast services and positioning methods. It requires software update, mobility and service continuity.

References

Feltrin, L, Tsoukaneri, G, Condoluci, M, Buratti, C, Mahmoodi, T, Dohler, M, and Verdone, R (2019). Narrowband IoT. A survey on downlink and uplink perspectives. *IEEE Wireless Communications*, 26(1),78–86.

Huawei (2015). NB-IoT-enabling new business opportunities.
Available from: https://www.huawei.com/minisite/4-5g/img/NBIOT.pdf

Kanj, M, Savaux, V, and Le Guen, M (2020). A tutorial on NB-IoT physical layer design. IEEE Communications Surveys & Tutorials, 22(4),2408–2446.

Liang, J M, Wu, K R, Chen, J J, Liu, P Y, and Tseng, Y C (2018). Energy efficient uplink resource units scheduling for ultra-reliable communications in NB-IoT networks. *Wireless Communications and Mobile Computing (WCMC 2018)*. pp 1–17.

Zhu, S, Wu, W, Feng, L, Zhao, P, Zhou, F, Yu, P, and Li, W (2019). Energy efficient joint power control and resource allocation for cluster-based NB-IoT cellular networks. *Transactions on Emerging Telecommunications Technologies*. 30(4), 3266.

42 Investigations of the physical and antibacterial properties of vitamin e-enriched lemongrass oil nano emulsions

Veda Prakash[a], Anil Kumar Murmu[b] and Dr Lipika Parida[c]

Assistant Professor, VSSUT, Burla, Odisha, India

Abstract

Oil in water nanoemulsions consisting of lemon grass oil, Tween 80 and vitamin E as the organic phase and water, citric acid and sodium benzoate as the aqueous phase were fabricated by the emulsion phase inversion method. All the nanoemulsions were fabricated with a surfactant-to-oil ratio (SOR) varying from 0.251.25. Properties of nanoemulsions such as droplet diameter, zeta potential (ZP) and poly dispersity index (PDI) were evaluated. The nanoemulsion formulated with a SOR of 0.25 was found to have the smallest droplet diameter of 143 nm, and the largest droplet size of 221.6 nm was exhibited by the nanoemulsion prepared with a SOR of 1.25. The formulated nanoemulsions exhibited an increasing trend of droplet size with SOR. The PDI of formulated nanoemulsions was less than 0.3, whereas it was 0.315 for the nanoemulsion prepared at SOR = 0.25. The PDI results indicated a decreasing trend with increasing SOR. The ZP measurements showed values ranging from -8.9 to -14 mV. The nanoemulsions showed antibacterial activity against *S. aureus and E. coli*. However, the antimicrobial property decreased with an increase in SOR. The nanoemulsions were found to be more resistant to *S. aureus* than to *E. coli*.

Keywords: Antibacterial, droplet diameter, lemon grass oil, nanoemulsion

Introduction

Nanoemulsions are colloidal dispersions with a size less than 200 nm that are thermo dynamically unstable. An emulsifier or surfactant is used to combine two immiscible liquids into a homogenous dispersion to create nanoemulsions. To reduce the surface tension between two liquids, an emulsifier contains both hydrophobic and hydrophilic groups. Two basic approaches, low energy (emulsion inversion point, spontaneous emulsification and phase inversion temperature) and high-energy methods (microfluidisation, ultrasonication and high-pressure homogenization) are employed to create nanoemulsions.

Tocopherols and tocotrienols commonly use vitamin E as a component (α, β, γ, δ); it is a component of numerous organic goods with both plants- and animal-derived origins (Gamna and Spriano, 2021). Due to its high oxidation susceptibility, vitamin E is typically added to foods, medications, and cosmetics in its esterified form (α-tocopherol acetate) (Gawrysiak-Witulska et al., 2009). In order to incorporate it into aqueous-based products, proper delivery mechanisms are required due to its limited water solubility.

Highly lipophilic components can be delivered and encapsulated using nanoemulsions. Nanoemulsions have quite a few advantages over conventional emulsions.

[a]vedaprakash_chemical@vssut.ac.in, [b]akmurmu_chemical@vssut.ac.in, [c]lparida_chemical@vssut.ac.in

DOI: 10.1201/9781003450924-42

Because they have sizes close to the wavelength of light and greatly scatter light waves, the droplets in conventional emulsions tend to be turbid. Conversely, due to the droplets' small size and poor light scattering, nanoemulsions are often clear or hardly turbid (McClements, 2012).

In this paper, lemongrass essential oil (EO) was utilized to formulate nanoemulsions. Lemongrass oil has antifungal, antibacterial, and antiviral properties that react against many kinds of yeasts and bacteria. In beverages and food, lemongrass is utilized as a flavoring. In manufacturing, lemongrass is used as a fragrance in deodorants, soaps, and cosmetics. In this study, the antibacterial activity of nanoemulsion of lemongrass EO and vitamin E mixture was examined against *S. aureus* and *E. coli* at different surfactant-to-oil ratio (SOR). The prepared nanoemulsions were first analyzed for their droplet diameter, poly dispersity index (PDI) and zeta potential (ZP). The fabricated nanoemulsions at various SOR were then examined for their antibacterial activity.

Materials and methods

Materials

We purchased vitamin E (DL-alpha-tocopherol acetate, 96%) and Tween 80 from LobaChemie Pvt. Ltd. located Maharashtra, India. We received 99% ultra-pure sodium benzoate and 1M citric acid solutions from LobaChemie Pvt. Ltd., Maharashtra, India. Lemongrass oil was bought from a nearby market. All the nanoemulsions mixtures were made using regular distilled water. The antibacterial activity of the created nanoemulsions was tested using the disc diffusion method against both Gram-positive (*Staphylococcus aureus*) and Gram-negative (*Escherichia coli*) bacteria.

Preparation and characterization of nanoemulsions

The formulations were prepared by the phase inversion technique, in which a water phase is titrated into an organic phase. Initially, the organic phase (20 wt%) was fabricated by adding lemongrass oil, surfactant Tween 80 and vitamin E acetate in a beaker and mixing the solution in a magnetic stirrer (500 rpm) for 30 min. The SORs were maintained at 0.25, 0.52, 0.75, 1.0, and 1.25. The aqueous phase (80 wt%) comprising distilled water, citric acid and sodium benzoate was then titrated to a stirring organic phase (500 rpm) at two drops per 1 second. The experiments were performed at ambient temperature (30°C).

Using a dynamic light scattering device (Zetasizer Ver. 7.12, Malvern Instruments, Malvern, UK), the droplet size of nanoemulsions was determined. Before each experiment, distilled water was used to dilute each mixture to eliminate scattering effects. The average size of dispersed phase droplets and PDI were determined. The value of PDI represents a measurement of the narrowness of the particle size distribution. A very narrow distribution occurs at PDI≤0.1. Additionally, ZP measurements were carried out and reported.

Disc diffusion-based antibacterial activity

E. coli and S. aureus were the bacterial strains used in the antibacterial testing. The typical disc diffusion method was used in the antimicrobial investigation. The nanoemulsion inhibited the formation and growth of the test microorganism, and the inhibiting growth zones' diameters were measured. The experiment was carried out three times.

Results and discussion

Characterization of lemongrass oil-vitamin E enriched nanoemulsions

The influence of the surfactant concentration on the nanoemulsions' droplet size was investigated by preparing emulsions at different SOR with a fixed aqueous phase composition. The mean droplet size of formulated nanoemulsion was 143 nm, prepared at a SOR of 0.25. The droplet size followed an increasing order with SOR (Figure 1(a)).

The nanoemulsion formulated at a SOR of 1.25 exhibited the largest droplet size of 221.6 nm. This finding has been associated with the physicochemical processes that are thought to be responsible for the system's spontaneous emulsification. It has been linked to a rise in surfactant adsorption at the interface, which lowers interfacial tension and encourages the formation of smaller droplets. Also, a greater rate of surfactant molecules moving from the organic phase to the aqueous phase has been associated with producing finer oil droplets at the border. We observed an increasing trend in droplet size with SOR. It has been postulated that the formation of a crystalline phase makes spontaneous breakage of the oil-water interface more challenging at high SOR (Anton and Vandamme, 2009). The value of PDI represents how narrowly distributed the droplet sizes are. A nanoemulsion can be considered stable for PDI<0.3. All the formulations prepared at SOR≥0.5 showed a uniform-sized dispersed phase (Figure 42.1(b)).

It was observed that the PDI values followed a decreasing trend with increasing SOR, with a narrower distribution occurring at PDI = 0.17, which was prepared at a SOR of 1.25 (Figure 42.1(b)). These results suggest that vitamin E-lemon grass oil nanoemulsions can be fabricated by the emulsion phase inversion approach at relatively low SOR.

The determination of ZP is crucial, as it influences the stability of nanoemulsion. The higher the ZP, the stronger the electrostatic repulsion between the dispersed particles, favoring long-term stability. Results of ZP ranged from -13.6 mV to -10.3 mV as the SOR of nanoemulsions varied from 0.25 to 1.25, with the lowest ZP (ZP = -8.9 mV) observed at a SOR of 0.50 and the highest (ZP = -14.0 mV) at SOR of 0.75 (Table 42.1). The negative charge on oil droplets causes negative ZP measurements, which appear during emulsification.

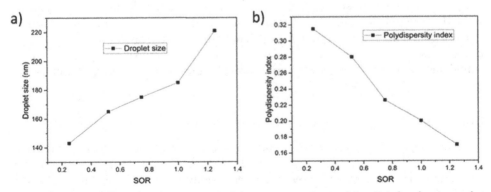

Figure 42.1 Effect of SOR on vitamin E-lemongrass nanoemulsions' a) droplet size and b) PDI

Source: Author

Table 42.1: Zeta potential of formulated nanoemulsions at different SOR.

Surfactant-to-oil ratio	Zetapotential, mV
0.25	-13.6
0.50	-8.9
0.75	-14.0
1.00	-10.0
1.25	-10.3

Source: Author

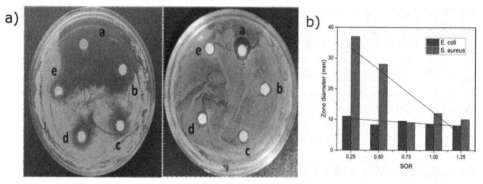

Figure 42.2 a) Antibacterial activity of nanoemulsions against *S. aureus* (left) and *E. coli* (right); a-e represent SOR values from 0.25 to 1.25, b) Effect of SOR of nanoemulsions on inhibition zone diameter

Source: Author

Antibacterial activity

In this work, the effectiveness of the prepared nanoemulsions with different SOR was tested against *E. coli* and *S. aureus* (Figure 42.2). The inhibition zone depended on the percentage of surfactant (or SOR) dispersed in the nanoemulsions. The antibacterial study revealed that nanoemulsion with a SOR of 0.25 produced the largest inhibition zone of diameter 37 mm against *S. aureus* (Figure 2(b)). Conversely, the nanoemulsion exhibited the lowest inhibition zone of size 8 mm with a SOR of 1.25 when tested against *E. coli*. These results revealed that lemongrass oil-encapsulated nanoemulsions displayed significantly higher resistance against *S. aureus* than *E. coli*. This is in accordance with the study that revealed that compared to using an essential oil alone, encapsulating them in colloidal systems such as nanoemulsions increases their antibacterial activity (Liang et al., 2012, Anwer et al., 2014).

Conclusions

In this study, Emulsions of different surfactant-to-oil ratio (SORs) were formulated using the emulsion phase inversion method. All formulations (SOR ≤ 1.00) showed the nanometric size and exhibited stable emulsions (poly dispersity index, PDI < 0.315). In addition, they all showed negative zeta potential (ZP). The droplet size depended mainly on the SOR, which exhibited an increasing trend of droplet size with SOR. Typically, the quantity of antibacterial activity in the nanoemulsion correlates with the extent of the zone of inhibition. The nanoemulsions prepared at different SORs showed resistance against *S. aureus* and *E. coli*.

References

Anton, N and Vandamme, T F (2009). The universality of low-energy nanoemulsification. *International Journal of Pharmaceutics*, 377(12), 142147.

Anwer, M K, Jamil, S, Ibnouf, E O, and Shakeel, F (2014). Enhanced antibacterial effects of clove essential oil by nanoemulsion. *Journal of Oleo Science*, 63(4),347–354.

Gamna, F and Spriano, S (2021). Vitamin E .A review of its application and methods of detection when combined with implant biomaterials. *Materials*, 14(13), 3691, 119. https://doi.org/10.3390/ma14133691.

Gawrysiak-Witulska, M, Siger, A, and Nogala-Kalucka, M (2009). Degradation of tocopherols during near ambient rapeseed drying.Journal. Food Lipids16(4), 524-539.

Liang, R, Xu, S, Shoemaker, C F, Li, Y, Zhong, F, and Huang, Q (2012). Physical and antimicrobial properties of peppermint oil nanoemulsions. *Journal of Agricultural and Food Chemistry*, 60, 30, 75487555. doi.org/10.1021/jf301129k.

McClements, D J (2012). Edible delivery systems for neutraceuticals: designing functional foods for improved health.Therapeutic Delivery. 3(7), 80-803.

43 Artificial intelligence based double fed induction machine controlled pumped storage turbine – A Critical Review

T.K. Swain[1,a], Meera Murali[2b] and Raghu Chandra Garimella[3,c]

[1]CoEP (Tech) University & Central Water and Power Research Station (CW&PRS) Ministry of Jal Shakti, Government of India, Pune, Maharashtra, India

[2]CoEP (Tech) University, Pune, Maharashtra, India

[3]Central Water and Power Research Station (CW&PRS), Ministry of Jal Shakti, Government of India, Pune, Maharashtra, India.

Abstract

Applications of artificial intelligence (AI) in the control of energy systems has gained much attention in recent years. One such application of AI in energy systems is the control of double fed induction machine (DFIM) controlled pumped storage system (PSS). This paper gives a comprehensive review of the current situation of AI-based DFIM controlled PSS and its mathematical modelling. Besides, a DFIM controlled PSS is one such system that may get benefitted from the implementation of AI control strategies. A DFIM based PSS consists of a high-speed rotor-side converter, a low-speed stator-side converter, and a generator, connected to a network of pumps and turbines. The high-speed rotor-side converter acts as the primary control device, while the low-speed stator-side converter acts as the secondary control device. This paper reviews mathematical modelling equations for achieving a novel AI based DFIM controlled PSS.

Keywords: Artificial intelligence, control theory, modelling, future of energy storage

Introduction

Pumped storage systems (PSS) are used for energy storage and have significant importance in ensuring grid stability. The traditional PSS control systems have limitations in terms of their ability to handle the dynamic changes in the power system. In addition, the traditional control methods used in these systems have limitations such as slow response time, low efficiency, and limited adaptability. To overcome these limitations, artificial intelligence (AI) techniques like neural networks, fuzzy logic, and expert systems may be utilized. With the advancements in AI, the control of PSS has become more efficient and flexible. The double fed induction machine (DFIM) controlled PSS has the potential to improve overall system performance by reducing costs and increasing efficiency. Double fed induction machine (DFIM) has been widely utilized in various power system applications due to the high efficiency and reliability. The integration of AI techniques in DFIM-based systems can further improve their performance and adaptability. The application of AI in DFIM controlled pumped storage systems is one such example. Figure 43.1 shows the typical control diagram of hydro-electric power plant and Figure 43.2 depicts the typical functional block diagram of pumped storage turbine. The major objective of this paper is to study the application

---

[a]tks15.elec@coep.ac.in/swain_tk@cwprs.gov.in, [b]mm.elec@coep.ac.in, [c]raghuchandra.g@cwprs.gov.in

DOI: 10.1201/9781003450924-43

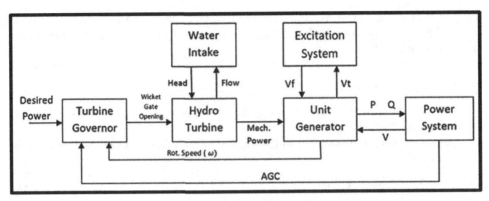

Figure 43.1 Typical layout of a hydropower plant with two surge tanks
Source: Author

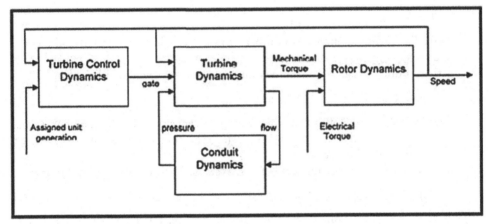

Figure 43.2 Functional block diagram of the pumped storage turbine
Source: Author

of AI techniques in DFIM controlled pumped storage systems, with a focus on mathematical modelling equations.

Literature review

In recent years, the use of AI in the control of DFIM controlled PSS has gained much attention. Researchers have applied various AI techniques, which include artificial neural networks (ANNs), fuzzy logic (FL), and reinforcement learning (RL) to control the DFIM in PSS (Garud, Jayaraj, and Lee 2021). ANNs have been used to model the dynamic behavior of DFIM and to control the power flow in PSS(Kumar and Saini 2022). FL has been applied to improve the stability of the DFIM controlled PSS and to control power flow in the system (Gang et al. 2006). RL has been used to optimize the performance of DFIM controlled PSS and to control the energy storage and retrieval in the system (Zhao et al. 2022).One of the main challenges in designing an AI-based control system for a DFIM based PSS is to develop a mathematical model that accurately represents the system behavior (Behara and Saha 2022). Mathematical modelling equations for a DFIM-based pumped storage system can be developed using state-space representations, transfer functions, and other mathematical tools

(Huang et al. 2022). However, the development of accurate mathematical models is essential for the successful implementation of these control strategies (Murray-Smith and Johansen 2020).

Mathematical modelling

The DFIM controlled pumped storage system may be modelled using the following mathematical equations:

1. Electric power balance equation for the system may be expressed as illustrated in Equation (1):

$$P\_in = P\_out + P\_loss \tag{1}$$

   where, P_in is input power, P_out is output power, and P_loss is the power loss.

2. Similarly, torque equation for the DFIM may be expressed as given in Equation (2):

$$T = 3/2 * p * L * i\_d * i\_q \tag{2}$$

   where, T is torque, p indicates number of poles, L is inductance, i_d and i_q are the d-axis and q-axis currents, respectively.

3. Further, stator voltage equation may be expressed as provided in Equation (3):

$$V\_s = R\_s * I\_s + L\_s * di\_s/dt + omega * L\_m * i\_q \tag{3}$$

   where, V_s is stator voltage, R_s is stator resistance, I_s is stator current, L_s is stator inductance, di_s/dt is derivative of stator current, omega is angular velocity, and L_m is mutual inductance.

4. Likewise, rotor voltage equation may be expressed as presented in Equation (4):

$$V\_r = R\_r * I\_r + L\_r * di\_r/dt - omega * L\_m * i\_d \tag{4}$$

   where, V_r is rotor voltage, R_r is rotor resistance, I_r is rotor current, L_r is rotor inductance, and di_r/dt is derivative of rotor current.

5. Subsequently, mathematical equations for the control algorithm are given in Equations (5) to (7) below.
   Load prediction equation:

$$y = f(x) \tag{5}$$

   where, x = [E_prod, E_cons] is input vector, and y is predicted load.

6. Speed control equation:

$$w\_DFIM = k\_1 * (y - y\_0) \tag{6}$$

   where, w_DFIM is speed of DFIM, k_1 is a gain constant, and y_0 is desired load.

7. Energy storage control equation:

$$E\_PSS = k\_2 * (w\_DFIM - w\_DFIM\_0) \tag{7}$$

   where, E_PSS is energy stored in PSS, k_2 is a gain constant, and w_DFIM_0 is desired speed of DFIM.

Accurate modelling is essential for the effective implementation of AI based control strategies and for optimization of system performance, reliability, and cost-effectiveness.

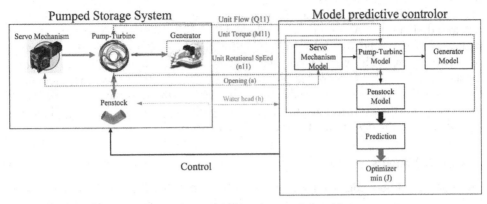

Figure 43.3 Typical methodology of AI based control algorithm
Source: https://doi.org/10.3389/fenrg.2021.757507

AI based control algorithm

Figure 43.3 depicts the typical method for the implementation of AI based double fed induction machine controlled pumped storage system.

In general, researchers may follow the steps provided below for developing an AI based control algorithm:

i.   Modelling the DFIM: Develop a mathematical model of DFIM using equivalent circuit equations, as described in a previous section.
ii. Modelling the PSS: Develop a mathematical model of PSS using non-linear differential equations, as described in a previous section.
iii. Combine the models: Combine the models of DFIM and PSS to create a single, comprehensive model of the system.
iv. Develop a control algorithm: Design and develop a control algorithm that uses the model of DFIM controlled PSS to regulate the system and maintain desired operating conditions.
v.   Implementation of control algorithm: Implement the control algorithm in code, using a programming language such as Python. This may involve adjusting the algorithm parameters to optimize performance and accuracy.
vi. Testing the algorithm: Test the control algorithm on simulated data to evaluate its performance and accuracy.
vii. Refinement of the algorithm: Refine the control algorithm based on results of the tests, making adjustments as necessary to optimize performance and accuracy.
viii. Deployment of the algorithm: Deploy the control algorithm in actual DFIM controlled PSS and monitor its performance over time to ensure it continues to meet desired operating conditions.

This gives major outline of steps involved in developing a control algorithm for achieving AI based DFIM controlled PSS. Moreover, the implementation requires a detailed and comprehensive approach, and may likely require a significant amount of programming and testing skills.

Conclusions

Double fed induction machine (DFIM) controlled pumped storage systems have the potential to provide effective energy storage and grid stability services. Artificial intelligence (AI) By developing a comprehensive mathematical model of a DFIM system and designing an AI based control algorithm, it is possible to evaluate the performance of these systems and determine the optimal control strategies for improving their performance. This paper has provided an overview of steps involved in developing an AI based) DFIM controlled pumped storage system (PSS), including modelling the system, developing an AI based control algorithm, and evaluating performance of the system. Further, research in this area may involve exploring other AI techniques and optimizing the design of AI based control algorithm.

Acknowledgments

The authors acknowledge Dr. R.S. Kankara, Director, CW&PRS, and Government of India, who has provided a greater support during the preparation of this manuscript.

References

Kumar, B R and Saha, A K (2022). Artificial intelligence control system applied in smart grid integrated dou*bly* fed induction generator-based wind turbine: a review. *Energies,* 15(17), 6488.

Gang, L I, Shijie Cheng, W E N Jinyu, P A N Yuan, and Jia, M A (2006). Power system stabilisation by a double-fed induction machine with a flywheel energy storage system. *IU-Journal of Electrical & Electronics Engineering,* 6(1), 69–76.

Garud, K S, Simon J, and Moo-Y L (2021). A review on modeling of solar photovoltaic systems using artificial neural networks, fuzzy logic, genetic algorithm and hybrid models. *International Journal of Energy Research,* 45(1), 6–35.

Huang, J, Yujie X, Huan G, Xiaoqian G, and Haisheng C. (2022). Dynamic performance and control scheme of variable-speed compressed air energy storage. *Applied Energy* 325, 119338.

Kumar, K and Saini,R P (2022). A Review on operation and maintenance of hydropower plants. *Sustainable Energy Technologies and Assessments,* 49, 101704.

Roderick, M and Johansen, T (2020). *Multiple Model Approaches to Nonlinear Modelling and Control.* CRC press.

Zhao, Y, Li, X, Hou, N, Yuan, T, Huang, S, Li, L, Li, X, and Zhang, W (2022). Self-powered sensor integration system based on thorn-like polyaniline composites for smart home applications. *Nano Energy,* 104, 107966.

44 Understanding the link between above ground biomass and vegetative indices: an artificial neural network and regression modelling approach

Kumari Anandita[1,a], Anand Kumar Sinha [2,b] and C Jeganathan[3,c]

[1]Research scholar, Department of Civil and Environmental Engineering. Birla Institute of Technology, Mesra, Ranchi, Jharkhand, India

[2]Professor, Department of Civil and Environmental Engineering. Birla Institute of Technology, Mesra, Ranchi, Jharkhand, India

[3]Professor, Department of Remote Sensing. Birla Institute of Technology, Mesra, Ranchi, Jharkhand, India

Abstract

Understanding the relationship between above ground biomass (AGB) and vegetative indices is important for many applications, such as monitoring and managing natural resources, assessing land use, and predicting climate change impacts. However, this complex relationship may vary depending on several factors, including the specific ecosystem and methods used to measure AGB and vegetative indices. An artificial neural network (ANN) is a very powerful tool for analyzing complex relationships. It can be used to analyze the relationship between AGB and vegetative indices. There are several vegetative indices, but only a few correlates with aboveground biomass estimation. As principle composite analysis (PCA) showed, support vector machine (SVR) may be better for forecasting aboveground biomass if the connection between predictors and response variable is complicated and non-linear.

Keywords: Above ground biomass, artificial neural network , Forest Health, MATLAB, vegetative indices

Introduction

The ecosystem carbon stocks are affected by anthropogenic land-use change, climate change (Keenan and Williams, 2018), and alterations to the nitrogen cycle (Fowler et al., 2013). The interaction between the source (resource acquisition), sink (metabolic tissue production), and regulatory systems (phenology, hormones) determines how much biomass is produced (Bahguna et al., 2015). This study used an AGB and vegetative indices dataset to train and test an artificial neural network (ANN) model and evaluate the model's performance using correlation analysis (Liang et al., 2022). The results of this present study provide insights into the relationship between AGB and vegetative indices.

Material and methods

Study area

BIT Mesra is located at 23.4123°N and 85.4399°E. Figure 1 shows the study area, the map was created using ArcGIS 10.3.A 980-acre (4 Km²) campus for the Institute is

[a]anandita1247@gmail.com, [b]aksinha@bitmesra.ac.in, [c]jeganathanc@bitmesra.ac.in

DOI: 10.1201/9781003450924-44

near the confluence of the Jumar and Swarnarekha rivers. It is bordered by a lush Sal (*Shorea Robusta*) forest that covers a major portion of the BIT Mesra campus. This tree species belongs to the *Dipterocarpaceae* family.

Data acquisition

Seventy plots (10 m × 10 m) were selected based on feasibility. The average number of trees in each plot was used to upgrade data. Tree height was calculated using the NASA Globe and diameter at breast height (DBH) was observed. All inventory data were collected during April-May 2022. Equation 1 was proposed by (Chave et al., 2014).

$$AGTB \text{ (kg)} = 0.0509 * \rho \, D2 \, H \tag{1}$$

Where,
D = Diameter at Breast Height (DBH) (cm);
ρ = The wood-specific gravity (gm/cm^3);
H = The tree height (m);
Sal tree wood-specific gravity = 0.73 gm/cm^3.

Optical images from Sentinel 2 were used in this study. Data was acquired for 02/04/2022 from Copernicus open access hub with a cloud cover of less than 10%. Several vegetative indices were calculated, which are given in Table 44.1.

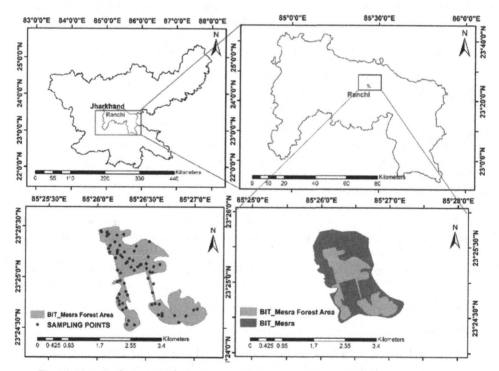

Figure 44.1 Study area (Birla Institute of Technology, Mesra. Ranchi)
Source: Author

Artificial neuron networking and regression analysis

MATLAB 2018a was used to perform ANN on the prepared dataset (Wang et al., 2017). The data were divided into testing (15%), training (70%), and validation (15%) sets randomly. 10 hidden layers were applied for testing. Levenberg-Marquardt training algorithm was used to train the data set. Two case scenarios were tested 1) all the parameters mentioned in Table 44.1 and 44.2) selected parameters after curve fitting analysis to filter out vegetative indices having the best direct correlation with Above-ground biomass Table 44.2.

The regression models were run once with principle composite analysis (PCA) enabled and again with PCA disabled. Finally, maps were created using observed values and predicted values. Inverse distance weighting (IDW) technique was used to prepare the above ground biomass (AGB) map using ArcGIS 10.3 software.

Result and discussion

The linear correlation between the predicted and actual AGB indicated by the R-value for the test set of the ANN model for case 1 is -0.77311, and for case 1 is 0.96384

Table 44.1: Details of vegetative indices calculated.

Abbreviation	Vegetative indices	Formula
MCARI	Modified chlorophyll absorption in reflectance index	$((B05 - B04) - 0.2 * (B05 - B03)) * (B05 / B04)$
CCCI	Canopy chlorophyll content index	$((B08 - B05) / (B08 + B05)) / ((B08 - B04) / (B08 + B04))$
GVMI	Global vegetation moisture index	$((B08 + 0.1) - (B12 + 0.02)) / ((B08 + 0.1) + (B12 + 0.02))$
WDRVI	Wide dynamic range vegetation index	$(0.1 * B08 - B04) / (0.1 * B08 + B04)$
GLI	Green leaf index	$(2.0 * B03 - B04 - B02) / (2.0 * B03 + B04 + B02)$
PPR	Plant pigment ratio	$(B03 - B01) / (B03 + B01)$
PVR	Photosynthetic vigor ratio	$(B03 - B04) / (B03 + B04)$
SIPI1	Structure intensive pigment index 1	$(B08 - B01) / (B08 - B04)$
SIPI3	Structure intensive pigment index 3	$(B08 - B02) / (B08 - B04)$
GNDVI	Green normalized difference vegetation index	$(B08 - B03) / (B08 + B03)$
NDVI	Normalized difference vegetation index	$(B08 - B04) / (B08 + B04)$
LCI	Leaf chlorophyll index	$(B08 - B05) / (B08 + B04)$
EVI	Enhanced vegetation index	$2.5 * (B08 - B04) / ((B08 + 6.0 * B04 - 7.5 * B02) + 1.0)$
GDVI	Green difference vegetation index	$B08 - B03$
SLAVI	Specific leaf area vegetation index	$B08 / (B04 + B12)$
GI	Greenness index	$B08 / B03$
PBI	Plant biochemical index	$B08 / B03$
CIGreen	Chlorophyll index green	$B08 / B03 - 1.0$
DVI	Difference vegetation index	$B08 / B04$
RVI	Ratio vegetation index	$B08 / B04$

Source: Literature review

Table 44.2: Significant linear correlations between vegetation indices and AGB.

Vegetative indices	Linear correlation with AGB
DVI	$R^2 = 0.970$
EVI	$R^2 = 0.761$
GI	$R^2 = 0.920$
GLI	$R^2 = 0.806$
MCARI	$R^2 = 0.869$
NDVI	$R^2 = 0.723$
PVR	$R^2 = 0.835$
RVI	$R^2 = 0.970$
SLAVI	$R^2 = 0.871$
WDRVI	$R^2 = 0.905$

Source: Author

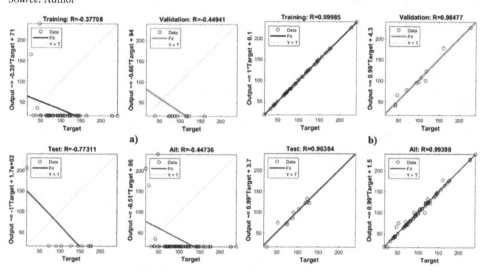

Figure 44.2 AGB results of different data sources (nntool). (a) Case 1; and (b) Case 2
Source: Author

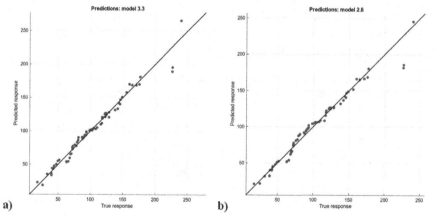

Figure 44.3 Best regression model output for prediction of AGB for Case 2 a) linear regression model b) support vector machine regression model
Source: Author

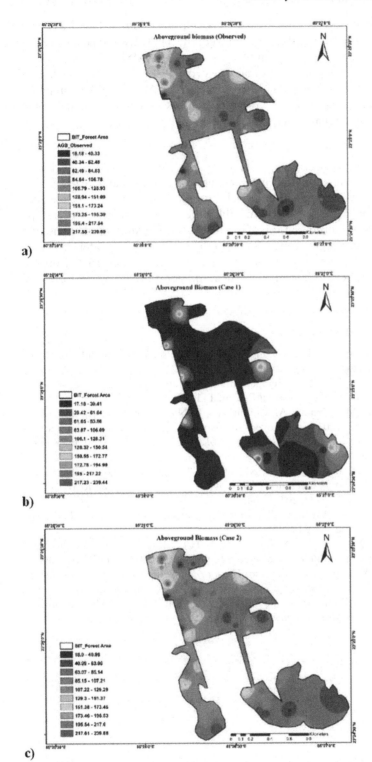

Figure 44.4 AGB map a) observed values b) predicted values Case 1 c) predicted values Case 2

Source: Author

(Figure 44.2). Filtration of vegetation indices through curve fitting has drastically improved the correlation. Two sets of regression were run 1) PCA was enabled, and 2) PCA was disabled. The linear regression model gave the best RMSE value of 8.00659 without PCA, whereas the Support vector machine proved to be the best model, with PCA giving us an RMSE value of 8.66735 (Figure 3). The AGB map was prepared from the predicted values (Figure 44.4). Less variation is seen in case 1, whereas the case 2 map shows greater similarity with the observed value.

Conclusion

The neural network tool suggested that machine learning with suitable feature engineering minimizes the AGB prediction model's training time while considerably improving its accuracy and predictive power. Table 44.2 gives us an idea of what type of vegetative indices to be considered for AGB estimation. The findings demonstrate that both SVM and linear regression (LR) can efficiently and accurately enhance the representativeness of spectral analysis feature data. Model constructed on the screened feature variables exhibiting high stamina and good predictive performance AGB prediction.

References

Alvaro, D et al. (2014). Improved allometric models to estimate the aboveground biomass of tropical trees. *Global change biology,* 20(10), 31773190.

Bahuguna, R N and Krishna S V J (2015). Temperature regulation of plant phenological development. *Environmental and Experimental Botany,* 111, 8390.

Chave, J., Réjou-Méchain, M., Búrquez, A., Chidumayo, E. N., Colgan, M. S., Delitti, W. B. C., ... & Vieilledent, G. (2014). Improved allometric models to estimate the aboveground biomass of tropical trees. *Global Change Biology,* 20(10), 3177–3190. https://doi.org/10.1111/gcb.12629

Fowler, D, Mhairi C, Ute S, Mark, A S, Cape, J N, Stefan, R, Lucy, J S et al. (2013). The global nitrogen cycle in the twenty-first century. *Philosophical Transactions of the Royal Society B: Biological Sciences,* 368(1621), 20130164.

Keenan, T F and Williams, C A (2018). The terrestrial carbon sink. *Annual Review of Environment and Resource,* 43, 219243.

Liang, R -T, Wang, Y -F, Qiu, S -Y, Sun, Y -J, and Xie, Y -H (2022). Comparison of artificial neural network with compatible biomass model for predicting aboveground biomass of individual tree. *Ying Yong Sheng tai Xue Bao. The Journal of Applied Ecology,* 33(1), 916.

Liang, W, Liu, L, Xu, S, Dong, J, and Yang, Y (2017). Forest above ground biomass estimation from remotely sensed imagery in the mount tai area using the RBF ANN algorithm. *Intelligent Automation & Soft Computing,* 18.

45 Design and fabrication of BLDC motor operated portable grass cutter

Kamala Kant Sahoo[a], Baisalini Sethi, Shambhu Kumar Mahato, Sibasis Harihar Sahu, Saroj Kumar Padhi and Sovan Prasad Behera

Assistant Professor, Parala Maharaja Engineering College Berhampur, Odisha, India

Abstract

Currently, a portable lawn cutter is powered by an expensive, high maintenance traditional SI engine or induction motor. Therefore, a portable grass cutter powered by a brushless direct current (BLDC) motor is designed and made as part of this study. The design goal is to create a lawn cutter that is lightweight, robust, simple to use, and low-maintenance. This grass-cutting device is manufactured for gardening and other uses. Its construction is simple, and operating it is simple. The parts used are a 48-volt BLDC motor, controller, 48-volt battery pack, motor mounting head assembly, and extended cylindrical frame with driving head. It is used to maintain and care for lawns in gardens, schools, colleges and other locations. The operation is comparable to that of a traditional portable grass cutter.

Keywords: BLDC motor, controller, conventional, mounting head, portable grass cutting machine.

Introduction

Grass-cutting machines have emerged as an essential part of our daily lives since it enable us to manage our yards. Furthermore, people are really quite concerned in how utilizing lawn mowers would affect the environment. People are thus seeking for ways to reduce and eventually get rid of their own carbon footprints. Environmental pollution is another issue that we deal with on a daily basis, particularly in our homes. A study found that 70% of Malaysians who live at home regularly cut their grass with fuel lawn mowers. Consequently, a lawn mower requires a lot of maintenance. For instance, one should periodically change the oil or fuel in the lawn mower to ensure that it functions efficiently when cutting the grass. Further variable expenses will be incurred as a result of the recent increase in fuel prices. A green lawnmower must be developed in order to support green innovation initiatives in order to solve these issues.

Literature review

Numerous publications are cited for the construction and design of a lawn cutter powered by a brushless direct current (BLDC) motor. Following is a review of the previously employed method by many researchers such as BB et al. 2021, Khan Mohammad Majedul Hasan et al. (2022) and Rahul et al. (2021) who worked in the field related to making an automatic lawn mower and tried to make an improvement in its efficiency. The majority of this lawn mower is powered by solar energy, which is more convenient and cost-effective than other energy sources,

[a]kamalakanta6553@gmail.com

DOI: 10.1201/9781003450924-45

particularly gas-powered ones. But because this energy is a renewable source and is secure, effective, and ecologically benign, our lawn mower is powered by a BLDC motor. We thus designed a lawn cutter without any power source derived from SI engines. Ulhe et al. (2016) constructed a residential lawnmower driven by solar energy is intended to overcome a variety of concerns that conventional mowers powered by internal combustion engines do not. Garg et al. (2017) has worked on a comprehensive study in quantification of response characteristics of incremental sheet forming process and given its importance related to the fabrication and material selection.

In this study, a new portable grass cutter driven by BLDC motor and rechargeable batteries was built. Additionally, the cost of producing the grass cutter was kept cheap while important considerations like durability, environmental friendliness, and light weight were made. A lawn mower is a machine that uses blades or strings to trim grass into uniform lengths in yards or gardens.

Design variables

Overall Design of the Lawn Mower body structure before the actual fabrication process was carried out; the lawn mower with different parts was designed using Solidworks 2016 software which is shown in Figure 45.1.

Fabricated grass cutter

Figure 45.2 shows the complete fabricated working model of grass cutter.

Block diagram for electrical connections

Figure 45.3 depicts the block diagram of electrical connections for the above fabricated grass cutter.

Empirical results

Cost analysis and comparison

Based on the data obtained from Honda Motors Ltd., the lead manufacturer of conventional portable grass cutter which worked on 4-stroke engine consumed approximately 0.5 ltr/hr fuel. As per the above condition, if the machine can run for 6 hr/day, then total fuel consumption is 0.5*6 = 3 ltr/day.

Based on India's current 105 rupee per litre gasoline price. Thus, the machine's daily running cost in rupees is equal to 105*3 = 315. So, assuming the grass cutter is utilized once per week throughout the year, the total running cost in Indian rupees would be 50*315 = 15,750. Original investment in the machine, including the frame, was Rs. 20,000, while maintenance costs were Rs. 5000.

Based on above data the total cost for the construction of fuel operated grass cutter is approximately 40,750 Rs. But in the case of current project work i.e., BLDC motor operated portable grass cutter, the total operating cost are as follows:

BLDC motor in Rs. = 9000, Controller in Rs. = 5000, Frame in Rs. = 5000, Maintenance cost in Rs. = 2000 and Battery (30 Ah) cost in Rs. = 10000.

So, the total cost for the construction of BLDC motor operated portable grass cutter is approximate 30,000 Rs. Therefore, from the above analysis on the basis of cost analysis, we can save 40% cost in BLDC operated grass cutter compared to the fuel powered grass cutter.

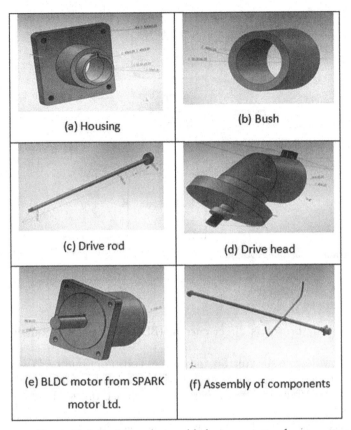

Figure 45.1 Design and assembled components of grass cutter
Source: Author

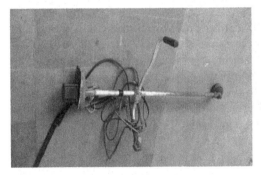

Figure 45.2 Fabricated components
Source: Author

Specifications

According to the available data sheet for engine specification fuel consumption is 500 ml/hr whose power output is 350 W/4000 rpm; torque at no load is 1 N-m and weight is 3 kg. Based on this data, we have selected a BLDC motor having the following specifications such as motor power output required is 220 W; speed is 3000-4000 rpm; torque is 0.7 N-m; voltage is 48 volt and weight is 1.85 kg.

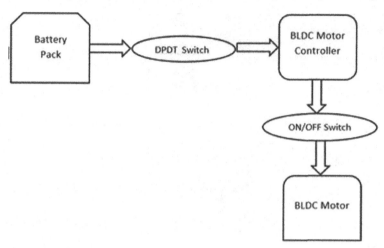

Figure 45.3 Block diagram
Source: Author

Calculations

Battery power consumption by brushless direct current (BLDC) motor per hour is calculated as follows:

Power required is 220 W; voltage is 48-volt. So, the current consumption/hr = P/V = 220/48 = 4.58 ampere/hr. From this calculation it is found that if the above motor run for 1 hr, it can consume approximately 4.58 ampere current. Based on this requirement, battery pack is selected.

But for this current project work, 30 Ah battery pack is needed. So that motor can be run for 30/4.58 = 6.55 hr approximately.

Conclusion

The outcomes of our project, "Design, Fabrication, and Modification of Conventional Portable Grass Cutter," were good. The project takers who intend to make more alterations will find it simpler. In the current project, a grass cutter that is both manually and motor-driven is used effectively. A string blade increases cutting efficiency, and it is simple to collect the grass cutter. Due to the modest weight and small footprint of this lawn cutter, operating expenses are lower. Compared to market lawn cutters that are powered by gasoline, the cost of the system is inexpensive.

Future scope

The manufactured portable lawn cutter may be powered by solar energy, paving the way for a more environmentally friendly future.

References

Abraham, D, and Divyashree, S (2021). Design and implementation of solar grass cutter. *I-Manager's Journal on Electrical Engineering*. 14(4), 47.

Arora, H, Sagor, J A, Panwar, V, Sharma, P, Mishra, S K, Arora, P G, and Singh, P K (2019). Design and fabrication of autonomous lawn mower with water sprinkler. *Think India Journal*. 22(12), 472–484.

Arunesh, R S, Arunesh, S, and Nivetha, N (2016). Design and implementation of automatic lawn cutter. *IJSTE–International Journal of Science Technology and Engineering*. 2(11), 202–207.

Bb, A, Abraham, D, Ms, H, and Dk, M (2021). Design and implementation of solar grass cutter. *Journal on Electrical Engineering*. 14(4), 47–52.

Garg, A, Gao, L, Panda, B N, and Mishra, S (2017). A comprehensive study in quantification of response characteristics of incremental sheet forming process. *The International Journal of Advanced Manufacturing Technology*. 89, 1353–1365.

Khan Mohammad Majedul Hasan, B, Shemeoun, M D, Islam, S M, and Debnath, D (2022). Design and fabrication of smart auto-grass cutter for analyzing its peformance (Doctoral dissertation, Sonargoan University).

Nagarajan, N, Sivakumar, N S, and Saravanan, R (2017). Design and fabrication of lawn mower. *Asian Journal of Applied Science and Technology (AJAST)*. 1(4), 50–4.

Rahul, J K, Chakraborty, S, Khayer, N, Hriday, A R, Mia, M L, Islam, M Z, and Ahmed, T (2021). Design and re-engineering of an automated solar lawn mower. In 2021 International Conference on Automation, Control and Mechatronics for Industry 4.0 (ACMI). IEEE. pp. 1–6.

Ulhe, P P, Inwate, M D, Wankhede, F D, and Dhakte, K S (2016). Modification of solar grass cutting machine. *International Journal of Innovative Science and Research Technology*. 2(11), 711–714.

46 Rheological and mechanical properties of different types of lightweight aggregate concrete

Snigdhajit Mukherjee[1,a], Rajesh Kumar[2,b], A Sofi[3,c] and Monalisa Behera[4,d]

[1]Department of Structural and Geotechnical Engineering, School of Civil Engineering Vellore Institute of Technology, Vellore, India

[2]Senior Scientist & Head, Organic Building Materials (OBM) Group CSIR-Central Building Research Institute, Roorkee, India

[3]Department of Structural and Geotechnical Engineering, School of Civil Engineering Vellore Institute of Technology Vellore, India

[4]CSIR-National Aerospace Laboratories Bangalore, India

Abstract

Numerous studies have discussed lightweight aggregate concrete because of its reduced self-weight and better dynamic responses, which significantly reduce the density of concrete. Basic fresh properties are used to determine the variable performance of this type of concrete under various conditions, including water/powder ratio and types of aggregates. Experimental demonstrations are being used to study the fresh and mechanical characteristics of conventional concrete and self-compacting lightweight concrete. Major mixtures were studied in terms of workability such as slump flow, V-funnel, viscosity, and mechanical properties viz. crushing strength under various curing environments. As varying super plasticizer dosages cause the dispersion and dissolution of the particles in concrete over time, a thorough examination of numerous experiments is taken into consideration. The study identifies the impact of inter particle behavior in a fresh state, which also tends to have a significant effect on mechanical properties. The paper focuses on the rheological values of various concrete mix types, which show how the structure is built up and how thixotropic the concrete is. Following several models that describe flow behavior and mathematically validate the rheological model, the linear and nonlinear behavior of self-compacting concrete have been analyzed.

Keywords: Lightweight concrete, rheology, thixotropy, workability

Introduction

With an annual global demand of about 26.8 billion tons, concrete consumption by the construction industry is at its peak. Low self-weight adversely affects the self-compacting concrete (SCC) features, such as its high fluidity which can cause aggregate to float on top of the concrete (Li et al., 2021). The self-weight of the SCC provides a driving force to maintain the concrete's flow property, which is related to the rheological parameters. Lightweight self-compacting concrete (LWSCC) serves both the attributes of SCC and lightweight concrete (LWC). The workability of lightweight aggregate concrete (LWAC) is mainly determined by proper gradation and basic properties of the lightweight aggregate (LWA) particles. The behavior of each form of LWC has thus been advised to be studied to draw a more precise conclusion (Mouli

[a]snigdhajit1997@gmail.com, [b]rajeshkumar@cbri.res.in, [c]sofime@gmail.com, [c]monalisabehera7@gmail.com

DOI: 10.1201/9781003450924-46

and Khelafi, 2007). This paper investigates the rheological properties of (LWC) with varying properties and compositions. Moreover, it specifies the thixotropic properties as well, which defines how the concrete will behave under dynamic conditions. The article throws significant light on the futuristic approach that must be considered for optimizing the mix design from the rheology point of view.

Mix proportion

LC^3 was used as a material for binding which contains 55% OPC (grade 43) with 30% calcined clay and limestone slurry (15%). The natural aggregates have been replaced by 50% and 100% LECA. The target strength was assumed to be 25 MPa after 28 days of curing. Table 46.1 represents some of the material components per cubic meter of LWC designed as per the EFNARC code.

Fresh properties

Slump flow test

The flow property of self-compacting lightweight concrete increased when the w/b ratio gets increased. This is happening because of the decrement in friction between the fine particles as well as water content regardless of the types of aggregate used (Figure 46.1). The flow value reached the maximum value when the w/b ratio was 0.35 for diatomite (Uygunoglu and Topcu, 2010). Sometimes the metakaolin which was used as supplementary cementitious material (SCM) leads to an increase in the slump flow because of the fineness and plate-like particle (Bogas et al., 2012).

V- funnel and J ring test

It has been observed that only the pumice at a w/b ratio of 0.32 gave a suitable value (Figure 46.2) of the V- funnel following the EFNARC code. The other LWA used such as pumice, diatomite was seen to have very low value (Uygunoglu and Topcu, 2010). According to Guneyisi et al., 2016, there was a little increase in the viscosity when the nano-silica dosage used as SCM varied between 0 and 0.5 %. The height difference in J-ring depends on the aggregate and w/b ratio same as that of the V-funnel as shown in (Figure 46.2).

Rheological properties

The term rheology refers to the flow rate. The absolute parameters calculated are yield stress and plastic viscosity which are not particular at a constant time. Bingham

Table 46.1: Material component per cubic meter.

Mixes	LC^3 (kg/m^3)	Natural CA (kg/m^3)	Natural FA (kg/m^3)	LECA %	w/c ratio	SP %	VMA %
M 1	420	600	1200	0	0.55	1.5	0.50
M 2	490	-	-	100	0.5	1	0.35
M 3	490	600	1200	50	0.45	1	0.35
M 4	490	600	1200	50	0.5	1.5	0.50

Source: EFNARC, Lotfy et al., 2014

Model follows a linear stress and strain relationship that leads to the concept of the thixotropy of a material. Thixotropy defines the reduction of viscosity of a colloidal substance at a shear rate (Aydin et al., 2022). The Bingham equation follows as $\tau = \tau_0 + \mu\dot{\gamma}$ where τ = shear stress (Pa), τ_0 = yield stress (Pa), μ = plastic viscosity (Pa.s) and $\dot{\gamma}$ = shear rate (1/sec). According to Gesoglu et al., 2014; the flow curve has been drawn to calculate the absolute parameters that we get from rheometers (Figure 46.3) which follows like T = Y + VN where T = Torque (Nm), Y is the intercept of the axis of torque related to yield stress, V = slope of the line, and N = speed. It is stated that the mixes having low yield stress must be ignored because the minimum limit is 10 Pa (Singh et al., 2017). The surface charge of the metakaolin particles when used as SCM increases the plastic viscosity at rest (Aydin et al., 2022).

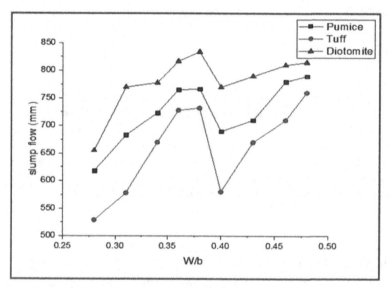

Figure 46.1 Relation between slump flow and w/b ratio
Source: Topcu and Uygunoglu, 2010.

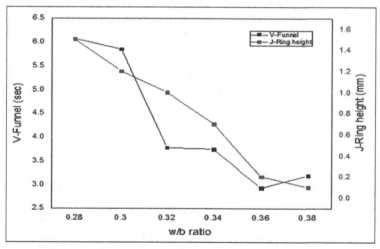

Figure 46.2 Relation between J-ring and V-funnel for different LWA for different w/b ratio
Source: Topcu and Uygunoglu, 2010

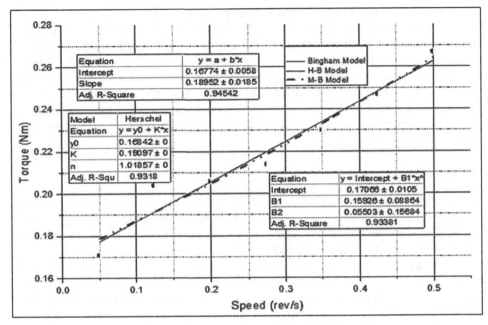

Figure 46.3 Relation between torque and speed for mix M 2
Source: Laboratory experiment

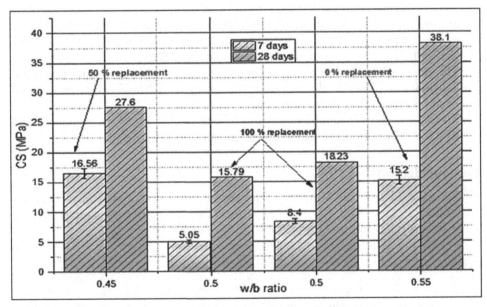

Figure 46.4 Relation between CS and w/b ratio for mix M 2 different replacement percentage of LWA
Source: Experiment

The shear rate and the thixotropy property can be well defined by particle cluster theory which leads to the shear thickening property which happened for mix M 1. The power law was evaluated for mix M 2 to find that mixes are having n value less than 1 under the same w/b ratio which leads to the breakdown of the cluster. This is known as the shear thinning condition. The lightweight fine and coarse aggregates reduce the plastic viscosity and tend to have a shear thinning property with a dynamic yield stress of 28.35 Pa calculated from the Torque vs Speed curve (Figure 46.3) for mix M 2. The change in stress is due to the aggregate proportion and force induced to shear the material at a constant rate being less. Moreover, the viscosity reduces under dynamic motion produced by the vane of the rheometer. The study has been made for the nonlinear power model and Modified Bingham (M-B) model where the flow index (n) varied from 0.9 to 1.02.

Mechanical properties

The crushing strength of a specimen is known as the compressive strength (CS) which is calculated at 7, 28 and 90 days of curing. Mix M 2 having the partial replacement of LECA in concrete (nearly 50 %) gives a satisfactory result (27 MPa) as shown in Figure 46.4 so that the target strength of 25 MPa can be achieved. When crushed limestone aggregate is used as LWA there is an improvement in CS of about 68% with the age of concrete and w/b ratio (Uygunoglu and Topcu, 2010).The strength reduces with an increase in LWA and the w/b ratio due to the stress concentration at the inter-face of LWA and cement paste keep on increasing (Gesoglu et al., 2014).

Conclusions

- Lightweight aggregate (LWA) minimizes yield stress and plastic viscosity as less force is required to provide a dynamic flow in concrete due to its surface texture and bulk density. The mathematical interpretation of linear and nonlinear behav-ior of concrete has been discussed through different rheological models and this change in flow is due to the high-water demand of LWA.
- The effect of metakaolin dosage as supplementary cementitious material (SCM) has been also presented through shear thickening (H-B Model) as the metakaolin particles increase the plastic viscosity.
- The studies show a significant approach such as the volume of aggregate used which will be considered for optimizing various mixes to get a satisfactory CS.

Acknowledgments

We are grateful to 'The Ministry of Environment, Forest and Climate Change (MoEF&CC), New Delhi, Government of India' for the sustained financial support to the project (File Number: 19/44/2018/RE; Project No.: GAP0120).

References

Aydin, E, Kara, B, Bundur, Z, Ozyurt, N, Bebek, O, and Gulgun, A (2022). A comparative evaluation of sepiolite and nano-montmorillonite on the rheology of cementitious materi-als for 3D printing. *Construction and Building Materials*, 350, 128935. doi: 10.1016/j. conbuildmat..128935.

Bogas, J A, Gomes, A, and Pereira, M F C (2012). Self-compacting lightweight concrete pro-duced with expanded clay aggregate. *Construction and Building Materials*, 35, 10131022. doi: 10.1016/j.conbuildmat.2012.04.111.

Gesoglu, M, Güneyisi, E, Ozturan, T, Oz, H, and Asaad, D (2015). Shear thickening intensity of self-compacting concretes containing rounded lightweight aggregates. *Construction and Building Materials*, 79, 4047. doi: 10.1016/j.conbuildmat.01.012.

Kumar, R and Srivastava, A (2022). Influence of Lightweight Aggregates and Supplementary Cementitious Materials on the Properties of Lightweight Aggregate Concretes. *Iranian Journal of Science and Technology, Transactions of Civil Engineering*. doi:10.1007/s40996-022-00935-5.

Lotfy, A, Hossain, K, and Lachemi, M (2014). Application of statistical models in proportioning lightweight self-consolidating concrete with expanded clay aggregates. *Construction and Building Materials*, 65, 450469. doi: 10.1016/j.conbuildmat.2014.05.027.

Topcu, İ and Uygunoğlu, T (2010). Effect of aggregate type on properties of hardened self-consolidating lightweight concrete (SCLC). *Construction and Building Materials*, 24(7), 12861295. doi: 10.1016/j.conbuildmat.2009.12.007.

47 On game theory integrated particle swarm optimization for crack detection in cantilever beams

Prabir Kumar Jena[1,a], Rabinarayan Sethi[2,b] and Manoj Kumar Muni[3,c]

[1,2,3]Department of Mechanical Engineering, VSSUT, Burla, Odisha, India

[2,3]IGIT, Sarang, Odisha, India

Abstract

An inverse technique has been discussed in the current work to simplify the procedure of estimating the crack details of the beam model. The Euler-Bernoulli beam with a crack has been chosen for current investigation. The proposed approach modified the particle swarm optimization (PSO) equation with the concept of game theory called as game theory integrated particle swarm optimization (GTPSO). The GTPSO modifies the equation of particles velocity of PSO as per the cooperative and defective strategy to improve the convergence speed and thus, forecast the crack details with accuracy by minimizing the function based on frequency information obtained from theoretical investigation. A fair comparison has been made among the results obtained through GTPSO and PSO. The results attained reveal that the proposed GTPSO is associated with good accuracy compared to PSO.

Keywords: Crack detection, natural frequency, inverse approach, particle swarm optimization, game theory integrated particle swarm optimization

Introduction

The location and intensity of crack are mainly the two parameters that bring considerable changes in beam vibration characteristics. The problem of crack detection can be organized into two categories. The first category of problem mainly deals with the theoretical and finite element methods and are discussed below. Yokoyama et al. (1998) employed a linear spring model to investigate variations of bending vibration characteristics of Euler-Bernoulli beam with crack of unlike sizes. Considering the above model, Loya et al. (2022) carried out the vibration analysis investigation of Timoshenko beam containing a crack in an elastic medium. The second category of problems mainly deals with the inverse way to find the crack details from the known values of natural frequencies. Some of the research work contributed to these directions are as follows: Chinchalkar (2001) and Kim and Stubbs (2003). To make the above process easy to implement and reliable, the crack identification problem could be modelled as an optimization problem to predict crack details by defining an objective function based on frequency records. Few studies related to the use of GA in the problem of crack detection are made by Horibe and Watanabe (2006) and Baghmisheh et al. (2008). Complexities involved in the use of operators of GA has attracted many towards the use of swarm intelligence techniques to the discussed problem domain. PSO, that utilizes swarm intellect has proved better than GA in terms of accuracy of results obtained as described by the work carried out by Baghmisheh et al. (2012), and Jena and Parhi (2015).

[a]pkjena_me@vssut.ac.in, [b]rabinsethi@igitsarang.ac.in, [c]manoj1986nitr@gmail.com

DOI: 10.1201/9781003450924-47

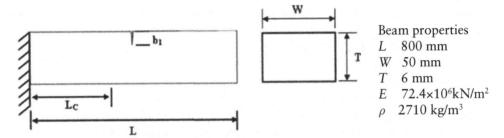

Beam properties
L   800 mm
W   50 mm
T   6 mm
E   72.4×10⁶kN/m²
ρ   2710 kg/m³

Figure 47.1 Geometry of cantilever beam
Source: Author

In the present paper care has been taken to tune the controlling parameters of PSO and the concept of game theory is used to modify the equations of PSO (GTPSO) to attain good accuracy of results.

Cracked beam model

The beam geometry under investigation is presented in Figure 47.1. The beam undergoes both longitudinal and transverse motion under the action of P_1, the axial force and P_2, the bending moment. The established relations among the additional strain energy, J and stress intensity factors (SIFs), (K_1) at the crack site expressed by Tada et al. (1973) as:

$$J(b) = \frac{W}{E^*} \int_0^{b_1} [K_1(P_1) + K_1(P_2)]^2 db \tag{1}$$

$$K_{11} = \frac{P_1}{WT} \sqrt{\pi b} \left(F_1 \left(\frac{b}{T} \right) \right) \text{and} K_{12} = \frac{6P_2}{WT^2} \sqrt{\pi b} \left(F_2 \left(\frac{b}{T} \right) \right) \tag{2}$$

F_1 (b/T) and F_2 (b/T) are the correction factors used to obtain the stress intensity factors (Tada et al., 1973).

Formulation for bending vibration

The equations of beam model under investigation undergoing free longitudinal and transverse vibrations are described as follows:

$$\frac{\partial^2 s_i}{\partial x^2} = \frac{1}{H_u^2} \frac{\partial^2 s_i}{\partial t^2} \text{ and } \frac{\partial^4 w_i}{\partial x^4} + \frac{1}{H_y^2} \frac{\partial^2 w_i}{\partial t^2} = 0 \tag{3}$$

where i = 1 for $0 \leq x \leq L_c$ and i = 2 for $L_c \leq x \leq L$.

The solution has been obtained by solving Equation (3) and the boundary conditions are:

$$\underbrace{S_1 = 0;\ W_1 = 0;\ W_1' = 0;}_{\text{Fixed end boundary conditions}}\ \underbrace{S_2' = 0;\ W_2'' = 0;\ W_2''' = 0}_{\text{Free end boundary conditions}} \tag{4}$$

The other related and appropriate end conditions at the cracked section are as follows:

$$S_1'(a=\tfrac{L_c}{L}) = S_2'(\alpha = \tfrac{L_c}{L}); W_1(\alpha) = W_2(\alpha); W_1''(\alpha) = W_2''(\alpha); W_1'''(\alpha) = W_2'''(\alpha) \tag{5}$$

$$AES_1'\big|_{L_c} = K_{11}(S_2|_{L_c} - S_1|_{L_c}) + K_{12}\left(W_2'\big|_{L_c} - W_1'\big|_{L_c}\right) \tag{6}$$

$$EIW_1''\big|_{L_c} = K_{21}(S_2|_{L_c} - S_1|_{L_c}) + K_{22}\left(W_2'\big|_{L_c} - W_1'\big|_{L_c}\right) \tag{7}$$

The natural frequencies obtained from the above forward approach is given to the inverse approach as an input to forecast the crack place and its extent.

Inverse analysis of crack detection

The present section explains the inverse way of finding the crack details from frequencies information employing GTPSO. Firstly, the underlying concept of PSO is summarized and subsequently, the way equation of PSO has been modified as per game theory is discussed.

Revisit of PSO

The thought of revised version of the classical PSO was instigated by Shi and Eberhart (1998) by suggesting an inertia weight parameter to the velocity revise equation of PSO. The governing equations of PSO that update the position of particles is as follows:

$$\left.\begin{aligned} x_{ij}(k+1) &= x_{ij}(k) + v_{ij}(k+1) \\ v_{ij}(k+1) &= wv_{ij}(k) + C_1R_1(pbest_{ij}(k) - x_{ij}(k)) + C_2R_2(gbest_j(k) - x_{ij}(k)) \end{aligned}\right\} \tag{8}$$

where R_1 and R_2 are two different random numbers generated in the range [0, 1]. The cognitive learning coefficients C_1 and social learning coefficients C_2 are changed with iterations (Jena and Parhi), (2015).

Game theory integrated particle swarm optimization

In the proposed approach of GTPSO, the PSO equation is modified as per cooperative and defective strategies that are employed in the evolutionary game theory. The particles for cooperative strategy update its velocity by the following equation:

$$V_{ij}^{k+1} = w \times V_{ij}^k + C_2 \times R_2 \times (gbest_j(k) - x_{ij}(k)) \tag{9}$$

Whereas, for defective strategy the particle updates its velocity by the following equation:

$$V_{ij}^{k+1} = w \times V_{ij}^k + C_1 \times R_1 \times (pbest_{ij}(k) - x_{ij}(k)) \tag{10}$$

The particle updates its position as per the Equation (8) and the inertia weight has been modified as:

$$w_i(k) = w_{min} + \frac{rank_i}{N}(w_{max} - w_{min}) \tag{11}$$

where *rank*$_i$ represents a rank assigned to each particle depending on the cost function value. The discussed algorithms minimize the cost function of the present problem based on frequency information. The iterative process continues until it satisfies the stopping criterion.

Cost function

The cost function established for the optimization problem is presented by the following equation:

$$\text{minimize } f = \sqrt{\sum_{i=1}^{n} \left(f_{norm,i}^{d} - f_{norm,i}^{e} \right)^2} \tag{12}$$

the following problem constraints: $0 < b_1 < T$ and $0" < L_c < L$
where $f_{norm,i}^{d}$ and $f_{norm,i}^{e}$ are the normalized beam natural frequency supplied from the desired source and attained from the algorithm. From the calculation point of view, the current problem considers the value of n equals to 3.

Results and discussion

The variations of the cracked beam natural frequencies for known value of crack intensity located at distinct crack site are plotted and shown in Figure 47.2. The

Table 47.1: Setting of control parameters for PSO and GTPSO algorithms.

Parameter	NP	D	max_iter	w_{min}	w_{max}	C_{1i}	C_{1f}	C_{2i}	C_{2f}
Value	10	2	100	0.4	0.9	2.0	0.5	0.5	2.0

Source: Author

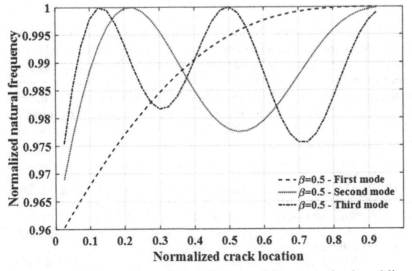

Figure 47.2 Comparison of normalizednatural frequencies for three different modes for a definite non-dimensional crack depth
Source: Author

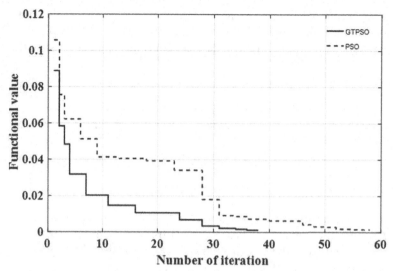

Figure 47.3 Comparison of convergence between GTPSO and PSO for crack location 600 and crack depth 2.4 mm
Source: Author

second and third natural frequencies follow a definite pattern, but not monotonic as detected in the case of the first natural frequency. From the convergence trend as depicted in Figure 47.3 it is obvious that GTPSO converges faster compared to PSO.

The minimum and maximum value algorithm tuning parameters that are taken to estimate the objective function value is presented in Table 47.1. The number of particles generated, and the dimension of the problem are presented by the term NP and D.

The results revealed that GTPSO predicts results with error of (0.16%, 0.16%) for crack location of 600 and depth of 2.4 mm compared to (0.28%,0.33%) from PSO.

Conclusion

The inverse analysis employing game theory integrated particle swarm optimization (GTPSO) and particle swarm optimization (PSO) algorithms have been carried out to predict crack details from the available frequency information. The regulating parameters of algorithms have been adjusted to improve the precision of results attained. The results attained from inverse analysis reveal that the GTPSO outperforms PSO in terms of predicting crack details. The future scope of the present work is to authenticate the outcomes of the discussed algorithm applied to different beam configurations.

References

Baghmisheh, M T V, Peimani, M, Sadeghi, M H, and Ettefagh, M M (2008). Crack detection in beam-like structures using genetic algorithms. *Applied Soft Computing.* 8, 1150–1160.
Baghmisheh, M T V, Peimani, M, Sadeghi, M H, Ettefagh, M M., and Tabrizi, A.F. (2012). A hybrid particle swarm–Nelder–Mead optimization method for crack detection in cantilever beams. *Applied Soft Computing.* 12, 2217–2226.
Chinchalkar, S (2001). Determination of crack location in beams using natural frequencies. *Journal of Sound and Vibration.* 247, 417429.
Horibe, T and Watanabe, K (2006). Crack identification of plates using genetic algorithm. *Journal of Jpn Society of Mechanical Engineering.* 49(3), 403410.

Jena, P K and Parhi, D R K (2015). A modified particle swarm optimization technique for crack detection in cantilever beams. *Arabian Journal for Science and Engineering*. 40, 32633272.

Kim, J T and Stubbs, N (2003). Crack detection in beam-type structures using frequency data. *Journal of Sound and Vibration*. 259, 145–160.

Loya, J A, Ruiz-Aranda, J, and Zaera, R (2022). Natural frequencies of vibration in cracked Timoshenko beams within an elastic medium. *Journal of Theoretical and Applied Fracture Mechanics*. 118, 103257.

Shi, Y and Eberhart, R C (1998). Parameter selection in particle swarm optimization. In Proceedings of 7th International Conference on Computation Programming VII.

Tada, H, Paris, P C, and Irwin, G R (1973). The stress analysis of cracks hand book. Pennsylvania, USA: Del Research Corp. Yokoyama, T and Chen, M C (1998). Vibration analysis of edge-cracked beams using a line spring model. *Engineering Fracture Mechanics*. 59(3), 403–409.

48 Evaluating chloride resistance of high strength recycled aggregate concrete integrating metakaolin

Uma Shankar Biswal[a] and DinakarPasla[b]

School of Infrastructure, Indian Institute of Technology Bhubaneswar, Bhubaneswar, India

Abstract

The current study examines the chloride diffusion performance of recycled aggregate concrete that entirely replaces coarse natural aggregate with recycled aggregate. In total, nine types of concrete mixes were prepared, three of which were mixed with w/b values of 0.25, 0.30, and 0.35; in the remaining six mixes, in order to make high-strength recycled aggregate concretes, metakaolin was substituted by 10, and 15% by weight of Portland cement, for each w/b. The diffusion characteristics of chloride was investigated using rapid chloride penetration test. Using analysis of variance, a further statistical significance of metakaolin on chloride diffusion property was determined. In this connection, the comparative investigation demonstrated promisingbenefits for the preparation of recycled aggregate concrete (RAC) with metakaolin.

Keywords: Recycled aggregate concrete, metakaolin, chloride diffusion, sustainability

Introduction

Rapid urban growth, limited land-fill sites, and excessive exploitation of natural resources have made recycling a crucial aspect of meeting the growing demand for construction materials (Biswal et al. 2022). Due to the lower density, water absorption, and grading of recycled aggregates (RA), they weaken concrete's strength and durability. Previous investigations attributed the cause to porous adhered mortar. However, uses of supplementary cementitious materials like flyash, GGBS (Shankar Biswal and Dinakar 2023), metakaolin (MK), and silicafume (Biswal and Dinakar 2022) seem to an easy, and effective way to improve the porous nature of RAand thus make recycled aggregate concrete (RAC).

There is currently very little data available regarding the impact of recycled coarse aggregate (RCA) on high-strength concrete (Gonzalez-Corominas and Etxeberria 2016). In addition, there is little data about MK's effect on the behavior of RAC of high strength. Our previous studyBiswal and Dinakar (2022) explored the impact of MK on the mechanical and durability attributes of high-strength RAC. This research investigates the chloride permeability of high-strength RAC manufactured with 100% RCA and two different replacement levels (10% and 15%)of MK so that it can be used with greater confidence in coastal and marine environments.

Material and method

Ordinary Portland cement 53 grade was used as the main cementing material, whereas MK as the supplementary cementitious material.As coarse aggregates, two size fractions (20 mm RCA and 10 mm RCA) of RCA were obtained from a recycling facility,

---

[a]usb10@iitbbs.ac.in, [b]pdinakar@rediffmail.com

DOI: 10.1201/9781003450924-48

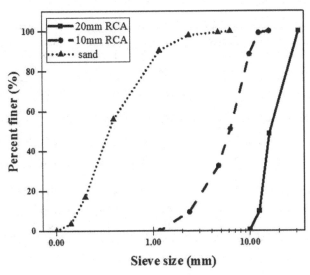

Figure 48.1 Grading of aggregates
Source: Author

Table 48.1: Properties of Materials considered for mix design.

Material	Specific gravity	Water absorption
Cement	3.14	-
Metakaolin	2.59	-
Sand	2.63	0.8 %
10 mm RCA	2.30	5.5 %
20 mm RCA	2.33	3.06 %

Source: Author

while as fine aggregate, locally sourced sand was used. Individual aggregate grading of aggregates is presented in Figure 48.1. The combined aggregate grading was carried out using the DIN standard to achieve the particles' maximum packing Biswal and Dinakar (2021a). The specific gravity along with water absorption of all the materials considered for the mix design are displayed in Table 48.1. The corresponding mix design details, following the absolute volume method Biswal and Dinakar (2021b) are shown in Table 48.2. The ASTM C 1202 standard was followed to perform the Rapid chloride penetration test (RCPT). Disc specimens with a diameter of 100 mm and thickness of 50 mm were utilized for the test. The samples were kept in water curing for 28 and 90 days, followed by one day of drying in the laboratory. Figure 48.2 depicts the experimental setup. The concrete sample's chloride ion diffusion resistance was measured by total charge passed (TCP).

Results and discussion

Rapid chloride penetration test

The RCPT can give a good estimate of the ionic permeability of concrete over a broad spectrum of concrete quality. The TCP observed for all the developed concrete

Table 48.2: Mix design details.

Mix ID	Portland cement (kg/m³)	Metakaolin (kg/m³)	Water (kg/m³)	20 mm RCA (kg/m³)	10 mm RCA (kg/m³)	Sand (kg/m³)
25RC00	680	0	170	402	501	537
25RM10	612	68	170	398	497	533
25RM15	578	102	170	397	495	531
30RC00	567	0	170	426	531	570
30RM10	510	57	170	423	528	566
30RM15	482	85	170	422	526	565
35RC00	486	0	170	443	553	593
35RM10	437	49	170	441	550	590
35RM15	413	73	170	440	549	589

Source: Author

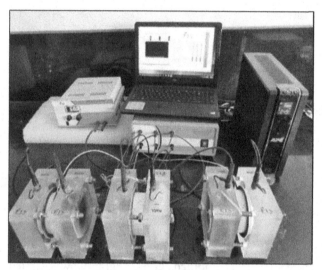

Figure 48.2 Test setup for RCPT test
Source: Author

is shown in Figure 48.3. It can be observed that with increased w/b, TCP through the sample increased both at 28 and 90days.For instance, when w/b was raised from 0.25 to 0.35, the TCPat 28 and 90 days of curingwas increased by 6% and 16%, respectively. This increase is due to increased porosity in the cement paste with the rise in w/b (Wee et al, 1999). Additionally, it was noticed that the inclusion of MK is more crucial at lower w/b. For instance,10 and 15% replacement levels reduced the TCP by 17.64 and 21.17%, respectively, for 0.35 w/b, whereas TCP was reduced by 24.65 and 30.13% for 0.25 w/b at 90 days of curing. The reduction in TCP is due to the pore structure refinement due to the inclusion of finer MK particles, which hinders the path of chloride ion migration (Panesar, 2019). Also, it can be noticed that the effect of MK is more crucial at 90 days. For instance, in 25RC00, a 15% reduction in TCP was observed between 28 days and 90 days, whereas for 25RM15, the reduction is 19%.This is due to the pozzolanic reactivity properties of MK, which facilitate a

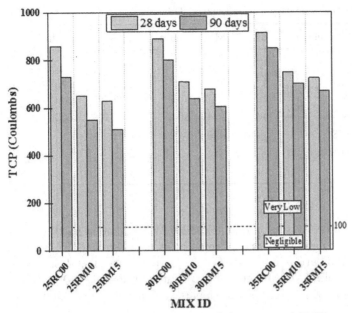

Figure 48.3 Influence of MK on total charge passed (TCP)
Source: Author

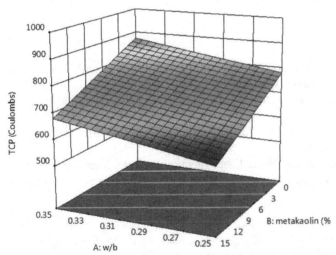

Figure 48.4 3D response surface plot of TCP for different levels of w/b and metakaolin (curing of 60 days)
Source: Author

chemical reaction with hydrated cement paste and water. This causes the formation of stronger phases that causes the microstructure to become more compact (Panesar, 2019).

Statistical analysis

ANOVA was done using Design Expert software (Version 13) to determine the significance of w/b, metakaolin level, and curing time on TCP, with all input parameters

Table 48.3: ANOVA on total charge passed of MK-based RACs.

Source	Sum of squares	df	Mean square	F-value	p-value
Model	1.626E+05	3	54211.61	79.93	< 0.0001
A-w/b	26226.75	1	26226.75	38.67	< 0.0001
B-metakaolin	1.170E+05	1	1.170E+05	172.50	< 0.0001
C-curing period	19404.50	1	19404.50	28.61	0.0001
Residual	9495.68	14	678.26		
Cor total	1.721E+05	17			
R^2=0.945					

Source: Author

showing significance ($p < 0.05$) (Biswal and Dinakar 2022). Metakaolin had the most impact on TCP for high-strength RAC, as confirmed by its high F-value (172.5) compared to w/b (38.67%) and curing time (28.61%). Equation 1 describes TCP's relationship with w/b, metakaolin level, and curing time, and Figure 48.4's 3D response surfaces provide a practical way to optimize variables, superior to a single-point statistical test.

TCP (Coulombs) = 612.227 + 935* w/b + −12.9286* Metakaolin

+ −1.05914* curing period   (1)

Conclusion

The following deductions can be made by replacing Portland cement with metakaolin in high-strength recycled aggregate concrete:

The inclusion of metakaolin has decreased the chloride ion penetration at both 28 and 90 days.

It is possible to achieve 'very low' chloride ion permeability even with 100% coarse recycled aggregate in the system.

The statistical analysis firmly established that metakaolin has a significant impact on the total charge passed for high-strength concrete made from recycled aggregates.

References

Biswal, U S and Dinakar, P (2021a). Effect of aggregate grading on the fresh and mechanical performance of recycled aggregate self compacting concrete. *The Indian Concrete Journal.* 95(5), 1–11.

Biswal, U S and Dinakar, P (2021b). A mix design procedure for fly ash and ground granulated blast furnace slag based treated recycled aggregate concrete. *Cleaner Engineering and Technology.* 5, 100314. https://doi.org/10.1016/j.clet.2021.100314.

Biswal, U S and Dinakar, P (2022). Influence of metakaolin and silica fume on the mechanical and durability performance of high-strength concrete made with 100% coarse recycled aggregate. *Journal of Hazardous, Toxic, and Radioactive Waste.* 26(2). https://doi.org/10.1061/(ASCE)HZ.2153-5515.0000687.

Biswal, U S, Mishra, M, Singh, M K, and Dinakar P (2022). Experimental investigation and comparative machine learning prediction of the compressive strength of recycled aggregate concrete incorporated with fly ash, ggbs, and metakaolin.*Innovative Infrastructure Solutions.* 7(4), 242. https://doi.org/10.1007/s41062-022-00844-6.

Biswal, U S and Dinakar P (2023). Evaluating corrosion resistance of recycled aggregate concrete integrating ground granulated blast furnace slag. *Construction and Building Materials* 370 (March): 130676. https://doi.org/10.1016/j.conbuildmat.2023.130676.

Gonzalez-Corominas, A and Miren, E (2016). Effects of using recycled concrete aggregates on the shrinkage of high performance concrete. *Construction and Building Materials*. https://doi.org/10.1016/j.conbuildmat.2016.04.031.

Panesar, D K (2019). Supplementary cementing materials. In *Developments in the Formulation and Reinforcement of Concrete*. 55–85. Elsevier. https://doi.org/10.1016/B978-0-08-102616-8.00003-4.

Wee, T H, Suryavanshi, A K, and Tin, S S (1999). Influence of aggregate fraction in the mix on the reliability of the rapid chloride permeability test. *Cement and Concrete Composites*. 21(1), 59–72. https://doi.org/10.1016/S0958-9465(98)00039-0.

49 Hydration kinetics and stability study of pure phases of cement clinker with the addition of SCMs and dopants

Dipendra Kumar Das[1,a], Rajesh Kumar[2,b] and A Sofi[1,c]

[1]Department of Structural and Geotechnical Engineering, School of Civil Engineering Vellore Institute of Technology (VIT), Vellore, India

[2]Senior Scientist & Head, Organic Building Materials (OBM) Group CSIR-Central Building Research Institute, Roorkee, India

Abstract

Cement is the most important binding material that is used in the development of the construction industry. Although the main component of modern cement is Portland cement clinker, there is potential for alternate development. The lower emission of carbon dioxide during the calcination of raw materials is the focus of research work because of anthropogenically induced climate change. Additionally, the goal is to utilize more locally available raw materials and alternative energy. Pure phases of cement clinker can be used to study the complicated cement hydration, which is due to the heterogeneous cement clinker composition. The most important compounds that are responsible for strength are tricalcium silicate and dicalcium silicate, which together make about 7080% of cement. The factors that affect the pure phases of cement clinker are summarized in this paper. The effects of mineral admixtures, dopants, elements, and chemical admixtures composites on reaction chemistry and the stability of the hydrated phases formed, hydration kinetics, and macroscopic properties are discussed by studying their hydration mechanism. The addition of supplementary cementitious materials effects the amount and the types of hydrates that are formed when reacted with individual phases of cement. The increase in the uptake of silica rich SCMs effects the ratio of calcium/silicate in the hydrated phase of calcium silicate and the pH of the matrix with the consumption of portlandite, which is formed during the hydration of tricalcium and dicalcium silicate. SCMs reaction begins later and is accelerated by temperature and pH.

Keywords: Alite, belite, dopants, supplementary cementitious material, Low carbon cement

Introduction

Cement is one of the largest construction materials. With a global output of about 2 billion tons per year, it is still the most used material today. The cement production industry is mainly responsible for around 7% of all worldwide emissions of CO_2. The primary reason this percentage is rising so quickly is that cement production is expanding faster than emissions are currently being decreased. The development of the cement production industry is one of the major sources of emissions of greenhouse gases, particularly CO_2 emissions. This is a result of the calcination of raw materials required for cement production and the combustion of fuels required to sustain high temperatures in a kiln. To understand the chemical and physical characteristics of cement, basic research is still needed to develop efficient characterization techniques. With the development of new cement, it is possible to replace more than 50% of the clinker mass without compromising quality. Since the clinker phases of cement are impure compounds that form some solid solutions with other minor oxides in

[a]dipendradkd@gmail.com, [b]rajeshkumar@cbri.res.in, [c]sofime@gmail.com

DOI: 10.1201/9781003450924-49

the clinker, their composition has a significant impact on several cement properties, including setting behavior and strength development. Although OPC has a variety of clinker phases that interact with water, the hydration process of C_3A and C_3S is very crucial for the early characteristics of cementitious products Stephan and Wistuba (2006). In the presented paper, the studies regarding the cementitious pure phases i.e., Alite and Belite with alkalis, SCMs, etc. are being discussed.

Chemical background

Alite phase (C_3S)

Alite is a solid solution of C_3S with minor oxides that determines the early strength. It contains monoclinic or trigonal forms in the clinker. The reaction of C_3S with the water produces calcium silicate hydrate, and portlandite. C-S-H constitutes 50–60% of the total solids by volume in a fully hydrated paste of cement, which determines the strength properties of the concrete. $Ca(OH)_2$ is naturally alkaline, which keeps the pH of concrete at 13 and prevents reinforcement corrosion.

Belite phase (C_2S)

Dicalcium silicate is another major crystalline component that makes about 15–30 wt.% of typical Portland cement clinkers. The different allotropic forms of C_2S are α, $\alpha'H$, $\alpha'L$, β, and γ. In industrial clinkers, the main form of doped belite is β -C_2S, which is formed from the ionic substitution of Ca^{2+} or Si^{4+} ions. The contribution on strength of this phase is small up to 28 days as it reacts slowly with water, but the strength increases significantly at later stages.

Composition and synthesis of pure phases

Tricalcium Silicate Synthesis(C_3S)

The solid reaction utilizing the calcium carbonate and silica ratio (4.695:1) is more efficient and frequently employed for large-scale synthesis of C_3S (Li et al., 2018). It has been seen that most of the study employs a heating rate of 5°C/min, however, a quicker heating speed of 7°C/min shows no detrimental effects on the quality of the C_3S when C_3S is heated at 1600°C for 3 hrs (Li et al., 2018). The mix composition for C_3S synthesis is given in Table 49.1.

Dicalcium Silicate Synthesis(C_2S)

The solid reaction is more efficient and frequently employed for large-scale synthesis of C_2S. At room temperature, γ-C_2S form is stable. The other forms of C_2S are stabilized by atmospheric quenching or the substitution of chemicals by foreign elements at room temperature. The sintering process is 1 hour and 45 minutes of ramping up to 1450°C followed by 4 hours of holding (Wesselsky and Jensen, 2009). The mix composition for 100 g of the final C_2S synthesis is given in Table 49.2.

Table 49.1: Mix constituents for the synthesis of C_3S.

Components	$CaCO_3(g)$	$SiO_2(g)$
C_3S	625.00	133.11

Source: Li et al., 2018

Table 49.2: Mix constituents for the synthesis of C₂S.

Components	Calcium carbonate	Silica	Boric Acid	Borax
β-C_2S	115.752	34.745	0.354	0
α'-C_2S	111.851	33.574	0	7.108

Source: Wesselsky and Jensen, 2009

Results and discussion

Effect of SCMs on C_3S

The induction period of hydration can be slightly delayed with the addition of gypsum when compared with plain alite, but the acceleration period and peak hydration rates are increased (Quennoz et al., 2013). The hydration of C_3S can be enhanced by the addition of sulphate. When compared to pure C_3S, the DoH of C_3S in C_3S/C_3A systems is greater, which can be explained by the increased C_3S hydration caused by the addition of gypsum (Zunino and Scrivener, 2020). $CaCO_3$ and $CaSO_4.2H_2O$ both have the potential to accelerate C_3S hydration. Up to the hydration age of 10–12 hours, the accelerating impact of the former is stronger than that of the latter, and subsequently, the reverse result is observed. Compared to pure C_3S and C_3S-$CaCO_3$ paste, the calcium/silica ratio of the C-S-H gel formed in C_3S- $CaSO_4.2H_2O$ paste is greater. Consequently, $CaSO_4.2H_2O$ can reach into the C-S-H gel and increase the strength of C_3S(Zhang, 2007). Metakaolin and limestone powder have little impact on the acceleration of C_3S, however, they do improve hydration. Chloride can alter the C_3S hydration kinetics and has the capacity to improve and accelerate the primary hydration peak, which results in a quicker rate of hydration (Sui et al., 2022). It is found that annealing ground and quenching C_3S do not alter the crystal structure. It may lead to a decrease in defect density that affects the hydration kinetics, extending the induction time, the disappearance of the peak at the end of the induction period, increasing the acceleration slope, and producing similar deceleration profiles (Bazzoni et al., 2013).

Alkalis and Acidic effect

Alkali additives on alite are supposed to increase the degree of hydration in a short duration period (the first few days), but on a longer duration period (28 days), the DoH is identical (Kumar et al., 2012). The addition of sodium hydroxide and sulphate causes the primary heat peak to be more intense, the induction time to be shorter, and the rate of deceleration to be quicker. It has been found that the increase in the alkalinity of the solution accelerates the precipitation of CH by lowering the number of Ca^{2+} ions needed to attain the required supersaturation concentration. Additionally, the concentration of silicate in the solution rises, thereby causing the C-S-H product to precipitate earlier. At the same DoH, alkalis boost the flexural and compressive strength of alite micro mortars (Mota et al., 2015). The addition of $CaCl_2$ to C_3S paste increases the rate and overall production of early hydration products.

Citric-acid-modified chitosan can greatly delay the rate of hydration of C_3S and hinder the C_3S dissolution and nucleation of portlandite and C-S-H phase. With the increase in doping dosage, CAMC shows an increase in inhibitory effect on the hydration of the C_3S (Wang et al.,2022).

Effect of oxides and dopants on C_2S

The stabilization of β-C_2S by B_2O_3 is independent of quenching temperature, while K_2O and SO_3 stabilize β-C_2S when the quenching temperature is over 1200°C. The β-C_2S stabilizers also affect the temperature of hydration. As compared to two other stabilizers, SO_3 delays the hydration process at an early stage of hydration but shows a high rate of heat release at later stage. The specimens stabilized by B_2O_3 and SO_3 exhibit disintegrated foil, whereas those stabilized by K_2O exhibit needle form. In comparison to B_2O_3 and K_2O, SO_3 produces C-S-H with a lower crystallinity (PARK, 2001). A partial breakdown of C_3S into the two distinct phases of α'H -C_2S and CaO can be produced by the addition of phosphorus at 0.1 wt% P_2O_5. This provides the concept that stabilization of α'H -C_2S polymorph can be obtained by phosphorus impurity and suggests that the phosphorus solubility limit in T1 C_3S is low (≤ 0.1 wt.% P_2O_5). The α'H phase of $C_2S(P)$ contains more than 1 wt.% P_2O_5 of phosphorus (De Noirfontaine et al., 2009). The doping of Fe and P with dicalcium silicate hydration studies suggest that the initial rate of the hydration of α'-C_2S form is greater than that of the β-C_2S form (Benarchid et al., 2005).

Conclusions

- The research in this field is increasing every year so as to know the effect of addition of SCMs on the individual clinker phase of cement in terms of reactivity, DoH, hydration kinetics, and strength of hydrated compounds formed at different environmental conditions like temperature, pH, and nature of the solution.
- The use of raw materials and industrial wastes that are harmful to the environment can be used with the aim of replacing cement clinker with optimized dosing of SCMs and dopants to reduce CO_2 emissions from the environment.

Acknowledgement

We are grateful to 'The Ministry of Environment, Forest and Climate Change (MoEF&CC), New Delhi, Government of India' for the sustained financial support to the project (File Number: 19/45/2018/RE; Project No.: GAP0090).

References

De Noirfontaine, M, Tusseau-Nenez, S, Signes-Frehel, M, Gasecki, G and , Girod-Labianca, C (2009). Effect of phosphorus impurity on tricalcium silicate t1: from synthesis to structural characterization. *Journal of the American Ceramic Society*. 92(10), 2337-2344. doi:10.1111/j.1551-2916.2009.03092.x.

Li, X, Ouzia, A, and Scrivener, K (2018). Laboratory synthesis of C_3S on the kilogram scale. *Cement and Concrete Research*. 108, 201207. doi: 10.1016/j.cemconres.2018.03.019.

Mota, B, Matschei, T, and Scrivener, K (2015). The influence of sodium salts and gypsum on alite hydration. *Cement and Concrete Research*. 75, 5365. doi: 10.1016/j.cemconres.2015.04.015.

Wang, L, Zhang, Y, Guo, L, Wang, F, Ju, S, Sui, S, Liu, Z, Chu, H, and Jiang, J (2022). Effect of citric-acid-modified chitosan (CAMC) on hydration kinetics of tricalcium silicate (C_3S). *Journal of Materials Research and Technology* 21, 36043616. doi: 10.1016/j.jmrt.2022.10.118.

Wesselsky, A and Jensen, O. M (2009). Synthesis of pure Portland cement phases. Cement and Concrete Research 39(11), 973-980. doi: 10.1016/j.cemconres.2009.07.013.

Zhang, Y and Zhang, X (2008). Research on effect of limestone and gypsum on C_3A, C_3S and PC clinker system. *Construction and Building Materials* 22(8), 1634-1642. doi: 10.1016/j.conbuildmat.2007.06.01

Zunino, F and Scrivener, K (2020). Factors influencing the sulfate balance in pure phase C_3S/C_3A systems. *Cement and Concrete Research* 133 (April), 106085. doi: 10.1016/j.cemconres.2020.106085.

50 Performance evaluation of quality of service parameters of proposed hybrid TLPD-ALB-RASA scheduling algorithm in cloud computing environment

Vijay Mohan Shrimal[1,a], Yogesh Chandra Bhatt[2,b] and Yadvendra Singh Shishodia[3,c]

[1]Computer Science Department, Jagannath University, Jaipur, India

[2]Emeritus Professor, Jagannath University, Jaipur, India

[3]Former Pro_Vice-Chancellor, Jagannath University, Jaipur, India

Abstract

Cloud computing enables centralized access to resources via a cluster, facilitating the dynamic allocation of resources in response to user demand. To achieve enhanced system performance during peak demand, cloud computing schedulers use task scheduling and resource allocation algorithms to balance workloads across cloud resources. However, the hybrid task length, task priority and task deadline merged adaptive load balancing with deadline (hybrid TLPD-ALB) scheduling algorithm does not meet the multi-objective requirements of load balancing. The resource aware scheduling algorithm (RASA) is added to hybrid TLPD-ALB to optimize resource utilization and minimize scheduling processing time. To ensure concurrent and fair scheduling and allocation of users without task starvation, a good scheduler performs these tasks quickly.

Keyword: Cloudsim, hybrid TLPD-ALB-RASA, adaptive load balancing, hybrid task length and deadline, shortest job first

Introduction

The process of job scheduling in cloud computing requires the development and implementation of new scheduling algorithms to allocate resources such as CPU, memory, and network bandwidth to tasks in a cost-effective and efficient way (Sanjay (2014). This study aims to improve the performance by using load balancing techniques to decrease the load on virtual machines (VMs) (Al-Faris et al., 2013). The study recommends the use of RASA, which combines Min-Min and Max-Min strategies, to optimize resource utilization (Ali and Alam, 2018).

This paper evaluates the performance of the hybrid TLPD-ALB-resource aware scheduling algorithm (RASA) algorithm using different combinations of cloudlets and virtual machines Kaur and Arora (2015). Five scenarios are considered, and the quality of service (QoS) metric, including total processing time, is measured (Arslan and Buyukkokten, 2010). The study compares and analyzes the results of traditional algorithms, previous hybrid algorithms (hybrid TLPD) and hybrid TLPD-ALB), and the newly proposed hybrid TLPD-ALB-RASA algorithm (Shrimal et al., 2022). With the above in mind, this study concentrates on designing and evaluating performance, with the goal of optimized the utilization of computing resources Kaur and Singh Dhindsa (2018) in cloud environments through traditional and hybrid algorithms.

[a]vijay2007shrim@gmail.com, [b]b_yogeshchandra@yahoo.co.in, [c]yad.shi@gmail.com

DOI: 10.1201/9781003450924-50

Literature review

An evaluation of the prevailing written works was conducted to compare various techniques for resource allocation. Kanani and Maniyar (2015) discussed how work scheduling algorithms can meet user expectations by maintaining high resource utilization and enhancing performance with rapid response time, while highlighting the challenges of scheduling with the Max-Min method. The RASA algorithm combines the benefits of both algorithms while minimizing their drawbacks (Akhilandeswari et al., 2017). In the Min-Min algorithm, larger jobs are required to wait for VM allocation, whereas in the Max-Min algorithm, smaller tasks are deprived of VM allocation. This is a limitation of both algorithms (Kaur and Kaur, 2022). This study identified gaps in previous research on load balancing optimization in cloud environments and developed a hybrid framework that includes resource provisioning and load balancing. This framework aims to optimize VM utilization by distributing the load evenly Kaur and Kaur (2019). Load balancing aims to improve both user satisfaction and resource utilization by ensuring the efficient and appropriate distribution of computer resources (Singh et al., 2017).

Research gap

Upon reviewing multiple research papers, it has become apparent that a small proportion of researchers have directed their efforts toward devising efficient scheduling algorithms for optimizing resource utilization in cloud computing. Thus, further investigation is necessary to enhance resource utilization in the field of cloud computing. As a result, it is imperative to consider a wider range of scenarios and QoS parameters to improve system performance and optimized resource utilization.

Methodology of proposed algorithm

The hybrid TLPD-ALB-RASA algorithm aims to tackle load balancing challenges, optimize resource utilization, and minimize overall processing time. The hybrid algorithm performs effectively in multiple optimization metrics, including reducing total processing time, optimizing resource utilization, and distributing workloads evenly among resources (Akhilandeswari et al., 2017). The algorithm first calculates the average task length (AvgTaskLen) and then creates two counters, one for tasks with length less than or equal to the average (TaskLenCountMIN) and the other for tasks with length greater than the average (TaskLenCountMAX). It selects a task (Ti) from the MyQ queue, compares its length to the average, and increments the corresponding counter.

Algorithm: Hybrid TLPD-ALB-RASA

Procedure 1: Credit based on length of task (Thomas and VP, 2015); Procedure 2: Priority credits assigning to task (Thomas and VP, 2015); Procedure 3: Deadline of the task; Procedure 4: Deadline time based adaptive load balancing

Procedure 5: Resource aware scheduling algorithm (RASA)

For all tasks Ti in the task set (MyQ),

Compute (AvgTaskLen) of all tasks in MyQ

If the length of task is less or equal to average task length then add task in the list of TaskLenCount$_{MIN}$

else

Add task in the list of TaskLenCount$_{MAX}$
End if condition
If TaskLenCount$_{MIN}$ is greater than or equal to TaskLenCount$_{MAX}$, then
Perform Max-Min scheduling
Else perform Min-Min scheduling
End if condition
End for

Total credits are calculated from hybrid TLPD algorithm and adaptive load balancing is added from hybrid TLPD-ALB so the flow chart of proposed hybrid TLPD-ALB-RASA is added after that previous flowchart:

Flowchart of hybrid TLPD-ALB-RASA:

The flowchart in Figure 50.1 categorizes tasks based on their length and then decides whether to use Max-Min scheduling (allocating resources to longer tasks first) or

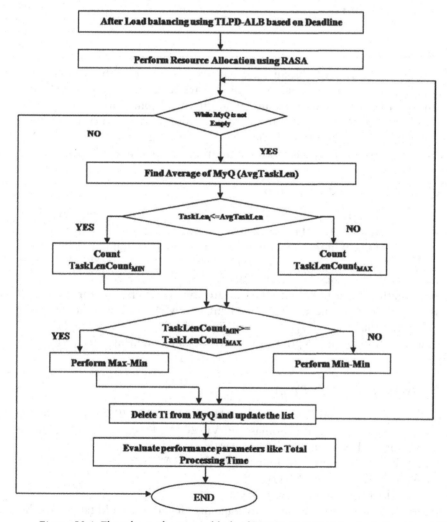

Figure 50.1 Flowchart of proposed hybrid TLPD-ALB-RASA scheduling algorithm
Source: Author

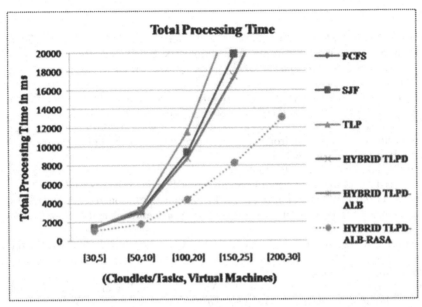

Figure 50.2 Graphical representation of total processing time in different scenarios
Source: Author

Min-Min scheduling (allocating resources to shorter tasks first) based on whether the number of tasks with minimum length count is greater than or equal to the number of tasks with maximum length count.

Simulation setup and result analysis

The system is composed of five hosts, each with 5000 MIPS processing speed and 5048 MB of RAM (Saini et al. 2019). Virtual machine configurations vary from 5 to 30, each hosting a different number of cloudlets (30, 50, 100, 150, and 200).

In Figure 50.2 the graph compares the minimum total processing time achieved by proposed and traditional scheduling algorithms in different scenarios. It shows that incorporating the RASA algorithm in the hybrid TLPD-ALB scheduling algorithm improves performance indicating optimal resource utilization and allocation. Overall, the graph indicates that the proposed hybrid TLPD-ALB-RASA scheduling algorithm outperforms traditional scheduling algorithms in terms of minimum total processing time in various scenarios.

Conclusion and future work

This paper evaluates the performance of a proposed algorithm called hybrid TLPD-ALB-RASA using combinations of varying cloudlets and varying virtual machines. The algorithm's quality of service (QoS) is measured by the total processing time metric in five different scenarios. The proposed TLPD-ALB-RASA algorithm outperforms compared with traditional and previous proposed algorithms in all scenarios, with improved QoS results. Future research exploring additional QoS parameters could improve resource utilization in cloud computing, leading to greater efficiency in resource usage. Designing and comparing additional hybrid algorithms based on the results of this study has the potential to yield even better results.

References

Akhilandeswari, P, Nymisha, K K, Gandavaram, J, and Srimathi, H (2017). CRASA: Cloud Resource Aware Scheduling Algorithm A Hybrid Task Scheduling Algorithm Using Resource Awareness. *ARPN Journal of Engineering and Applied Sciences*. 12(12), 3706-3710.

Al-Faris, M, Al-Dubai, A, and Al-Fuqaha, A (2013). Load balancing algorithms in cloud computing: A survey. *Journal of Computer Networks and Communications*. 1-13.

Ali, S A and Alam, M (2018). Resource Aware Min-Min (RAMM) Algorithm for resource allocation in cloud computing environment. International Conference on Information and Communication Technologies (ICICT).

Arslan, M and Buyukkokten, O (2010). Quality of service support for cloud computing systems. In Proceedings of the ACM Symposium on Applied Computing. pp. 1369-1374.

Euniza, J, Nor, M H, Jaya, P R, and Haron, Z (2014). Double layer concrete paving blocks using waste tyre rubber as aggregate replacement. *Applied Mechanics and Materials*, 554, 128132.

Gaikwad, S, Nalage, S, Nazare, N, and Joshi, R (2019). Use of waste rubber chips for the production of concrete paver block. *International Research Journal of Engineering and Technology*, 6(3), 48294832.

Kanani, B and Maniyar, B (2015). Review on Max-Min task scheduling algorithm for cloud computing. *Journal of Emerging Technologies and Innovative Research*. 2(3), 781-784

Kaur, P and Arora, V (2015). Cloud computing: A review of open research challenges. *Journal of Network and Computer Applications*. 52, 1-10.

Kaur, A and Kaur, B (2019). Load balancing optimization based on hybrid Heuristic-Metaheuristic techniques in cloud environment. *Journal of King Saud University - Computer and Information Sciences*. 31(2), 148-159.

Kaur, R and Singh Dhindsa, D K (2018). Efficient task scheduling using load balancing in cloud computing. *International Journal of Advanced Network and Applications*. 10(3), 3888-3892. doi:10.35444/ijana.2018.10037

RamKumar, S, Vaithiyanathan, V, and Lavanya, M (2014). Towards efficient load balancing and green IT mechanisms in cloud environment. *World Applied Sciences Journal*. 29(3), 159-165.

Saini, H, Upadhyaya, A, and Khandelwal, M K (2019). Benefits of cloud computing for business enterprises: a review. Proceedings of International Conference on Advancements in Computing & Management 2019.

Sanjay, M (2014). Introduction to cloud computing. In Proceedings of the International Conference on Information Management and Engineering. pp. 73-77.

Satyanarayanan, M (2009). From cloud computing to cloudlets. *IEEE Pervasive Computing*. 8(4), 14-23.

Singh, A B., Bhat J, S, Raju, R, and D'Souza, R (2017). A comparative study of various scheduling algorithms in cloud computing. *Scientific & Academic Publishing*. 7(2), 28-34.

Shrimal, V M, Bhatt, Y C, and Shishodia, Y S (2022). Performance evaluation of QOS parameters of hybrid TLPD scheduling algorithm in cloud computing environment. *International Journal of Advanced Research in Computer and Communication Engineering*. 11(12), 183-190.

Thomas, A, G, K and VP, J R (2015). Credit Based Scheduling Algorithm in Cloud Computing Environment. In Proceedings of the International Conference on Information and Communication Technologies. pp. 913-920.

51 Design and development of concrete paver block using waste tyre rubber

Kali P Sethy[1,a], M. Srinivasula Reddy[2,b], Sundaram Jena[1,c], Debasmita Malla[1,d] and Arpan Pradhan[3,e]

[1]Department of Civil Engineering Government College of Engineering Kalahandi, Odisha, India

[2]Civil Engineering Department, G Pulla Reddy Engineering College, Kurnool, AP, India

[3]Department of Civil Engg. Chirst Deemed University, Bangalore, India

Abstract

Waste rubber tyre is one of the critical ecological issues around the world. The removal of used rubber tyres is currently becoming a serious global problem for trash management. According to estimates, 1.2 billion discarded tires are shipped around the world each year. As a result of the expansion in auto portable creation, there is a need to appropriately arrange the tremendous measures of utilized rubber tyres. Waste materials change between strong, fluid, and vaporous materials. Rubber tyres have a place with strong waste materials that cannot be arranged in landfills without unsafe impacts on climate. In the event of removal of tyres in landfills, enormous area of land will be consumed with no advantage. Rubber tyre in concrete is precisely sliced to the necessary sizes. And a mixture proportion of rubberized concrete developed with the fractional substitution of coarse aggregate. Conventional concrete along with 5%, 10%, 15%, and 20% replacement were made to verify different concrete parameters.

Keywords: Waste tyre, concrete, rubber, environmental hazard

Introduction

Use of rubber concrete is reasonable and financially effective. It opposes the high strain, impact, and temperature. The rubberized concrete has also great water opposition with low retention, worked on corrosive obstruction, low shrinkage, high effect opposition, and phenomenal sound and warm protection (Gaikwad et al., 2019). An enormous piece of disposed of tyres winds up amassed in garbage removal destinations without having been exposed to a particular treatment before capacity (Thomas et al., 2015). The tyres maintain water for an extensive stretch, since it is impermeable and has a unique shape, it can reproduce territory to mosquitoes and unlike irritations (Mohammed et al., 2012). Throughout the long term, the removal of discarded rubber tyres has been a basic natural worry. Tyres is intended to have a generally excellent protection from climate and waste tires do not debase in landfill (Martinez et al., 2013). Double sheet concrete paving block by substituting aggregates with unused tyre with balance among mechanical characteristics and different characteristics like the thickness (Euniza et al., 2014).

Materials used

The basic concrete materials such as, cement OPC 53 grades, fine aggregate, coarse aggregate, water along with waste rubber is used.

[a]kaliprasanna87@gmail.com, [b]sm26@iitbbs.ac.in, [c]sundaramjena75@gmail.com, [d]malladebasmita20@gmail.com, [e]dr.arpanpradhan@gmail.com

DOI: 10.1201/9781003450924-51

Cement (OPC 53 grade)

The manufacturing of paver block 53 grade OPC are used conforming to Indian Standards (IS): 8112- 1989 was utilized. The cement's specific gravity was viewed as 3.15 and the normal consistency of the cement was attained as 33%.

Fine Aggregate and Coarse Aggregate

Confirming natural river sand to Zone II according to IS: 383-1970 is considered here in this current study. Coarse aggregate of 20 mm sizes is utilised for the development of paver block samples.

Water

Water used in the experimental work is normal drinking water and is confirmed to IS 105001983. The water is utilized for the arrangement of fresh concrete and curing of concrete samples throughout the test duration.

Rubber waste

Waste rubber tyre are collected from the nearby local garages and cut into small pieces according to the requirement. The rubber chips separated from the tyre are sieved through 12 mm and retained in 10 mm and are utilised for the substitution of coarse aggregate in the present research work.

Mix details

Concrete of M_{20} grade mix was prepared utilizing waste tyre rubber with coarse aggregate substitution in part at 0%, 5%, 10%, 15%, and 20%. Then the mix design

Table 51.1: Specification details of aggregate (fine and coarse).

Sl. No.	Test	Fine Agg.	Coarse Agg.
		Result	
1	Specific Gravity	2.46	2.73
2	Water Absorption	1.1%	0.25%
3	Fineness Modulus	5.5	23.61%
4	Size	Passing through 4.75mm	-
5	Crushing value	-	2.91

Source: Author

Table 51.2: Mix details for 1 m³ of rubber concrete.

Sl.No	Conc. name	Grade	Cement (kg\m³)	Rubber (%)	C. A (kg\m³)	F. A (kg/m³)	Water (kg)
1	RCP0	20MPa	315.00	0	1110	821	157.6
2	RCP5	20MPa	299.25	5	1110	821	157.6
3	RCP10	20MPa	283.50	10	1110	821	157.6
4	RCP15	20MPa	267.75	15	1110	821	157.6
5	RCP20	20MPa	252.00	20	1110	821	157.6

Source: Author

is according to the Indian Standard code of IS 10262:2009. The mix identification is given in the mix details Table 51.2.

Preparation procedure of test specimens

In the present experimental investigation, the paver block sample of height 75 mm is used. Outer dimension of the paver block is considered as 250 mm length and 200 mm width. Individually, 36 numbers of specimens were casted for each mix for all the tests. In the conventional mix, 0% rubber is added whereas in the mix abbreviated with Rubberized Concrete Paver (RCP):AQ5 contains 5%, RCP10 contains 10%, RCP15 contains 15% and RCP20 contains 20% of waste rubber tyre. Total three major tests, like workability, compressive and bending forces tests are performed for the developed samples.

Results and discussion

Workability test

Workability of the developed fresh concrete mixture is examined using slump cone apparatus. The workability parameter of fresh concrete is conducted as per the IS 1199: 1959.Immediately the mixture of fresh concrete is taken, and the slump test was conducted. The results of the workability parameter are plotted in the Figure 51.1.

Compressive strength test

The cast specimens were left for curing till the test duration. For each duration, minimum of three samples were considered.

According to the obtained strength parameters, graph is plotted and shown in Figure 51.2. The most common test is compressive strength test, which gives an idea in regard to all the traits of concrete (Loganathan et al., 2021).

Flexural strength test

Flexural stiffness of the developed samples was tested by using the CTM testing machine. The paver block samples are tested for the to check the bending strength

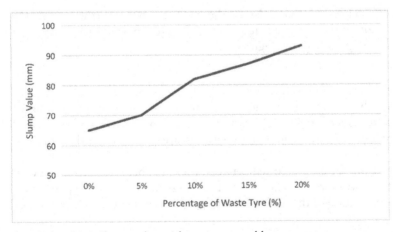

Figure 51.1 Slump value with waste tyre rubber percentage
Source: Author

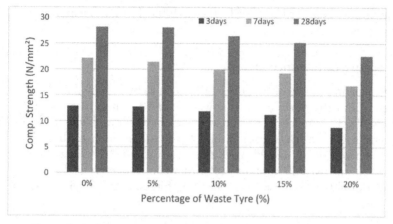

Figure 51.2 Comp. strength value with replaced rubber percentage
Source: Author

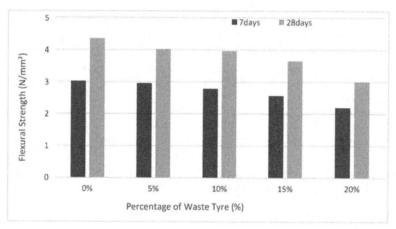

Figure 51.3 Flexural strength value with replaced rubber percentage
Source: Author

and is described as the pressure that exists at the most significant furthest either the strain or stress portion of the fibre. Average of three concrete paver block results at 7 days and 28 days were plotted in the Figure 51.3.

Conclusions

This present research work is about the development of a mix proportioning of M20 grade concrete paver block samples with waste rubber tyre. The developed samples are termed as RCP and following decisions are drawn from the obtained outcomes.

It can be noticed from the graph plotted between waste rubber tyre percentage with slump values that, the workability is keeps on increasing with the increase percentage of rubber. It may be because of the less or insignificant water absorption of the waste tyres.

With the rise percentage of waste rubber tyre, the compressive strength is also gets reduced. And a target compressive strength of 25.21 N/mm^2 is achieved at a rubber tyre replacement of 15%. And at 20% replacement, the target strength is slightly missed.

Overall, this present study may conclude that, the waste rubber tyre paver block is more environment friendly sustainable blocks.

References

IS 1199. (1959). Methods of sampling and analysis of concrete. New Delhi: Bureau of Indian Standards.

IS 383. (1970). Specification for coarse and fine aggregates. New Delhi: Bureau of Indian Standards.

IS 8112. (1989). Specification for 43 grades ordinary Portland cement. New Delhi: Bureau of Indian Standards.

IS 10262. (2009). Guidelines for concrete mix design proportioning. New Delhi: Bureau of Indian Standards.

Jusli, U, Nor, M H, Jaya, P R, and Haron, Z (2014). Double layer concrete paving blocks using waste tyre rubber as aggregate replacement. *Applied Mechanics and Materials*, 554, 128132.

Loganathan, M, Manohar, B, and Manishankar, K (2021). Comparative study on behaviour of concrete pavement block using coconut fibre and coconut shell. *ARSCB*, 25(4), 79087914.

Martinez J D, Puy, N, Murillo, R, García, T, Navarro, A M, and Mastral, A M (2013). Waste tyre pyrolysis – a review. *Renewable and Sustainable Energy Reviews*, 23, 179213.

Mohammed, B S, Khandaker, M, Anwar, H, Jackson, T E S, Grace, W, and Abdullahi, M (2012). Properties of crumb rubber hollow concrete block. *Journal of Cleaner Product*, 23, 5767.

Shivradnyi, G, Sandesh, N, Namdev, N, and Rajendra, J (2019). Use of waste rubber chips for the production of concrete paver block. *International Research Journal of Engineering and Technology*, 6(3), 48294832.

Thomas, B S, Gupta, R C, Meher, A P, and Kumar, S (2015). Performance of high strength rubberized concrete in aggressive environment. *Construction and Building Materials*, 83, 320326.

52 Effect of permeability property of micro concrete on durability of concrete repair

Dipti Ranjan Nayak[1], Rashmi R. Pattnaik[2] and Bikash Chandra Panda[1]

[1]Indira Gandhi Institute of Technology, Odisha, India

[2]Orissa University of Agriculture & Technology, Odisha, India

Abstract

The permeability property of concrete can be considered as an important property while study the durability aspect of any structure. The serviceability life of any concrete structure depends on various properties such as physical, chemical, and durability. Similarly, the repair of any concrete structure needs special attention to the selection of the repair materials based on their properties. This article mainly focuses on the selection of the right kind of repair materials. Six different micro- concrete was chosen having dissimilar properties. The grain size, compressive strength of each micro concrete sample was calculated. Then the initial and secondary rate of absorption of each micro concrete specimen was evaluated. The sorptivity coefficients (K_i and K_s) of micro concrete were not the only factor determining permeability properties; the compressive strength was also found to affect permeability properties, lower sorptivity coefficient observed with higher compressive strength and vice versa.

Keywords: Micro-concrete, absorption, sorptivity, durability

Introduction

The durability of any concrete structure can be evaluated by its mass transportation properties. Studies have already shown that concretes durability is largely influenced by its mass transport properties, such as sorptivity and permeability. Less durability of concrete found at higher sorptivity or permeability (Elawady et al., 2014). Particularly the durability of a reinforced concrete structure can be determined by the resistance property of water penetration as it speeds up the course of corrosive processes inside the elements. Hence, the factors liable for the longevity of concrete structures as well as their confrontation with the reason for corrosion influence the sorptivity property for describing the speed of water penetration into concrete (Choudhary et al. 2020). Based on the above a test has developed by the researcher which is derived from Darcy's law of unsaturated flow (Hall 1989). The transport property is also influenced by aggregates, but broadly, it contains very small openings which are irregular and do not allow the movement of water by capillary action, hence doesn't influence to sorptivity.

Objectives of the research

It is very difficult to evaluate the water transport mechanism of concrete having a porous network. This is because, numerous dissimilar types of transport mechanisms combined with different types of pores which usually appear in the same porous system. Study on water transport mechanisms for different designed mixes is highly essential to evaluate the sorptivity coefficient of a concrete structure. This experimental study tried to understand the behaviour of the micro-concrete.

DOI: 10.1201/9781003450924-52

Materials and methods

Materials used for experiments

Six different cementations micro-concrete such as M1, M2, M3, M4, M5, and M6were selected and used in this study. In order to maintain flow of 120mm on a flow table, sufficient care has been given while mixing the quantity of water to maintain the flow 120mm on the flow table. Specimen M1, M2, and M3weretaken from the laboratory prepared micro-concrete, mixed withOPC-53 having 53 MPa compressive strength at 28 days, PSC cement and aggregate of size 4.75mm. The ratio of the mix (cement and aggregate) was 1:1added with, powdered Poly-carboxylate ether as a superplasticizer. Similarly, specimens M4, M5, and M6 were taken from the micro-concrete prepared from OPC with silica fume, fly ash, and smaller aggregate in different proportions.

Test program and specimen preparation

Sieve analysis

Each sample of 400 g was sieved by using a set of Sieves of different size from 2.36 mm, to 25 microns according to IS:2386 (Part-I). The samples were allowed to pass through the series of sieves by setting 20 minutes of sieving duration in the sieve shaker. After the sieving duration the weight retained, percentage of passing, and cumulative percentage of passing were determined. The cumulative percentage of micro-concrete passing through a 150- micron screen as per Table 52.1 was taken into consideration to find out the grade of micro-concrete from coarser to finer.

Compressive strength

Micro-concretes of smaller particle 4.75 mm or downgraded has been used to evaluate the compressive strength of the samples, (ASTM C109, 2020) test technique was employed and 50 mm cubes were prepared for testing.Table 52.2 shows the results of the compressive strengths of the specimens, M1 to M6 prepared from the micro concretes calculated at 3, 14, and 28 days after the day they were casted.

Table 52.1: Grading of micro-concretes from M1 to M6.

Size of sieves	Cumulative passing (%) of repair materials					
in mm	M1	M2	M3	M4	M5	M6
2.36	100	100	100	99.9	100	99.6
1.18	85	85	85	84.7	88.8	86
0.6	70	70	70	71.5	70.4	70.8
0.3	55	55	55	57.5	52.8	52.6
0.15	48.25	48.63	33	47.5	39.2	43
0.075	10.75	11.23	4.4	11.9	8.8	31
0.045	1.25	0.94	1.2	1.5	2	8.2
0.025	0.25	0.2	0.2	0.3	0.8	3
pan	0	0	0	0	0	0

Source: Author

Table 52.2: Compressive strength of the micro-concretes specimens from M1 to M6.

Micro-Concrete	Compressive strength in MPa		
	3 days	14 days	28 days
M1	35.74	47.28	56.76
M2	31.63	36.37	45.27
M3	35.69	37.35	41.71
M4	39.96	43.02	55.93
M5	39.08	46.53	65.28
M6	17.96	25.01	42.39

Source: Author

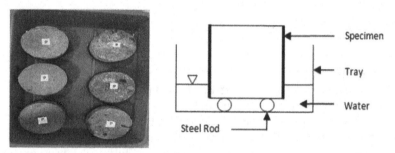

Figure 52.1 Typical arrangement for sorptivity test
Source: Author

Sorptivity

Figure 52.1 shows the typical arrangement of the sorptivity test of micro concretes. The sorptivity of micro concrete specimens M1, M2, M3, M4, M5, and M6 were conducted as per (ASTM C1585, 2004) with cylinders' size 100 mm dia and the heights were 50 mm. The samples were oven dried at a temp of 50^0 with 80% relative humidity for 3 days. Later on, the samples were kept in a sealed container separately. After 15 days, the weight of the specimens was taken. The sides of the sample were wrapped with adhesive tape leaving one surface free for absorption. Then the specimens were placed in a plastic tray as shown in Fig.1 over the support and the level of tap water was maintained at 3mm above the top of the support during the period of the test. The initial timing and date were recorded. The mass of the specimens was taken at an interval of 5 min, 10 min, 20 min, 30 min, 60 min, 6 hrs, 3 days, 7 days, and 9 days.

Results and discussion

Grading of micro concretes

Table 52.1 shows the cumulative percentage of the small aggregates present in micro-concrete sample M1 to M5The sieving data indicates that the particle size of micro-concrete samples was below 2.36 mm. It was found that the cumulative percentage of aggregate fleeting through a 150μ sieve is minimum in M3 sample and maximum in case of M2. The order from coarser to finer micro-concrete is found in the order of M3, M5, M6, M4, M1, and M2.

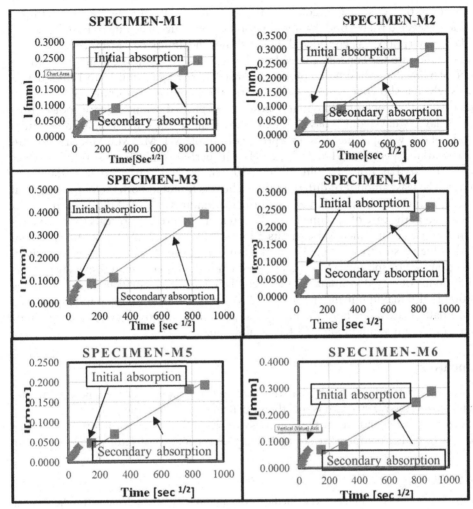

Figure 52.2 Initial and secondary absorption of specimen M1 to M6
Source: Author

Compressive strength

The compressive strength of six different micro-concretes specimens M1, M2, M3, M4, M5, and M6 were measured at 3, 14, and 28 days are indicated inTable 52.2. The mean of the three samples were taken to evaluate the outcomes of each period at 3, 14, and 28 days. From the experiment, maximum compressive strength of 65.28 MPa was found in case of specimen M5 at 28 days and a minimum value of 41.71 MPa was found for specimen M3 under similar curing conditions. Dissimilar strength gained by different micro-concretes specimens obtained at 3, 14, and 28 days.

Sorptivity of micro-concrete

Two phases of tests such as initial absorption and secondary absorption were conducted to determine the rate of water absorption in concrete. The 1st phase tests were conducted to evaluate resistance property of concrete against the water infiltration and the first six hours of the test was termed as initial absorption. In the secondary absorption, the resistance of concrete against water infiltration was evaluated for a period of seven days beyond the initial period of six hours. The rate of initial and secondary absorption was considered which shows the ability of resistance to penetration of fluid into the concrete pores. Initial water absorption was calculated from the slope of a line that is best fit to *l* plotted against square root of time from 1 minute to 6 hours which is shown in figure 52.2. Similarly, the secondary absorption of water was found from the slope of a line that is best fit to *l* plotted against square root of time from3 days, 7 days, and 9 days.Table.3. Shows the average dry weight, differential weight of dry, wet specimens, and sorptivity of sample M1, M2, M3, M4, M5, andM6 with a variable period (ASTM 1585, 2004). The period at which the sorptivity values were calculated at 5min, 10 min, 20 min, 30 min, 60 min, 6 hrs, 3 days, 7 days, and 9 days. The least amount of differential weight was found in both the initial and final reading in the case of specimen M5,and the maximum amount of differential weight was found in the case of specimen M3. The trend in Figure 52.2 reveals that at 9 days of time interval, specimen M5 exhibits a low sorptivity of 0.2242 mm and specimen M3 has high a sorptivity of 0.3478 mm. Figure 52.2 shows that the initial sorptivity coefficient (k_i) of specimen M3 is 0.0014 which is the maximum and specimen E has a minimum of 0.0006. Similarly, the secondary sorptivity coefficient (k_s) of M3 is 0.0004 is also maximum and in the case of M5, it is 0.0002 and found minimum. The above data reveals that both the initial and secondary sorptivity coefficient of specimen M3 has a maximum and M5 has a minimum.

Conclusion

It is observed from this study that the mechanical property of the micro-concrete, such as compressive strength is mostly influencing the permeable property.

Sorptivity is dependent on the compressive strength. Lesser sorptivity is found in the case of higher compressive strength and vice versa at a constant slump.

References

ASTM C109 (2004). Standard test method for compressive strength of hydraulic cement mortar.

ASTM C1585 (2004). Standard test method for measurement of rate of absorption of water by hydraulic-cement concretes.

Choudhary, R, Gupta, R, Nagar, R, Jain, A (2020). Sorptivity characteristics of high strength self-consolidating concrete produced by marble waste powder, fly ash, and micro silica. *Materials Today: Proceedings*, 32(4), 531535.

Elawady, E, Hefnawy, A A, Ibrahim, R A F (2014). Comparative study on strength, permeability and sorptivity of concrete and their relation with concrete durability. *International Journal of Engineering and Innovative Technology*, 4(4), 132139.

Hall, C (1989) Water sorptivity of mortars and concretes: a review. *Magazine of Concrete Research*, 41(147), 51–61.

53 Effect of wood dust filler on mechanical properties of polyester composite

Chandrakanta Mishra, Harish Chandra Baskey, Chitta Ranjan Deo[a], Deepak Kumar Mohapatra and Punyapriya Mishra

Department of Mechanical Engineering, Veer Surendra Sai University of Technology (VSSUT) Sambalpur 768018, India

Abstract

In the current work, the focus has been made to fabricate a noble eco-friendly composite by reinforcing waste wood dust (WD) to the polyester matrix. The consequence of the different proportions of WD filler on the mechanical performance of composites has also been studied. An optimal value of tensile strength and hardness is found at 15wt% of WD reinforcement, whereas the 10 wt% WD reinforcement exhibited higher flexural and impact strengths. However, a reduction in all the above properties was observed at higher loading of WD filler i.e., at 20 wt%. Again, an improvement in thermal stability was also observed due to incorporation of WD filler. Based on the experimental findings, the fabricated composite can suitably be used for structural applications in the automobile and construction fields.

Keywords: Wood dust filler, polyester resin, mechanical properties, composite

Introduction

Over the last few decades, polymeric composite materials have replaced the use of conventional metals and become one of the most promising areas of interest for researchers. Again, due to their recyclable nature, high stiffness, low cost, low density, eco-friendliness, and favorable mechanical properties, wood-dust filled composites (WDCs) are becoming popular in construction, automotive, and infrastructure applications. Pérez et al. (2012), examined the mechanical performance of polypropylene/wood-dust composites. A reduction in strength and strain along with reliable fracture toughness data was observed with an increase in fiber content. (Pradhan et al., 2022), reported that the tensile and thermal performances of the composites get better with an increase in the particle dimension. Again, the effect of a compatibilizer agent on the performance of composite materials was also studied (Dairi et al., 2017). The findings of the study reflected that, the mechanical strength of composite significantly improved with reduction of elongation at break due to addition of compatibilizer agent. Likewise, the influence of filler concentration and chemical modification of filler on mechanical performances of WD filled polystyrene composites was proposed by Adeniyi et al. (2022). The experimental findings reflected that, the sample reinforced with 4% NaOH treated WD fillers with 30% wt.% and particle size 841 μm possessed higher mechanical performances than that of other. Similarly, Jain and Gupta (2018), also examined the influence of hybridization of Teak and Sal wood floors on the mechanical and thermal performances composite along with water absorption characteristics. The hybrid composite with an equal percentage of Teak and Sal woods demonstrated the highest mechanical property amongst all combinations. In

---

[a]chittadeo@gmail.com

DOI: 10.1201/9781003450924-53

sight of the above context, this research mainly confined to determine the influence of the locally available WD content on mechanical performance of the WD polyester composite.

Experimental procedure

Materials and method

The matrix was prepared by combining the unsaturated polyester resin and catalyst methyl ethyl ketone (MEKP) in 10:1 ratio with 2% cobalt naphthenate accelerator. After that, the required amount of oven drayed WD as per Table 53.2 was thoroughly mixed with the polyester mix. Then the filler-matrix mix was poured into the steel mold and allowed the casting to cure for 48 hours at room temperature. The cast part was removed from the mold and test specimens were cut in accordance with the ASTM standard. The composition of composite samples is presented in Table 53.1.

Mechanical characterization

Tensile and flexural testing

The tensile and flexural tests were performed as per the ASTM D 638 and ASTM D 790 standard respectively with the help of a universal testing machine (UTM; model INSTRON 3382) having maximum capacity of 10 KN. During the test a crosshead speed of 2 mm/min was maintained.

Impact testing

A digital Izod impact tester was used to assess the impact energy of the composite's samples. The specimens were prepared as per the ASTM D 256 standard. The size of the notched specimen was taken as 63 × 12.7 × 5 mm.

Hardness test

A Micro-hardness tester of make: Matsuzawa, Japan Model: MMT-X7B was used to evaluate the micro-hardness of the fabricated composite as per ASTM E384 standard. In order to determine the micro-hardness, a load of 100 gf and dwell period of 15 s was employed.

Thermo-gravimetric test

A thermo-gravimetric analysis (TGA) was carried out by using Mettler-Toledo brand Thermo-gravimetric Analyzer as per ASTM D3850 standard. The analysis was done

Table 53.1: Combination of WD-polyester composites.

Sl. No.	Designation	Wt.% of WD	Wt.% of polyester	Thickness (mm)
1	WD-0	0	100	5.0
2	WD-5	5	95	5.0
3	WD-10	10	90	5.0
4	WD-15	15	85	5.0
5	WD-20	20	80	5.0

Source: Author

at temperature range 30-700°C with a heating rate of 30°C/min in nitrogen atmosphere. The samples were held in a crucible of 100 μL volume.

Results and discussion

Tensile strength

The variation of tensile strength with respect to content of WD filler is presented in Figure 53.1(a). A rising trend of tensile strength from 13.594 MPa to 47.396 MPa due to the increase of wood dust filler inclusion from 0-15 wt% is clearly noticed. However, the tensile strength is found to be reduced with further increase of WD filler content (20 wt %). A maximum value of 47.396 MPa is retained at 15 wt% of WD filler, which is about 239.66% higher than that of neat polyester. This may occurs due to the agglomeration of fillers at higher percentage. Similar growing trend was also observed by García et al. while dealing with wood-plastics composites (García et al., 2009; Agayev and Ozdemir, 2019).

Flexural Strength

Flexural strength of wood dust filled polymer composites is shown in Figure 53.1(b). It is seen that; the flexural strength is increased from 18.64 MPa to 57.39 MPa by the inclusion of WD up to 10 wt% which is about 207.88% higher than that of WD-0. This can be attributed to the higher flexural stiffness of wood dust filler. However, the subsequent addition of WD above 10% lead to decrease in flexural strength due to inadequate wetting of WD. A similar growing trend of variation of flexural characteristics was reported by García et al. while working with wood-plastics composites (García et al., 2009; Agayev and Ozdemir, 2019).

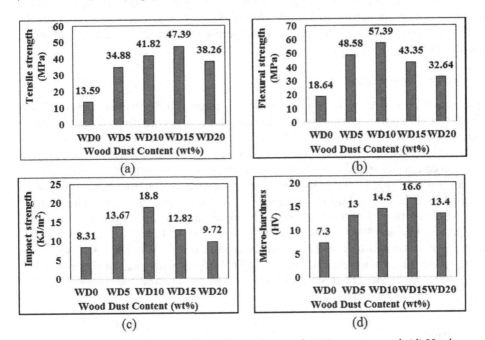

Figure 53.1 (a) Tensile strength (b) Flexural strength (c) Impact strength (d) Hardness strength

Source: Author

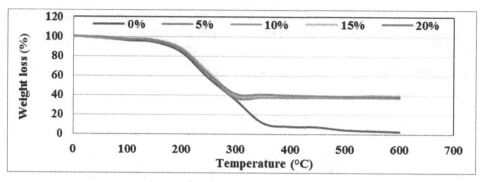

Figure 53.2 Analysis of TGA curves WD filled polyester composite
Source: Author

Impact strength

From Figure 1(c) it can be noticed that the impact strength of composite is improved with an increase of WD content up to10 wt.% and subsequently reduces with further increase of WD content. A maximum impact strength of 18.8 KJ/m² is found for WD-10 which corresponds to an increment of 126.23%. This attribute to the potential WD in impact strength as well as preventing fracture formation in the composite (Moreno and Saron, 2017).

Micro hardness

The results of the hardness test on the composite samples are shown in Figure 1(d). Due to the incorporation of wood dust filler varying from 5-15 wt.%, the micro-hardness is increased from 13 HV to 16 HV. Several researchers had reported the similar trend of increase in the hardness of composites with different wt.% of wood dust composition (Singh and Muniappan, 2020).

Thermo-gravimetric analysis

Figure 53.2 illustrates the TGA analysis of neat unsaturated polyester composite and wood dust filler polyester composite. The first phase of degradation of neat polyester sample occurs in the range of 80-180°C and weight loss is restricted to below 10% is noticed. However, in the second phase i.e., 180-265°C, major degradation is observed due to decomposition and degradation of polymer network structure. Again, at temperatures of 265-600°C, the polyester sample begins to decompose very slowly with weight loss below 5%. But, the addition of WD to polyester resin is significantly improving the thermal degradation rate. The maximum thermal degradation encountered at temperature 308°C, 316°C and 320°C for 5 wt%, 10 wt%, and 15 wt% WD filled polyester composite respectively. Whereas for 20 wt% WD filled polyester composite shows a maximum degradation at 310°C. This improvement of thermal stability is due to the barrier effect of decomposition products by the dispersion of the WD filler in the polyester matrix.

Conclusion

Based on the results of the current experimental investigation, the following conclusions are made based on the experimental study:

- The use of WD fillers in polyester composites leads to improvements in tensile strength, flexural strength, impact resistance, and micro hardness. The addition of 15% filler increases tensile strength by 239.66% over neat resin.
- Flexural strength is about 39 MPa at 10% filler, which is 207.88% greater than unreinforced polyester composite. Similarly, the composite sample with 10wt% reinforcement of WD composite exhibited a higher impact strength value of 18.8 KJ/m2, which is 126.23% more than neat matrix.
- Due to the greater weight percent of WD filler, the WD-15 composite (16 HV) exhibits a substantial improvement in micro-hardness.
- The inclusion of WD filler into polyester has significantly improved the thermal stability and maximum temperature of degradation was observed at 320°C for WD-15 composite.

In view of the above conclusion, WD can be effectively employed to prepare composites with polyester resin. As per the experimental result, this wood dust filled polyester composite can be used in different fields, such as the automobile and construction sectors.

Reference

Adeniyi, A G, Abdulkareem, S A, Adeoye, S A, and Ighalo, J O (2022). Preparation and properties of wood dust (isoberlinia doka) reinforced polystyrene composites. *Polymer Bulletin*. 79(6). 4361–4379.

Agayev, S, and Ozdemir, O (2019). Fabrication of high density polyethylene composites reinforced with pine cone powder: Mechanical and low velocity impact performances. *Materials Research Express*. 6(4).

Dairi, B, Djidjelli, H, Boukerrou, A, Migneault, S, and Koubaa, A (2017). Morphological, mechanical, and physical properties of composites made with wood flour-reinforced polypropylene/recycled poly(ethylene terephthalate) blends. *Polymer Composites*, 38(8). 1749–1755.

García, M, Hidalgo, J, Garmendia, I, and García-Jaca, J (2009). Wood-plastics composites with better fire retardancy and durability performance. *Composites Part A: Applied Science and Manufacturing*. 40(11), 1772–1776.

Jain, N K, and Gupta, M K (2018). Hybrid Teak/Sal wood flour reinforced composites: Mechanical, thermal and water absorption properties. *Materials Research Express*. 5(12).

Moreno, D D P and Saron, C (2017). Low-density polyethylene waste/recycled wood composites. *Composite Structures*. 176 1152–1157.

Pérez, E, Famá, L, Pardo, S G, Abad, M J and Bernal, C (2012). Tensile and fracture behaviour of PP/wood flour composites. *Composites Part B: Engineering*. 43(7), 2795–2800.

Pradhan, P, Purohit, A, Sangita Mohapatra, S, Subudhi, C, Das, M, Ku Singh, N and Bhusan, S B (2022). A computational investigation for the impact of particle size on the mechanical and thermal properties of Teak wood dust (TWD) filled polyester composites. *Materials Today: Proceedings*. 63 756–763.

Singh, A and Muniappan, A (2020). Experimental investigation on the mechanical properties of wood sawdust and plaster of paris reinforced composite. *IOP Conference Series: Materials Science and Engineering*. 992(1).

54 Assessment of conservation potential of sub-watersheds in the Baitarani River basin in Odisha by using morphometric analysis

Swagatika Sahoo[a] and Janhabi Meher[b]

Department of Civil Engineering, VSSUT, Burla, Odisha, India

Abstract

This research work examines the morphometry of Baitarani River basin in great depth to comprehend its morphological properties and hydrological behavior. The resulting morphometric parameters were then used for weighted sum analysis based on the sub-watershed priority for managing and conserving their natural resources. There have been sixteen parameters considered, including two linear, four form, six areal, and four relief characteristics. Using publicly available data with a resolution of 30 m SRTM DEM, streams, watersheds, and sub-watersheds were determined. The findings indicate that Baitarani River is a seventh order stream. The sub-watershed (SWs), SW4 and SW6 are the most vulnerable to soil erosion, followed by SW2 and SW5 are moderately susceptible. SW3 and SW1 are the least susceptible. Therefore, it can be concluded that weighted sum analysis is a reliable tool for developing policies for strategic utilization of available natural resources at the micro level.

Keywords: Morphometry, SRTM DEM, weighted sum analysis

Introduction

Geomorphology, hydrology, soil, and landscapes are all related. Geo-morphometry is used to measure and understand how a region's attributes relate to each other in terms of their sizes and shapes. Finally, morphometry can be used to assess drainage geometry, sedimentation, runoff, and soil erosion as reflections of hydrological and geomorphic processes. Thus, the morphometry of a basin contributes to the explanation of its hydrological characteristics. Prioritizing a watershed is crucial for planning and developing a watershed sustainably. It is the process of evaluating a watershed's sub-watersheds based on their importance for water and land preservation. Analyzing morphometric variables quantitatively can help prioritize soil or water conservation. Recently, remote sensing and geographic information systems have been developed for thematic morphometry mapping and quantitative assessments of drainage networks. Automated geospatial methods have demonstrated how powerful and reasonably priced they can be when compared to manually processed data. Morphometric parameters serve as the foundation for the prioritization procedure. Based on the compound factor value, the weighted-sum approach assigns weights to morphometric parameters and categorizes them based on their priority ranking (PR). Based on the watershed prioritization results; highly sensitive zones to soil erosion were identified as being critical for adequate water resource planning and management.

Morphometric analysis of the Naula watershed was carried out by Malik et al. (2019). The Dikrong watershed has undergone soil erosion susceptibility analysis using the weighted sum analysis (WSA) technique on a few significant morphometric factors by Chakravartty (2019). Machine learning (ML) techniques were used to

[a]swagatikasahoo527@gmail.com, [b]jmeher_ce@vssut.ac.in

DOI: 10.1201/9781003450924-54

predict and validate compound factors, and WSA was used to prioritize watersheds according to their vulnerability to soil erosion by Darji et al. (2021). Setiawan and Nandini (2021) conducted a study in the Sari Watershed on the island of Sumbawa by integrating principal component analysis (PCA) and WSA. The outcome demonstrated that the integrated method was a reliable tool for managing and planning irrigation, agriculture, and natural resources.

The current study uses a mix of weighted sum analysis and morphometric analysis to point out and prioritize the sub-watersheds (SWs) of the Baitarani river basin. This study's three main goals are parameter measurement, hydro-geomorphic characterization, and prioritizing sub-watersheds for natural resource preservation.

Material and methodology

In this study, Baitarani basin, drained by Baitarani River is located between east longitudes of 85°10' to 87°03' and between north latitudes of 20°35' to 22°15'. The area of Baitarani basin is about 14218 km^2 from which about 6.7% and 93.3% area situated in Jharkhand and Odisha respectively. The mean annual rainfall for the basin is about 1450 mm. The maximum temperature varies from 25°C to 38°C and the minimum temperature varies from 12°C to 24°C.

Sub-watershed delineation

The research area was extracted by mask from 30 m spatial resolution SRTM DEM data that was downloaded from http://earthexplorer.usgs.gov. The first stage in creating various thematic maps of morphometric parameters was geo-referencing with WGS_1984_UTM_Zone_45 using ArcGIS 10.3 data management tool. In the second step watershed delineation of the study area was carried out with spatial analyst tool. Next fill sink, flow direction and flow accumulation maps were created and pour points were snapped on the flow accumulation map for delineation of sub-watersheds. Following this process six sub-watersheds were identified in the study area as shown in Figure 54.1.

Morphometric analysis

A morphometric analysis provides insight into the characteristics of the stream network and basin. It is essentially carried out with parameter estimation of three aspects of the basin, i.e., linear, areal and relief, using standard mathematical expressions found in literatures.

Prioritization of sub-watersheds

It involves ranking sub-watersheds based on the degree of conservation treatment to be given to each sub-watershed. Once the sub-watersheds are prioritized, the quantitative assessment of the hydrological parameters of the watershed serves as necessary information for the adoption of appropriate conservation of soil and water in the watershed. In this study, arithmetic average method and weighted sum approach were used to prioritize the sub-watersheds.

Arithmetic average method

Here all morphometric parameters are assigned with equal priority but the simple arithmetic mean is the main drawback of this method.

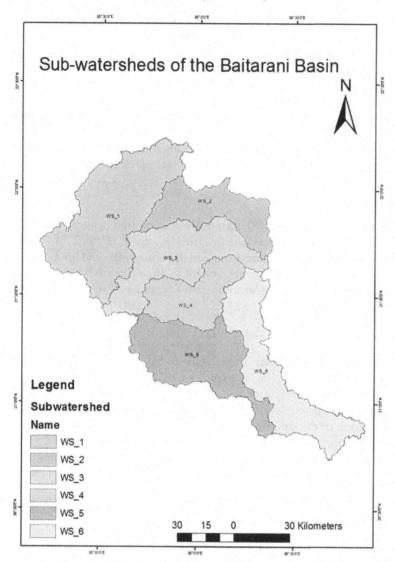

Figure 54.1 Delineation of sub-watersheds in Baitarani Basin
Source: Author

Weighted sum approach

This method is efficient and excellent for prioritizing watersheds. Here different morphometric parameters are ranked and correlation analyses are conducted to assign weights to each input parameter according to its importance. The linear and areal parameters are arranged from highest to lowest, with the highest value coming in first, second, and so on. The lowest value is ranked last. Erodibility is negatively correlated with shape factors, while relief parameters are directly related to soil and water quality degradation. The compound factor determines priority ranking and categorization. The compound factor's mathematical expression is as follows:

$$CF = PPR_{mp} * W_{mp} \qquad (1)$$

In cross correlation analysis, W_{mp} stands for the weight of the morphometric parameter, CF for the compound factor, and PPR_{mp} for the primary priority ranking based on the morphometric parameter.

$$W_{mp} = \frac{\text{Sum of correlation coefficient}}{\text{Grand total of correlation}} \quad (2)$$

Ultimately, the ranking will be determined by the compound factor in such a manner that the lowest value will be given priority rank 1, the next lowest will be given priority rank 2, and so on.

Average soil erosion

Universal Soil Loss Equation (USLE) model was used to estimate average soil erosion rate and prioritize sub-watersheds in Baitarani basin. Mathematically speaking, the USLE model is represented by the expression:

$$ASE = R * K * L * S * C * P \quad (3)$$

where ASE is the Average soil erosion (measured in ha-m/100 km²/year). Rainfall erosivity factor, soil erodibility factor, slope length factor, slope steepness factor, cover management factor, and support practice factor are all represented by R, K, L, S, C, and P, respectively.

Results and discussion

This research aimed to estimate morphometric parameters and rank sub-watersheds according to their vulnerability to soil erosion.

Arithmetic average method

Arithmetic average method-based compound factor and corresponding priority ranking is given in Table 54.2. The computed compound factor was validated with the average soil erosion (ASE) estimated using USLE method and a R^2 of 0.55 was obtained.

Weighted sum approach

WSA as one of the prioritization techniques entails preliminary ranking of parameters, development of correlation matrices, weightage computation, compound factor estimation, and final priority ranking.

Table 54.1: Importance of parameters on soil erosion susceptibility.

Parameter	D_d	L_g	F_s	R_t	I_f	C_c	C	R_f
Weightage	0.099	0.099	0.030	-0.137	0.088	0.063	-0.099	0.109

Parameter	R_c	R_e	R_{SL}	R_{bf}	B_h	R_r	R_h
Weightage	-0.048	0.109	0.118	0.014	0.154	0.104	0.151

Source: Author

Table 54.2: Compound factor based prioritization of sub-watersheds.

Sub-watersheds	1	2	3	4	5	6
Area (km²)	3054.06	1303.98	1946.77	1174.64	2033.68	2046.45
ASE (ton/ha/yr)	75.38757	89.27039	93.73296	115.7372	123.7399	110.7719
AAM based CF	3.75	3.25	3.875	2.375	3	3.125
Priority ranking	5	4	6	1	2	3
WSA based CF	5.189	2.719	3.776	1.677	2.633	2.064
Final priority ranking	6	4	5	1	3	2

Source: Author

Cross-correlation analysis

It was used to estimate the weightage of linear, shape, areal and relief morphometric parameters. Correlation coefficient values of all the columns were summed up, which were further added to compute the grand total. Each column total was then divided with the grand total to estimate weightage of each parameter as shown in Table 54.1.

Compound factor computation

All the parameters were ranked sub-watershed wise from 1 to 6 based on their estimated values. Equal parameter values were assigned with equal rank irrespective of sub-watershed ordering. Linear and relief parameters, directly related to soil erosion process were assigned with rank 1 for highest value and rank 6 for lowest value respectively. The reverse ranking with assigned to shape and areal parameters as they hold inverse relationship with soil erosion process. Highest priority ranking of watershed indicates greater erosion of soil from its surface. Weighted sum approach based compound factor and corresponding priority ranking is given in Table 54.2.

Soil erosion susceptibility zone

As the WSA was found to have higher correlation (0.66) than the Arithmatic Average Method (AAM) (0.55) with the estimated ASE, this approach was applied to divide the study area into various soil erosion vulnerability zones. Four separate soil erosion vulnerability zones i.e. very high, high, moderate and low have been established based on the compound factor values.

Conclusions

This study categorized and ranked sub-watersheds based on morphometric factors that are vulnerable to soil erosion. Compound factor (rank) was calculated using AAM and weighted sum analysis (WSA), and validated with average soil erosion (ASE). WSA was found to be a better prioritization method due to its wider range of CF values and more rational framework. Therefore, finally WSA was applied to prioritize the sub-watersheds based on the developed soil erosion susceptibility zones, which depicted that the very high soil erosion vulnerability zone is covered by sub-watersheds (SWs), SW4 and SW6. The high soil erosion vulnerability zone is covered by sub-watersheds, SW2 and SW5. The moderate soil erosion vulnerability zone is covered by sub-watershed SW3.And sub-watershed SW1 is included in the low soil erosion vulnerability zone.

References

Chakravartty, M (2019). Soil erosion susceptibility assessment for prioritization of hilly areas of a watershed in the Arunachala Himalayas in northeast India based on weighted sum analysis method on morphometric parameters. Bulletin of pure and applied sciences, 3f, 2. 10.5958/2320–3234.2019.00010.6

Darji, K R, Patel, D P, Vakharia, V, Panchal, J, Dubey, A K, Gupta, P and Singh, R P (2021). Watershed Prioritization and Decision Making Based On Weighted Sum Analysis, Feature Ranking and Machine Learning Techniques.

Malik, A, Kumar, A,Kushwaha, D P., Kisi, O, Salih, S Q, Al-Ansari, N and Yaseen, Z M (2019). The Implementation of a Hybrid Model for Hilly Sub-Watershed Prioritization Using Morphometric Variables: Case Study in India. Water 2019, 11, 1138; doi: 10.3390/w11061138.

Setiawan, O and Nandini, R (2021).Sub-watershed prioritization inferred from geomorphometric and landuse/landcover datasets in Sari Watershed, Sumbawa Island, Indonesia IOP Conf. Ser.: Earth Environ. Sci.747 012004. 10.1088/1755-1315/747/1/012004

55 Next generation communication network using NB-IoT

Lakhmi Priya Das[1,a], Srinivas Sethi[1,b], Ramesh Kumar Sahoo[1,c], Sunita Dalbisoi[2,d] and Ashima Rout[2,e]

[1]Department of CSEA, IGIT Sarang, Dhenkanal, India

[2]Department of ETC, IGIT Sarang, Dhenkanal, India

Abstract

The Internet of Things (IoT) provides an environment where data processing devices are interconnected that are labeled with unique identifiers (UIDs) and transmits data on a network without the involvement of humans and computers. An IoT system assembles web-enabled smart devices that have processors, sensors, and hardware for communication to collect, send, and process data. The cloud is where data is sent through the IoT gateway for analysis. NarrowBand Internet of Things (NB-IoT) based on low power wide area (LPWA) technology to empower IoT devices and services in a wide range. In deep coverage, NB-IoT enhances user device power dissipation, system capacity, and spectrum efficiency. In rural areas, new signals and physical layer channels are constructed to expedient the requirements. The NB-IoT technology can coincide with 2G, 3G, 4G, and 5G network technology. This research paper reviews the NB-IoT signal analysis in non-cellular network over cellular networks.

Keywords: Unique identifiers, internet of things, narrowband internet of things, low power wide area

Introduction

NarrowBand-Internet of Thing (NB-IoT or LTE Cat NB), designed to intensify the architecture of IoT using extant cellular networks (Nurlan et al., 2021) This narrowband and low power technology perform on bilateral data transmission in a very adequate, secure and impeccably. Cellular mobile communications were designed to focus more on voice service and mobile broadband services. After facing difficulties in communication links (Saputro et al., 2012), a new advancement was made in cellular networks as GSM, UMTS and LTE by 3GPP was introduced for machine type communication (MTC) services (Escolar et al., 2021). This MTC has become an important asset of 5G network (Veloso et al., 2020). The signals and channels are built in a way to understand the necessity of extended coverage in buildings and rural areas. This technology is easier than GSM/GPRS, and costs will decrease as demand increases.

Architecture of NB –IoT

A structured and understandable architecture of NB-IoT is concerned for its devising, diminishing, design, cost estimation and final formation. NB-IoT architecture is quite similar to the wireless sensor network (WSN) (Nurlan et al., 2021). In its advancement WSNs architecture and topologies play an important role. The LTE network

[a]lakhmipriyadas@gmail.com, [b]igitsethi@gmail.com, [c]ramesh0986@gmail.com, [d]sunita.muskan1@gmail.com, [e]ashimarout@igitsarang.ac.in

DOI: 10.1201/9781003450924-55

Figure 55.1 Layered structure of the NB-IoT architecture
Source: (Okorogu et al., 2019)

of cellular system forms a backbone system of NB-IoT. NB-IoT can be separated into 6 different layers in Release 13 these are described well. The physical link is similar to WSNs. The medium access control (MAC) layer is above it. For multiple access and medium access, it supports protocols. Another layer is the radio link layer which is placed in between the MAC layer and upper layer. There is another layer above it called Packet Data Convergence protocol. It distributes networking, routing, traffic scheduling and other tasks. Above it is the radio resource control layer. This layer contributes to the radio resources of packets in a channel carried out through user datagram protocol (UDP) and other cellular mechanism. The upper layer is non access stratum (NAS) which built a data transfer link or communication between the NB-IoT server and user equipment (UE) (Okorogu et al., 2019). Figure 1 shows the NB-IoT layered structure.

Low power consumption in NB-IoT

NB-IoT longer expandable time (Lauridsen et al., 2018) can be achieved through power saving mode (PSM) and expanded discontinuous reception (eDRX) (Chen et al., 2017). The PSM was first introduced in Rel-12, here the terminals were marked online and made deep sleep for an expandable time for power saving. Whereas in eDRX, was introduced in Rel-13. The deep sleep is extended to make the terminal idle which reduces extra usage of cells.

Modes of operations in NB-IoT deployment

There are three different modes of operation for NB-IoT deployment:

- *Stand-alone(independent) operation*: when cellular networks are not usable to serve narrowband spectrum as in GSM. By placing more GSM carriers to NB-IoT service, majority of machine communications can perform continues transition.
- *Guard band operation*: when cellular networks are usable and NB-IoT is in the guard band of LTE carriers to avoid difficulties.
- *In-band operation*: when cellular networks are present, NB-IoT is to be arranged in between LTE resources and bearer. This operation is preferred to be coherent

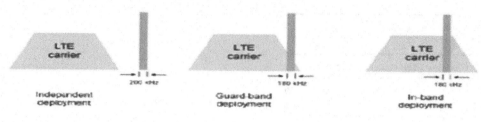

Figure 55.2 Deployment of NB-IoT
Source: (Chen et al., 2017)

for mobile operators. It required no change in hardware to the radio access network and allocates spectrum resources to LTE or NB-IoTservices. Figure 55.2 shows the operation modes (Chen et al., 2017).

NB-IoT Signal Analysis

For installation and maintenance work, the NB-IoT signal analysis is performed to ensure that proper communication and service delivery is achieved. Avoid consuming additional bandwidth, which also reduces the amount of IoT devices (Cao et al., 2019) that are connected to the network. Avoid the effects of inappropriate power levels, which could result in signals not reaching the IoT devices or multiple transmissions that reduce the uptime of the IoT (Talebkhah et al., 2021) devices batteries. Wireless signal analysis for installation and maintenance includes two main aspects: - The physical profile of the signal, including the power and frequency characteristics of RF, as well as verification of wireless carrier configuration and compliance with 3GPP standards. The data (modulation) quality of the signal, including modulation verification and error vector magnitude (EVM) distortion.

Non-cellular network solutions

There are different technologies used for non-cellular network solutions:

a) SigFox: It was designed to minimize the cost and energy accumulation of connected devices in an IoT network (Ramnath et al., 2017) of low-speed communication and telephony infrastructure. Moreover, to low speed it also works on narrow band that provide devices to invade obstacles, possible communication over long distances even in metropolitan areas. In structure of SigFox, the connection accomplished through antennas and base stations distributed throughout the area, which gather data from endpoints and transmit to the SigFox server, data is stored here.

b) LoRa: It is a leading radio modulation technology used for non-cellular networks. LoRaWAN is a standard networking protocol, which operates in industrial, scientific and medical bands. The main purpose of using LoRaWAN is that it is operated on battery devices to send small amounts of data at infrequent intervals to the internet. The LoRaWAN networks are accessible in public, private and community systems.

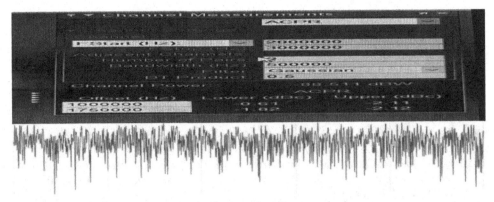

Figure 55.3 frequency spectrum using gaussian filters method
Source: Author

Cellular network solutions

Cellular IoT technology implements piggy backing on the existing dense and steady GSM infrastructure, this improves the performance and economy. Cellular network usage: 2G, 3G or LTE are high power consuming and this is the reason they are not supported by the IoT applications. Using the actual GSM, UMTS, LTE and 5G (Yu et al., 2016) infrastructures for connecting smart devices (Chen et al., 2017). It uses low-power and energy saving devices: sensors, actuators or microcontrollers. The IoT solution uses the best standards NB-IoT and LTE-M. In version 13, 3GPP (Hidayati et al., 2019) introduced a 4G network technology standard called LTE-M. The 4G wireless chips are designed for special power saving mode as specified in LPWA (Nashiruddin et al., 2020) specification. These chips are half-duplex in order consume less power. They work on maximum data rate of 100 kbps, which makes it slower than conventional 4G connection. Hence NB-IoT developed as wireless response to non-cellular LPWAN and cellular network. In general, the NB-IoT can be served in cellular as well as in LoRaWAN of non-cellular network.

Graphical analysis of frequency spectrum

The power density is observed here by comparing the two different frequencies in cellular network. Figure 55.3 showing different frequency spectrum channel measurements using the Gaussian filters method. The spectrum analyzer described the signal in the frequency domain and time domain. The range of frequency is from 23 MHz. The bandwidth is 0.5 MHz. The number of adjacent channels is two (consider). Gaussian filtration method is used. It will take very little power with less bandwidth. The spectrum analyzer power can display the power spectrum of the signal in three units, Watts, dBm, and dBW.

Conclusion

The NarrowBand Internet of Things (NB-IoT) operates within a globally recognized and unified ecosystem. Communication is based on low speed as well as on a very narrow band, which allows devices to invade obstacles, possible communication over long distances even in metropolitan areas. By integrating low speed and high speed, in future the 5G networks consolidate all mobile networks. In the case of NB-IoT it seems to be promising to overcome the short sights of the future development.

References

Cao, J, Yu, Maode, P, and Gao, W (2019). Fast Authentication and Data Transfer Scheme for Massive NB-IoT Devices in 3GPP 5G Network. *IEEE Internet of Things Journal*, 6(2), 15611575.

Chen, M, Miao, Y, Hao, Y, and Hwang, K (2017). Narrow Band Internet of Things. *IEEE Access*, 5, 2055720577. doi: 10.1109/ACCESS.2017.2751586.

Escolar, A M, Alcaraz-Calero, J M, Salva-Garcia, P, Bernabe, J B, and Wang, Q (2021). Adaptive network slicing in multi-tenant 5G iotnetworks. *IEEE Access*, 9, 1404814069. 10.1109/ACCESS.2021.3051940

Hidayati, A, Reza, M, Adriansyah, N M, and Nashiruddin, M I (2019). Techno-economic Analysis of Narrowband IoT (NB-IoT) Deployment for Smart Metering. *Asia Pacific Conference on Research in Industrial and Systems*, 16. doi: 10.1109/APCoRISE46197.2019.9318920

Lauridsen, M, Krigslund, R, Rohr, M, and Madueno, G (2018). An empirical NB-IoT power consumption model for battery lifetime estimation. *IEEE Conference on Vehicular Technology (VTC)*, 15. doi: 10.1109/VTCSpring.2018.8417653

Nashiruddin, M I and FaradilaPurnama, A A (2020). NB-IoT Network Planning for Advanced Metering Infrastructure in Surabaya, Sidoarjo, and Gresik. *International Conference on Information and Communication Technology (ICoICT)*, 16. doi: 10.1109/ICoICT49345.2020.9166315

Nurlan, Z, Othman, T Z M, Adamova, A, and Zhakiyev, N (2021). Wireless sensor network as a mesh: vision and challenges. IEEE Access.10, 46-67. 10.1109/ACCESS.2021.3137341

Okorogu, V N, Obioma, P C, Kanebi, U, and Okeke, N C (2019).Narrowband IoT and its implications in developing countries. *International Journal of Scientific & Engineering Research*, 10(12), 22295518

Ramnath, S, Javali, A, Narang, B, Mishra, P, and Routray, S K (2017). IoT Based Localization and Tracking. *International Conference on IoT and Applications of IEEE Access Top of FormBot*. 14. 10.1109/ICIOTA.2017.8073629

Saputro, N, Akkaya, K, and Uludag, S.(2012). A survey of routing protocols for smart grid communications. *Computer Networks*, 56(11), 27422771.

Talebkhah, M, Sali, A, Marjani, M, Gordan, M, Hash, S J, and Rokhani, F Z (2021). IoT and big data applications in smart cities: Recent Advances, Challenges, and Critical Issues. *IEEE Access*, 9, 5546555484. 10.1109/ACCESS.2021.3070905

Veloso, C, Pinto, L M, Tavares, V B, Batista, P, Lins, S, Linder, N, and Klautau, A (2020). Testbed for 5G connected artificial intelligence on virtualized networks. *IEEE Access*, 8, 223202223213. doi: 10.1109/ACCESS.2020.3043876

Yu, R, Bai, Z, Yang, L, Wang, P, Move, O A, and Liu, Y (2016). A location cloaking algorithm based on combinatorial optimization for location-based services in 5G networks. special section on green communications and networking for 5G wireless. *IEEE Access*. 4, 65156527. 10.1109/ACCESS.2016.2607766

56 Effect of agro waste ash on the mechanical properties of recycled coarse aggregate concrete

M. Srinivasula Reddy[1,a], Kaliprasanna Sethy[2], G. Nagesh Kumar[3] and K.V.S Gopala Krishna Sastry[4]

[1,3,4] Civil Engineering Department, G Pulla Reddy Engineering College (Autonomous), Kurnool, Andhra Pradesh, India

[2] Civil Engineering Department, Government College of Engineering, Kalahandi, Odisha, India

Abstract

The present research focuses on the study of the influence of agro waste ash on the mechanical strength properties of recycled coarse aggregate (RCA) concrete. The agro waste ashes employed in this study are obtained by burning rice straws and cotton stalks. To determine the optimal substitution levels of rice straw ash (RSA) and cotton stalk ash (CSA), preliminary experimental investigation is carried out on mortar mixes by replacing cement in the mortar in various percentages by weight. From the preliminary study the optimum replacement levels for RSA and CSA, individual mixes, are found as 5% and 10%, respectively. Then the mortar developed with 5% RSA, 10% CSA and 85% cement is shown the maximum strength over all the mixes. Keeping the cementitious content with 5% RSA, 10% CSA and 85% OPC cement, M30 grade concrete is developed using natural aggregates and is designated as Mix0. Different concrete mixes are then developed by replacing natural coarse aggregate with 10%, 20%, 30%, 40%, 50% and 60% by weight with RCA, and the mixes are designated as Mix1, Mix2, Mix3, Mix4, Mix5 and Mix6, respectively. The mechanical strength properties (compressive, tensile and flexural strengths) are evaluated and found that the concrete mix with 50% substitution of conventional coarse aggregates with RCA has maximum strength properties. From this research it is found that concrete with superior mechanical properties can be developed using RSA, CSA and RCA, over the conventional concrete.

Keywords: Recycled coarse aggregates (RCA), cement substitution, mechanical properties, agricultural waste

Introduction

Generally, agriculture wastes are burnt in the farmlands itself and the ash produced is disposed in landfills or water bodies, thereby causing environmental hazards. Several advancements in the research on the utilization of ashes that are produced from the combustion of agriculture waste has resulted in employing these waste materials in the civil engineering field and especially in concrete industry (Hilal et al., 2020; Hamada et al., 2021; Ordeiro and Sales, 2015; Zeyad et al., 2017). Globally, various agricultural waste ashes such as bamboo leaf ash, coconut shell ash, POFA, rice straw ash, wheat straw ash, rice husk ash, sugarcane bagasse ash, ground nut hulk ash, etc., (Agwa et al., 2020; Mosaberpanah and Umar, 2020; Memon et al., 2018) are being generated in large amounts. Utilizing these ashes from agro waste in the place of cement for producing concrete has gained so much attention because of its dual advantage. First is,

[a]sm26@iitbbs.ac.in

DOI: 10.1201/9781003450924-56

it decreases the environmental hazards from ineffective management of agricultural waste and the second is, the replacement of cement with these waste's ash helps in reducing the cement consumption which again is a huge concern for the construction industry. As cement production leads to large amounts of CO_2 emissions, replacement of cement with agro waste ash helps in reducing the carbon footprint (Pandey and Kumar, 2022; Raheem and Ikotun, 2020). India is a country where the rice, cotton crops farming is in large proportions with that the waste generation and hence, in the present study, RSA and CSA are employed as binder materials to replace the cement for developing concrete. Another major problem, the construction industry confronting with, is the disposal of demolished waste especially the concrete demolition waste. Utilization of the aggregate from the demolished concrete in the place of natural aggregates using for producing concrete is very beneficial to the construction industry (Abed et al., 2020; Akhtar and Sarmah, 2018). Therefore, to address both the issues, efforts are made in the present study to partially replace cement with agro waste ash and natural coarse aggregate with recycled coarse aggregate.

In the present study, at first, mix proportioning calculations are made conventional concrete of M30 grade is developed using OPC and natural aggregates and the mix is designated as M0. Later, in order to incorporate optimum percentage replacement levels of RSA and CSA in the place of cement, preliminary investigations are conducted on the mortar samples and the found as 5% RSA and 10% CSA, separately. For this, cement mortar cubes are prepared by replacing cement with RSA in 5%, 10%, and 15% by weight of cement and the same is followed with CSA. The percentage at which the maximum 28 day compressive strength shown by the mortar samples is chosen as optimum replacement level. Efforts are also made to determine the mortar strength when cement is substituted by 5% RSA and 10% CSA, at once, and found the maximum 28-day compressive strength. After determining the optimum replacement levels, M30 grade concrete is developed by using 5% RSA, 10% CSA and 85% OPC cement, natural fine and coarse aggregates. This mix is designated as Mix0. Various mixes are developed, keeping the cementitious content and fine aggregate content as constant, by replacing the natural coarse aggregates with RCA in various percentages i.e., 10 to 60%, at an increment of 10%, and the mixes are designated as Mix1, Mix2, Mix3, Mix4, Mix5 and Mix6, respectively. All the seven mix proportions are casted to determine the mechanical strength properties after 7 days and 28 days curing.

Materials methods and testing

The materials employed to develop the RCA concrete are OPC cement, river sand, crushed stone aggregate, RCA, RSA, CSA and water. Cement used in this study is Birla OPC 53 grade, RCA is obtained from the nearby demolished building, and RSA and CSA are obtained by burning the semi-dry rice straws and cotton stalks, respectively, which are obtained from the local region. Potable water is employed in the mix preparation. The specific gravity of cement is 3.04, fine aggregate is 2.62, natural coarse aggregate is 2.82, RCA is 2.74, RSA is 2.25 and CSA is 2.35. FM (fineness modulus) of fine, natural coarse and recycled coarse aggregates are obtained as 2.67, 4.77 and 4.72, respectively. The sand used in the present study falls under the medium sand category. The fineness of cement, RSA and CSA are 300 sq m/kg, 1846 sq m/kg and 1962 sq m/kg, respectively. Table 1 presents the chemical composition of the binder materials i.e., OPC, RSA and CSA.

For each mix proportion, a set of 9 cubes each of size 100 × 100 × 100 mm, 9 cylinders each of size 300 mm deep and 150 mm diameter, 9 beams each of size 750 × 150

Table 56.1: Binder materials oxide composition.

Material	Chemical oxide composition (%)									
	SiO_2	CaO	Al_2O_3	Fe_2O_3	K_2O	MgO	SO_3	Na_2O	Cu	Zn
OPC	21.7	62.3	0.30	2.8	5.2	3.5	2.85	0.50	–	–
RSA	75.6	2.6	18.5	–	–	0.10	–	–	1.20	0.95
CSA	30.1	31.2	36.4	0.20	–	0.30	–	–	1.0	0.60

Source: Author

Table 56.2: Mix proportioning of different mixes.

Mix ID	Mass of material required in kg per 1 m³ of concrete mix						
	Cement	RSA	CSA	W/C ratio	Fine aggregate	Natural coarse aggregate	RCA
Mix0	358.89	21.11	42.22	0.45	607.98	1163.36	0
Mix1	358.89	21.11	42.22	0.45	607.98	1047.02	116.34
Mix2	358.89	21.11	42.22	0.45	607.98	930.69	232.67
Mix3	358.89	21.11	42.22	0.45	607.98	814.35	349.01
Mix4	358.89	21.11	42.22	0.45	607.98	698.02	465.34
Mix5	358.89	21.11	42.22	0.45	607.98	581.68	581.68
Mix6	358.89	21.11	42.22	0.45	607.98	465.34	698.02

Source: Author

× 150 mm, are casted. At first, cementitious content i.e., OPC, RSA and CSA and fine aggregate are dry mixed for about 5 minutes followed by gradual addition of water. The mixture is then blended for about 3-4 minutes or until uniform mix is attained. After uniform mix is ensured, the fresh concrete is then transferred into the cube, cylinder, and beam specimens. The mix proportioning is done by adopting absolute volume method and mix proportioning of the seven mixes are specified in Table 56.2.

The specimens are then kept for water curing till the time of testing. All the cubes, cylinders and beams are casting and curing is done as per the procedure explained above and the mechanical strength tests are performed after 7 days and 28 days curing period. All the tests are carried out in accordance with IS 516-1959 and IS 5816-1999 standards.

Results and discussion

The 7 days and 28 days compressive, tensile, and flexural strength tests are conducted for all the specimens and the results are presented through Figures 13, correspondingly.

Figure 56.1 shows the compressive strength of concrete developed with OPC, RSA, CSA, fine aggregates, and regular coarse aggregates replaced in various percentages with RCA. It can be observed from the results that, with the replacement of regular coarse aggregate by RCA, the strength decreased slightly, initially, and then increased with further increase in the replacement levels till 40% RCA replacement. A further increase in the replacement levels beyond 40% RCA has shown a decline in the strength. The initial decrease in the compressive strength with increase in the RCA is

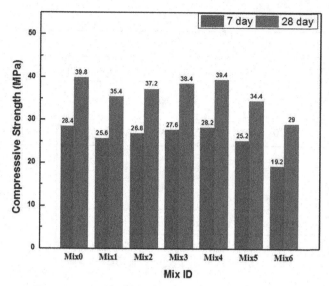

Figure 56.1 7 day and 28 day compressive strength of various mixes
Source: Author

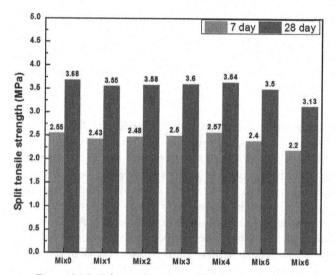

Figure 56.2 7 day and 28 day split tensile strength of various concrete mixes
Source: Author

possibly because of the enhancement in the voids created by the pores present on the surface of the RCA. With further increase in the RCA content, increase in the compressive strength could be the presence of different sizes of coarse aggregate in the mixture helping to form a dense matrix with the cement mortar consisting of cement, RSA, CSA and fine aggregate. At 40% RCA content, the mix showed the compressive strength, 28.20 N/sq.mm and 39.4 N/sq.mm at 7 and 28d, respectively, which is on par with the conventional mix and therefore, RCA content up to 40% can be incorporated in the mix to develop M30 grade concrete.

Similarly, Figure 56.2 shows the tensile strength of concretes developed in this study. From the figure, it can be noticed that the split tensile strength is initially decreased

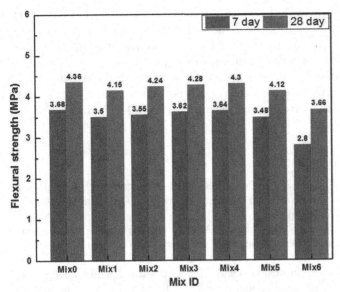

Figure 56.3 7 day and 28 day flexural strength of the developed mixes
Source: Author

with RCA content in the mix, a like compressive strength, and then increased with further addition of RCA. At 40% coarse aggregate replacement with RCA, the split tensile strength is showed as 2.57 N/sq.mm at 7 days and 3.64 N/sq.mm at 28 days. This strength obtained is on par with the strength of reference concrete mix. Therefore, 40% RCA can be used in the mixes. Generally, concrete strength lowers with the raise of the RCA volume in the conventional concrete, i.e., concrete made using ordinary cement alone as the binder material. However, in the current study, improvement in the strength is observed which could be due to the presence of RSA and CSA in the mix. The RSA and CSA might be helping in improving the strength by filling the voids present on the RCA surface.

Figure 56.3 depicts the 7 days and 28 days flexural strength of the developed concretes. From the figure it can be noticed that the flexural strength is following the same trend as that of other mechanical strength properties, where the strengths decreased initially, with the introduction of RCA, and then increased with further increase in RCA content. Similar reasons are expected for this behavior of strength variation. At 40% coarse aggregate replacement with RCA, the strength (flexural) is showed as 3.64 N/sq.mm after 7 days and 4.3 N/sq.mm after 28 days. Test results prove that the concrete developed in this research work, by incorporating 40% RCA in the place of natural coarse aggregate, 5% RSA and 10% CSA in the place of cement, have the mechanical properties on par with the conventional concrete.

Conclusion

The current study is planned to find the efficient means to make use of the agro wastes i.e. RSA, CSA and construction waste i.e. RCA. The following are the conclusions from this research:

Agricultural waste ashes such as RSA, CSA can be used successfully as a partial replacement to cement to develop M30 grade concrete. About 5% RSA and 10% CSA can be substituted in the place OPC to develop M30 grade concrete. Replacing natural coarse aggregate with RCA decreases the strength initially and then shows compressive strength on par with concrete made with natural coarse aggregates with further increase in RCA volume in the mixes. All the mechanical strength properties followed the same trend with the different mixes. At 40% RCA content, the mechanical strength properties of the concrete developed are on par with the one developed using natural coarse aggregate. The compressive, tensile, and flexural strengths at 28 days for the mix with 40% RCA are obtained as 39.4 MPa, 3.64 MPa and 4.30 MPa, respectively.

References

Abed, M, Nemes, R, and Tayeh, B A (2020). Properties of self-compacting high-strength concrete containing multiple use of recycled aggregate. *Journal of King Saud University - Engineering Sciences*, 32(2), 108–114.

Agwa, I S, Omar, O M, Tayeh, B A, and Abdelsalam, B A (2020). Effects of using rice straw and cotton stalk ashes on the properties of lightweight self-compacting concrete. *Construction and Building Materials*, 235, 117541.

Akhtar, A, and Sarmah, A K (2018). Construction and demolition waste generation and properties of recycled aggregate concrete: A global perspective. Journal of Cleaner Production, 186, 262–281.

Hamada, H, Alattar, A, Yahaya, F, Muthusamy, K, and Tayeh, B A (2021). Mechanical properties of semi-lightweight concrete containing nano-palm oil clinker powder. *Physics and Chemistry of the Earth*, 121, 102977.

Hilal, N, Ali, T K M, and Tayeh, B A (2020). Properties of environmental concrete that contains crushed walnut shell as partial replacement for aggregates. *Arabian Journal of Geosciences*, 13(16), 1–9.

IS 516 (1959): Method of tests for strength of concrete.

IS 5816 (1999): Method of test splitting tensile strength of concrete.

Memon, S A, Wahid, I, Khan, M K, Tanoli, M A, and Bimaganbetova, M J S (2018). Environmentally friendly utilization of wheat straw ash in cement-based composites. *Sustainability*, 10(5), 1322.

Mosaberpanah, M A and Umar, S A (2020). Utilizing rice husk ash as supplement to cementitious materials on performance of ultra high performance concrete:–a review. *Materials Today Sustainability*, 7, 100030.

Pandey, A and Kumar, B. (2022). Utilization of agricultural and industrial waste as replacement of cement in pavement quality concrete: a review. *Environmental Science and Pollution Research*, 29, 24504–24546.

Raheem, A A and Ikotun, B D (2020). Incorporation of agricultural residues as partial substitution for cement in concrete and mortar – A review. *Journal of Building Engineering*, 31, 101428.

Zeyad, A M, Johari, M, Tayeh, B A, and Saba, A M (2017). Ultrafine palm oil fuel ash: from an agro-industry by-product into a highly efficient mineral admixture for high strength green concrete. *Journal of Engineering and Applied Science*, 12(7).

57 Evaluation of expansive soil stabilized with lime and silica fume

Sushismita Tripathy[1,a] and Pragya Paramita[2,b]

[1]Department of Civil Engineering, IGIT, Sarang, Dhenkanal, Odisha, India

[2]Department of Civil Engineering, VSSUT, Burla, Sambalpur, Odisha, India

Abstract

Expansive soil has both expanding and shrinking nature with respect to water so it is an unsuitable soil for civil engineering works. In this experimental research an attempt is made to enhance the geotechnical properties of expansive soil using silica fume and lime as additives. This study highlights the potential of silica fume and lime to stabilize the soil. The influence of silica fume and lime on bearing capacity, compressive strength, permeability and swelling pressure of soil was investigated. From the study it is concluded that the strength and bearing capacity increased by more than three times also swelling pressure and permeability reduced by 50% and 25% when silica fume and lime are added in the proportion of 7.5% and 6% with the expansive soil. The above result confirmed the use of lime and silica fume enhance the engineering properties of expansive soil.

Keyword: Expansive soil, silica fume, lime, swelling pressure, permeability

Introduction

The soft soil which has low strength and is highly compressible, generally appears as a civil engineering problem when the highways and runways constructed over it. The soil having CBR (> 8KPa) and Unconfined compressive strength (UCS)(> 48KPa) taken as soft soil. If the soft soil is present in subgrade or considered as subgrade in pavement construction it should be stabilized. There is another way to do construction on soft soil by replacing soft soil by a strong material such as crushed rock. But the process is very expensive so it bothers the administration so we should require a new technique for soft soil which is stabilization and cost effective also. One method is to use lime as stabilizing agent in expansive soil. We use lime and silica fume for improving the geotechnical properties of expansive soil such as plasticity index, bearing capacity, swelling pressure, permeability, shear strength.

Iman et. al. (2003) researched on stabilization of fine-grained soils by adding micro silica and lime or micro silica and cement. From studies it was concluded that the micro silica and lime shows better result as compare to micro silica and cement. Aim of this research is to check the effect of nano silica and silica fume with a hydrating element applying it to the soil (Fynn et. al., 2004). It was found that nano-silica increases the water demand and decreases the dry density. Whereas a decrease is observed in liquid limit. Qassun et. al. (2015) experiment on properties of expansive soil using the mixture of lime silica-fume. The study shows that UCS increases with curing period and maximum strength occurs at 12% lime and 13% silica fume which is 12 times higher than the virgin soil. The swelling pressure also reduces up to 92.93% with 4% lime and 8% SF. Kolay et. al (2016) concluded that the reduction in the swelling pressure 92% and 63% when Class C fly ash and Ottawa sand is added to kaolinite clay and bentonite clay respectively.

[a]sushi.tripathy@gmail.com, [b]pragyapattanayak23@gamail.com

DOI: 10.1201/9781003450924-57

Experimental methodology

The aim of the investigation was to evaluate the different engineering properties of clay with different compositions of silica fume and lime. Collection of soil sample, silica fume and lime sample, characterization of properties through different geotechnical tests, development of soil-silica fume composites, development of soil-silica fume-lime composites, major laboratory tests (geotechnical properties) for prepared composites to investigate the results. The major laboratory tests are compaction test, unconfined compressive strength, California bearing ratio, permeability and swell pressure test, pH test, specific gravity test, Atterberg's limits. At first the optimum silica fume content was found out, and then different percentages of lime were added in the soil with optimum silica fume content to analyze the characteristics of stabilized soil.

Material used

The materials used in this study are expansive soil, Silica fume and lime. As per USCS Classification system, the soil was classified as clay having high plasticity (CH). Silica fume is light gray in color with a specific gravity of 2.2. The samples are powdered and stored in air tight container for further use. The Table 57.1 shows the geotechnical properties of the soil.

Engineering properties of soil

Table 57.1: Geotechnical properties of expansive soil.

Sl No.	Properties of clay		Value
1	Liquid Limit(%)		60
	Plastic limit(%)		24
	Shrinkage limit(%)		7.58
	Plasticity index(%)		36
2	Specific gravity		2.7
3	Differential free swell Index (%)		60
4	Unconfined compressive strength (Kgf/cm^2)		0.65
5	Grain size Distribution	Sand (%)	2
		Silt(%)	31.5
		Clay(%)	66.5
6	Soil Classification (As per IS1498)		CH
7	Maximum Dry Density(gm/cc)		1.7
	Optimum Moisture Content (%)		14.38
8	Un soaked CBR (%)		3.9
	Soaked CBR (%)		1.95
9	Swelling pressure		75Kpa
10	Coefficient of permeability(cm/sec)		3.859x10^{-5}

Source: Author

Result and discussion

In the present study mix of the soil with silica fume in the proportions of 2.5%, 5%. 7.5%, 10% has been done and then its optimum with different proportions of lime i.e., 2%, 4%, 6%, 8% has performed. Addition of 6% lime with 7.5% of silica fume shows the lowest plasticity index i.e., 14.21% (Figure 57.1). Hence this mix proportion can be considered as fixation point which gives maximum workability. From the plasticity chart it is confirmed that the untreated soil classified as CH class clay, falls in the class MI soil after addition of proportion of lime and silica fume. From the Figure 57.1 we get to know that the virgin soil has high swelling potential which is decreased to low by addition of silica fume. The shrinkage limit of virgin soil was found out to be 7.58% which is less than 10% which is considered as having critical degree of expansiveness. It is reduced to marginal by addition of 7.5% silica fume and it becomes noncritical by addition of 7.5% silica fume and 6% lime (Figure 57.2). Standard compaction test was conducted with both silica fume and lime mixed with soil. The observation from the test showed that the addition of SF resulted the optimum moisture content to be increased up to 7.5% and after that it decreased for further addition of SF. Up to 7.5% maximum dry density also (MDD) was also increased and then decreased (Figure 57.3). From the results we observed the maximum value of CBR of the virgin soil was 1.95% for soaked and 3.9%for un soaked condition respectively. For addition of 7.5% SF the maximum CBR value obtained 3.41% for soaked and 7.8% for un soaked condition (Figure 57.4). For combination of 7.5% SF and 6% lime CBR increased such that the highest CBR value was obtained to be 6.33% for soaked and 16.58% un soaked condition (Figure 57.4). This increase is 4.25 times compared to virgin soil for un soaked case and 3.24 times for soaked case. The increment in the maximum CBR value was produced from the

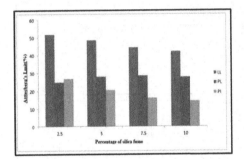 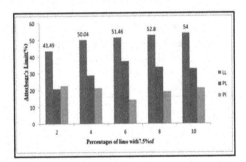

Figure 57.1 Atterberg's limit with different percentages of silica fume and life
Source: Author

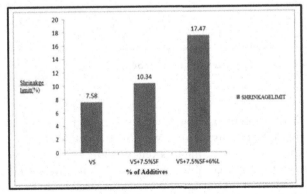

Figure 57.2 Variation of shrinkage limit with different percentages of lime and SF
Source: Author

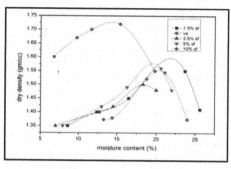
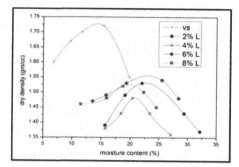

Figure 57.3 Compaction curve with different percentages of lime and SF
Source: Author

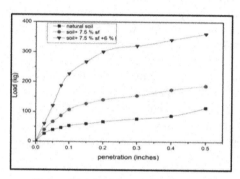
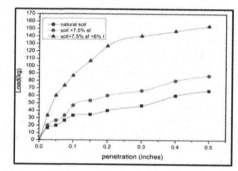

Figure 57.4 Bearing capacity curve with different percentages of lime and SF with unsoaked and soaked condition
Source: Author

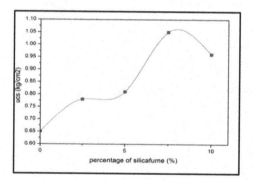
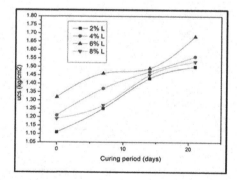

Figure 57.5 UCS curve with different percentages of lime and SF
Source: Author

soil due to improvement in compaction and pozzolanic reaction. For addition of silica fume from 2. 57.5% strength increased, and this increase was due to pozzolanic reaction of silica fume particle with the soil and resulting in strength gain over time. The soil provides higher compressive strength for combination of 7.5% silica fume and 6% lime at 21 days curing. The strength of mix proportion was increased by 61% of natural soil which is 2.6 times of natural soil strength (Figure 57.5). Swelling pressure of virgin soil decreased substantially from 75 kPa to 58 kPa with addition of silica

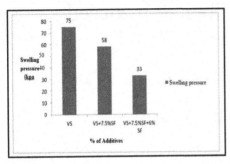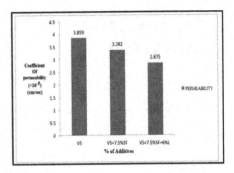

Figure 57.6 Variation of swelling pressure and permeability coefficient with additives
Source: Author

fume of 7.5% SF, further decreased with the combination of soil-7.5% SF-6% Limei. e.33kPa (Figure 57.6). The soft clayey soil used in the study is highly compressible and impervious in nature. The coefficient of permeability of mixed proportions was reduced 1.34 times than the virgin soil.

Conclusion

The plasticity index reduced by addition of silica fume but a significant variation occurs when test of mix proportion of soil-SF-lime takes place. The optimum plasticity index was occurred at soil with 7.5% SF and 6% lime i.e. 14.21. The shrinkage limit is also increased which reduces the degree of expansiveness of the soil. Increase in the silica fume percentage shows the increment in the OMC and maximum dry density. The optimum percentage of soil and SF mixture is 7.5% SF. The combination of silica fume and lime further enhanced the moisture content and the maximum dry density decreased. The grains used for composite with the soil that creates larger grains, which has some extra voids in it. For that reason, the dry density is reduced. The California bearing ratio for silica fume-lime blended clay found to be 4.25 times greater than that of untreated soil. The silica fume and lime has a remarkable effect on swelling pressure and permeability of soil. Swelling pressure and coefficient of permeability value reduces phenomenally with rise of SF and lime to the soil.

References

Bagherpour, I. and Choobbasti, A.J., (2003). Stabilization of fine-grained soils by adding microsilica and lime or microsilica and cement. *Electronic Journal of Geotechnical Engineering*, 8(B), 1–10.

Fynn, Rechard. (2004) Effect of nano silica and silica fume with a hydrating element. *Soils and Foundations*, 36(1), 97103.

Kolay, P.K. and Ramesh, K.C., (2016). Reduction of expansive index, swelling and compression behavior of kaolinite and bentonite clay with sand and class C fly ash. *Geotechnical and Geological Engineering*, 34(1), 87–101.

Qassun et. al. (2015). enhancement of expansive soil properties using lime silica-fume mixture. *E-Journal of Civil Engineering*, 1(2).

Shafiqu, Q.S.M., Ali, A. S., Abdul-Hussein, H.N., (2015). Enhancement of Expansive Soil Properties Using Lime Silica-Fume Mixture. 6(10):1239–1257.

58 Latex Modified Steel Fiber Reinforced Concrete

Tapas Ranjan Baral[1,a], Sujit Kumar Pradhan[2,b] and Debakinandan Naik[3,c]

Department of Civil Engineering, IGIT Sarang, Dhenkanal, Odisha, India

Abstract

All around the globe, concrete is the most extensively used construction material. The important ingredient in traditional concrete is Portland cement. However, traditional concrete has low compressive strength, flexural strength, split tensile strength, and modulus of elasticity. In the present day, there will be studies and experiments conducted on steel-fiber-reinforced concrete. Concrete that has been changed with latex is just regular concrete that had some of the water in the mix replaced with latex. Sikacim as an admixture can enhance the properties of concrete, like strength and water resistance, and have lower permeability than traditional concrete. Since concrete is vulnerable to tension, the addition of steel fiber improves the characteristics of concrete in tension. Many researchers have located that steel fiber can provide higher strength as compared to other fibers like plastic fiber, glass fiber, etc. As a result, steel fibers, both straight and crimped, of 1% and 0.5% in concrete modified with different percentages of 0.6% and 1% of M20 grade are cast in two different proportions. In this research, a comparison between concrete and fiber modified steel fiber reinforced concrete (FMSFRC) is found by different tests conducted after 7 days and 28 days.

Keywords: Steel fiber, mix design, split tensile strength, modulus of elasticity

Introduction

Scattered fibers in the concrete mixture were first lightened by Bernard (1874). But the application of steel fiber is limited (Lantsoght, 2023) Because of insufficient standard codes present for fiber reinforced concrete (FRC). Steel fiber uses in concrete has increased rapidly over the last decade. The tensile cracks of composite materials are delayed and controlled by adding steel fiber to concrete (Fattouh et al., 2022). Steel fiber extended the life of concrete under compressive fatigue, and the optimum fiber content results in the longest life span for concrete mesostructured and fiber characteristics (Gonzalez et al., 2022). FRC can control plastic shrinkage (Sukasawang and Yohannes, 2020).

In concrete, fiber has an important role. The compressive fatigue loading and fatigue life were not healthy in the case of the large specimen (Gonzalez et al., 2022). For shear capacity of SFRC (Lantsoght 2023) concluded that for improvement experimental research is important. Zhao et al. (2022) studied behavior of SFRC with vibration time and found that with increase in vibration time the steel fibers settled bottom of the specimen and due to high flowing SFRC began to segregate. According to Cho and Nam (2022). 0.1% of graphen oxide nonoflake (GONF) shows 12.614.6% of reduction in slab thickness.

---

[a]tapas1040baral@gmail.com, [b]sujitpradhan@igitsarang.ac.in, [c]debakinandananik@gmail.com

DOI: 10.1201/9781003450924-58

Steel fiber

Properties of steel fiber	Specification of Steel fiber in FRC
Aspect ratio = 60	Good adhesion
Tensile strength =1100 N/mm^2	Compatibility with binder
Density = 7850 Kg/m^3	Good flexural strength
Length = 30 cm	Good Impact factor strength and fat

Source: Author

Experimental program

Each group consists of six cubes (15 × 15 × 15cm), six cylinders (15 cm (dia) × 30 cm) and two beams of size 1. (10 × 10 × 50) cm^2. (15 × 15 ×) cm respectively.

Table 58.1: Materials quantity used in experimental program.

Sample	Cement content	Fine aggregate	Coarse aggregate	W/C ratio	Latex percentage	Steel fiber
PC	388	673	1241	0.52	0	0
LMC	388	673	1241	0.48	0.6	0
LMFRC1	388	673	1241	0.48	0.6	0.5
LMFRC2	388	673	1241	0.48	0.6	1
LMFRC3	388	673	1241	0.48	1	0.5
LMFRC4	388	673	1241	0.48	1	1
LMFRC5	388	673	1241	0.48	0.6	0.5

Source: Author

Conceptual framework

Steel fiber by weight of cement is appended to the grade 'A' mix design concrete, which has a 1:1.73:3.19 proportion. After the mixture is prepared, it is poured into moulds of various shapes such as bar, cube, and cylinder, and the compressive strength test procedure is followed. Molds are kept in a water bath for 7 and 28 days to cure. After their respective days, the moulds are repaired for testing.

Mix design for M20 grade concrete

The mix calculation per unit vol. of cement concrete given below Figure 58.1.

Salient features of mix design for LMFRC

Table 58.2: Mix proportion (Mix design data was decided as per Shetty 2006)

Materials	Quantity
Cement content	388 Kg/m^3
W/C ratio	0.48
sand	672.41 kg/m^3
20 mm CA	744.41 Kg
10 mm CA	496.25 Kg
Steel fiber	0.5% and 1% straight and crimped
Length of steel fiber	30 mm
Dia of steel fiber	0.5 mm

Source: Author

Results and discussion

Cube compression test

The test was conducted as per IS 516-1959. The size cube is 150x150x150 mm were used and the specimens are placed in the 200-ton capacity testing machine shown in Figure 58.1 and a uniform rate of 550 kg/cm² per minute is applied until the failure shown in Figure 58.2. The results are tabulated in Table 58.3.

Flexural test

For the flexure test, use 100*100*500 mm-sized LMSFRC beams. In the testing machine, two rollers are used to hold the specimen, and these rollers have a gap of 400 mm between each other and 50 mm of bearing on each support. The load is transformed to the beam by two rollers, which are kept on the beam, and they have a gap of 200 mm. The results are shown in Table 58.3.

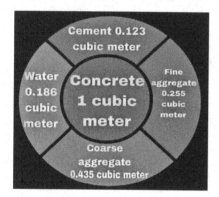

Figure 58.1 Volume of 1 m3 concrete
Source: Author

Table 58.3: Test results.

Sl No	Type of fiber	Aspect ratio	Fiber content %	Latex Sikacim %	Compressive strength (N/Mm2)		Split tensile strength (N/Mm2)		Flexural strength (N/Mm2) 28 Days
					7d	28d	7d	28d	28 days
1	.....	Conventional concrete	.....	.....	16.65	25.32	.....	.....	.....
2	.....	.....	.....	0.6	18.93	30.05	2.23	2.567	3.7
3	Straight and crimped	60	0.5	0.6	21.58	32.19	2.912	3.33	4.1
			1	0.6	23.932	33.851	3.19	3.747	5.3
4	Straight and crimped	60	0.5	1	21.732	32.57	3.058	3.357	4.8
			1	1	24.891	34.093	3.59	3.87	5.8
5	Straight	60	0.5	0.6	21.36	32.076	2.815	3.28	4.9
			1	0.6	22.44	34.75	3.16	3.51	5.9
6	Straight	60	0.5	1	21.87	33.34	3.042	3.405	4.8
			1	1	22.691	35.729	3.39	3.67	5.7

Source: Author

Conclusion

The aim of this study is to analyze the influence of steel fiber on normal concrete. The study investigates the impact of steel fiber and steel fiber with Sikacim on the flexural behavior and compressive strength of FRC cured in the lab. The compressive strength marginally increased by about 18.71% in Latex modified concrete (LMC) and 32.30% in Latex modified fiber reinforced concrete (LMFRC) as compared to the plane cement sample. LMFRC has 11.48% more compressive strength than LMC on average. Splitting strength gives about 5% to 10% more value than the direct tensile strength. In the last 10 years, knowledge on fibers has increased drastically; further knowledge is needed on design rules and standard values for fibers.

Acknowledgement

Authors would like to thank Department CIVIL Engineering, IGIT Sarang for providing the laboratory set ups to conduct the experiments.

References

Cho, B H and Nam, B H (2022). Concrete composites reinforced with graphene oxide nano-flake (GONF) and steel fiber for application in rigid pavement. https://doi.org/10.1016/j.cscm.2022.e01346

Fattouh, M S, Tayeh, B A, Agwa, I S, and Elsayed, E K (2022). Improvement in the flexural behaviour of road pavement slab concrete containing steel fiber and silica fume. https://doi.org/10.1016/j.cscm.2022.e01720

Gonzalez, D C, Ruiz, G, and Ortega, J J (2022). Size effect of steel fiber–reinforced concrete cylinders under compressive fatigue loading: Influence of the mesostructured. https://doi.org/10.1016/j.ijfatigue.2022.107353

Lantsoght, E O L (2023). Theoretical model of shear capacity of fiber reinforced concrete beams. https://doi.org/10.1016/j.engstruct.2023.115722.

Shetty, M S (2006). Concrete Technology: Theory and Practice. S. CHAND LTD: India.

Sukasawang, N and Yohannes, D (2020). Using fiber reinforced concrete to control early-age shrinkage in replacement concrete pavement. https://doi.org/10.1007/978-3-030-58482

Zhao, M., Li, J., and Xie, Y.M. (2021).Effect of vibration time on steel fibre distribution and flexural behaviors of steel fibre reinforced concrete with different flowability. https://doi.org/10.1016/j.cscm.2022.e01114

59 Study of indirect tensile strength and cyclic indirect tensile modulus of fiber reinforced cement stabilized fly ash, stone dust and aggregate mixture for pavement application

Sanjeeb Mohanty[a], Dipti Ranjan Biswal[b],
Brundaban Beriha[c], Ramchandra Pradhan[d] and
Benu Gopal Mohapatra[e]

KIIT Deemed to be University, Bhubaneswar, Odisha, India

Abstract

Owing to the huge quantity of fly ash (FA) generated in the thermal power plants, the researchers and road authorities across the globe are exploring the use of bulk amount of FA in road construction. The presence of high volume of fine particles in fly ash, makes it of lower strength and brittle when stabilized with cement. Hence, a mixture of stone dust (SD), aggregates (AG), and FA was prepared to enhance the gradation of fly ash prior to cement stabilization. In order to enhance the strength and brittle behavior of stabilized mix, randomly oriented polypropylene fiber (FI) has been used for reinforcement. Therefore, a laboratory study was carried out to observe the unconfined compressive strength (UCS), indirect tensile strength (IDT) and cyclic indirect tensile modulus of fiber reinforced cement stabilized FA, stone dust, aggregate mixture with fiber reinforcement. Cement dosage was varied from 4to 8% in 2% intervals and fiber percentage was varied as 0%, 0.25%, 0.35%, and 0.5%. Results shows that inclusion of fiber upto 0.25% enhances the compressive strength, however, any further addition of fiber have minimal effect on strength. However, IDT increases with increment in fiber dosage from 0 to 0.5%. Inclusion of fiber also results in ductile behavior of stabilized FA. An optimum fiber dosage of 0.25-0.35% was observed for the stabilized fly ash and fly ash-stone dust-aggregate mixture. A strong regression equation was developed between IDT and cyclic IDT modulus.

Keywords: Flyash, fibre, compressive strength, indirect tensile strength, cyclic IDT modulus

Introduction

The disposal of huge quantity of fly ash (FA) is a major concern throughout the globe. Further, paucity of crushed stones near the road construction site and strict environmental regulations inspires the researchers and authorities to explore the use of industrial waste like FA as structural layer of pavement as stabilized layer. Due to presence of high volume of fine particles in FA, it unfulfilled the specification requirement of subbase and base, hence it is used in stabilized form. Typically compressive strength is used to judge the suitability of stabilized material for pavement application. But study of tensile strength is also needed as the failure of the stabilized layer is largely due to tension (Biswal et al., 2020). Limited studies have focused on indirect tensile behavior of fiber reinforced cement stabilized materials. Many researchers observed enhancement of compressive strength behavior of stabilized materials with addition of fibers in the range of 0-1% (Kumar and Singh, 2008; Kaniraj and Havanagi, 2001; Chore

[a]sanjeeb2007.mohanty@gmail.com, [b]dipti.biswalfce@kiit.ac.in, [c]brundaban.berihafce@kiit.ac.in, [d]ramapradhan68@gmail.com, [e]bmohapatrafce@kiit.ac.in

DOI: 10.1201/9781003450924-59

and Vaidya, 2015). Past studies show that compressive strength decreases beyond a certain % of FA which varies from 0.3 to 1%. Elastic parameters such as resilient modulus (M_r) and Poisson's ratio are also vital inputs for Mechanistic -Empirical Pavement Design. Therefore, laboratory study was undertaken to study the UCS, IDT, M_r behavior of cement stabilized FA and FA –stone dust (SD)-aggregate(AG) mixture with without fiber. As the strength of stabilized fly ash do not meet the requirement of subbase or base due to presence of high volume of fine particles, there is a need to enhance the gradation of FA by adding stone dust or aggregates prior to stabilization.

Materials and methods

In this experimental program, FA, SD, AG, cement and fibers are the raw materials. The FA collected from National Aluminium Company Limited (NALCO) located in Angul (Odisha) can be classified as class F.Ordinary Portland Cement (OPC) of 43 grade was used as stabilizer. Stone dust was collected from nearby granite quarry crushing plant. The particle distribution curve of collected fly ash, FA and SD-AG (SA) mixtures is shown in Figure 59.1. Randomly oriented 100% virgin, non-corrosive 12mm length and 24-micron diameter polypropylene (PP) fiber has been used as reinforcing material in the mixture. FA, Cement, SD and AG and fibers were mixed with predetermined water corresponding to optimum moisture content. In this study, 70% FA with 30% SA and 60% FA with 40% SA were considered for strength study in order to meet the specification requirement of base and subbase. Cement percentage was taken as 4 and 6%, where fiber percentage was varied as 0.25%, 0.35% and 0.5%. Cylindrical samples (diameter 75 mm and height 150 mm) were prepared for UCS tests as per IS 2720:10. Cylindrical specimens(diameter100mm and height 60 mm) were prepared and cured for 28 days for IDT and M_r test. The M_r or cyclic IDT modulus was determined as per ASTM D4123-82 (Figure 59.2).

Result and analysis

Effect of fiber on compressive strength

The effect of dosage of fiber on cement stabilized 70FA-30SA and 60FA-40SA mixtures at various cement content are represented in Figure 59.3 (a) and Figure 59.3 (b)

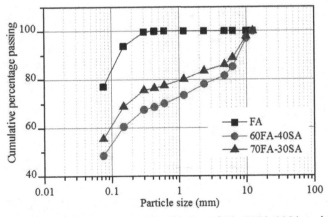

Figure 59.1 Grain size distribution of FA, 70FA-30SA and 60FA-40SA
Source: Author

Figure 59.2 Cyclic IDT modulus jig in servo hydraulic universal testing machine
Source: Author

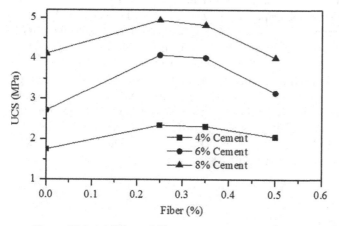

Figure 59.3 (a) Effect of fiber content on UCS of 70FA+30SA various cement content
Source: Author

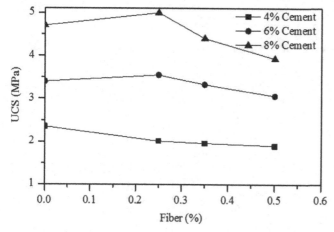

Figure 59.3 (b) Effect of fiber percentage on UCS of 60FA-40SA
Source: Author

respectively. It was found that the strength of 70FA-30SA mixture increases with increase in fiber content from 0 to 0.25%. The enhancement of strength of FA-SA mixtures may be attributed to the mobilization of tensile strength of randomly dispersed fibers due to inter-particle friction between fibers and surrounding fly ash, stone dust and aggregate mixtures (Ranjan et al., 1996, Zornberg, 2002).

Effect of fiber on indirect tensile strength

Indirect tensile strength of fiber reinforced cement stabilized FA and stone dust mixtures of 70FA-30SA and 60FA-40SA at 4% and 6% cement and 0% fiber, 0.25% fiber, 0.35% fiber and 0.5% fibers are presented in Figure 59.4. Increase in IDT of the specimens was observed with increase in cement. IDT was positively impacted by increase in fiber content from 0% to 0.5%. The enhancement of strength of FA-SA mixtures may be attributed to the inter particle friction between fibers and surrounding fly ash, stone dust and aggregate mixtures.

Effect of fiber on indirect tensile modulus

Cyclic IDT Modulus of 70FA-30S with 4% cement and at various percentages of fibers such as 0, 0.25, 0.35 and 0.5% are shown in Figure 59.5. It is observed that

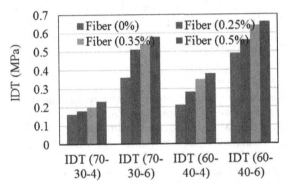

Figure 59.4 Indirect tensile strength of 70FA-30SA and 60FA-40SA at 4% and 6% cement and 0% fiber, 0.25% fiber, 0.35% fiber and 0.5% fibers
Source: Author

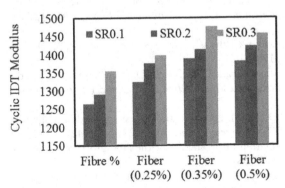

Figure 59.5 Variation of cyclic IDT modulus of 70FA-30SA and 4% cement at various fiber percentage and stress ratio
Source: Author

Cyclic IDT modulus increases with stress ratio which is in consistency with past literatures. The modulus at stress ratio 0.3 has been considered for further analysis based on the study by Yeo (2008) and Paul et al. (2015). A remarkable enhancement of modulus is observed upto 0.35% fiber content and any further addition of fibers have less effect on modulus. Though, IDT increases with increase in fiber content, IDT modulus gradually decreases at fiber percentage of 0.5 %. This may be due to the development a greater number of slip surface in the mixture matrix. Considering 0.35% as the optimum dosage, a cyclic IDT modulus of 1450 MPa can be used for the mechanistic design of pavement.

Conclusion

In the present study, a detailed experimental program was conducted to study unconfined compressive strength (UCS), indirect tensile strength (IDT), and cyclic IDT modulus of fiber reinforced cement stabilized FA, SD and AG mixture. UCS of fiber reinforced stabilized fly ash increases with addition of 0.25% fiber. Inclusion of fiber results in ductile behavior of cemented fly ash, and fly ash stone dust mixtures as compared to unreinforced materials, which shows brittle behavior. An optimum fiber dosage of 0.250.35% was observed in case of stabilized FA and FA-SD-AG mixture. However, IDT increases with increased fiber content from 0 to 0.5%. Cyclic IDT test was conducted to assess the cyclic IDT modulus. Considering 0.35% as the optimum dosage, a cyclic IDT modulus of 1450 MPa can be used for the mechanistic design of pavement.

References

ASTM, D4123-82. (2003). Standard test method for indirect tension test for resilient modulus of bituminous mixture. ASTM International, West Conshohocken, PA.

Biswal, D R, Sahoo, U C, and Dash, S R (2020). Mechanical characteristics of cement stabilised granular lateritic soils for use as structural layer of pavement. *Road Materials and Pavement Design*, 21(5), 12011223.

Chore, H S and Vaidya, M K (2015). Strength characterization of fiber reinforced cement–fly ash mixes. *International Journal of Geosynthetics and Ground Engineering*, 1, 1-8.

Ranjan, G, Vasan, R M, and Charan, H D (1996). *Journal of Geotechnical Engineering* 122, 419426. IS, 2720. 1991. Methods of test for soils–determination of unconfined compressive strength. Bureau of Indian Standard, New Delhi.

Kaniraj, S R and Havanagi, V G (2001). Behavior of cement-stabilized fiber-reinforced fly ash-soil mixtures. *Journal of Geotechnical and Geo Environmental Engineering*, 127(7), 574584.

Kumar, P., & Singh, S. P. (2008). Fiber-reinforced fly ash subbases in rural roads. *Journal of Transportation Engineering*, 134(4), 171180.

Paul, D K, Theivakularatnam, M, and Gnanendran, C T (2015). Damage study of a lightly stabilised granular material using flexural testing. *Indian Geotechnical Journal*, 45(4), 441448.

Ranjan, G, Vasan, R M, and Charan, H D (1996). Probabilistic analysis of randomly distributed fiber-reinforced soil. *Journal of Geotechnical Engineering*, 122(6), 419426.

Yeo, R (2008). The development and evaluation of protocols for the laboratory characterisation of cemented materials (No. AP-T101/08).

Zornberg, J G (2002). Discrete framework for limit equilibrium analysis of fibre-reinforced soil. *Géotechnique*, 52(8), 593604.

Printed in the United States
by Baker & Taylor Publisher Services